Evolutionary Computation with Intelligent Systems

Demystifying Technologies for Computational Excellence: Moving Towards Society 5.0

Series Editors:
Vikram Bali and Vishal Bhatnagar

This series encompasses research work in the fields of data science, edge computing, deep learning, distributed ledger technology, extended reality, quantum computing, artificial intelligence, and various other related areas, such as natural-language processing and technologies, high-level computer vision, cognitive robotics, automated reasoning, multivalent systems, symbolic learning theories and practice, knowledge representation and the semantic web, intelligent tutoring systems, and education.

The prime reason for developing and growing out this new book series is to focus on the latest technological advancements – their impact on society, the challenges faced in implementation, and the drawbacks or reverse impact on society due to technological innovations. With these technological advancements, every individual has personalized access to all services and all devices connected with each other communicating among themselves, thanks to the technology for making our life simpler and easier. These aspects will help us to overcome the drawbacks of the existing systems and help in building new systems with the latest technologies that will help society in various ways, proving Society 5.0 as one of the biggest revolutions in this era.

Healthcare and Knowledge Management for Society 5.0
Trends, Issues, and Innovations
Edited by Vineet Kansal, Raju Ranjan, Sapna Sinha, Rajdev Tiwari, and Nilmini Wickramasinghe

Society 5.0 and the Future of Emerging Computational Technologies
Practical Solutions, Examples, and Case Studies
Edited by Neeraj Mohan, Surbhi Gupta, and Chuan-Ming Liu

Evolutionary Computation with Intelligent Systems
A Multidisciplinary Approach to Society 5.0
Edited by R. S. Chauhan, Kavita Taneja, Rajiv Khanduja, Vishal Kamra, and Rahul Rattan

For more information on this series, please visit: https://www.routledge.com/Demystifying-Technologies-for-Computational-Excellence-Moving-Towards-Society-5.0/book-series/CRCDTCEMTS

Evolutionary Computation with Intelligent Systems

A Multidisciplinary Approach to Society 5.0

Edited by
R.S. Chauhan, Kavita Taneja, Rajiv Khanduja,
Vishal Kamra, and Rahul Rattan

CRC Press
Taylor & Francis Group
Boca Raton London New York

CRC Press is an imprint of the
Taylor & Francis Group, an **informa** business

First edition published 2022
by CRC Press
6000 Broken Sound Parkway NW, Suite 300, Boca Raton, FL 33487-2742

and by CRC Press
2 Park Square, Milton Park, Abingdon, Oxon, OX14 4RN

CRC Press is an imprint of Taylor & Francis Group, LLC

Library of Congress Cataloging-in-Publication Data
Names: Chauhan, R. S. (Ranjit Singh), editor.
Title: Evolutionary computation with intelligent systems : a
multidisciplinary approach to society 5.0 / edited by R.S. Chauhan,
Kavita Taneja, Rajiv Khanduja, Vihsal Kamra, and Rahul Rattan.
Description: First edition. | Boca Raton : CRC Press, [2022] |
Series: Demystifying technologies for computational excellence: moving towards
society 5.0 | Includes bibliographical references and index.
Identifiers: LCCN 2021046706 (print) | LCCN 2021046707 (ebook) |
ISBN 9780367744939 (hbk) | ISBN 9780367744946 (pbk) | ISBN 9781003158165 (ebk)
Subjects: LCSH: Artificial intelligence—Industrial applications. |
Society 5.0. | Evolutionary programming (Computer science) | Internet of things.
Classification: LCC TA347.A78 E96 2022 (print) | LCC TA347.A78 (ebook) |
DDC 006.3—dc23/eng/20211118
LC record available at https://lccn.loc.gov/2021046706
LC ebook record available at https://lccn.loc.gov/2021046707

ISBN: 978-0-367-74493-9 (hbk)
ISBN: 978-0-367-74494-6 (pbk)
ISBN: 978-1-003-15816-5 (ebk)

DOI: 10.1201/9781003158165

Typeset in Times
by codeMantra

Contents

Preface

Evolutionary computation is one of the most powerful problem-solving concepts inspired from natural evolution. This book provides the emerging and highly relevant topics, concepts, and their applications. It will be beneficial for academicians, researchers, engineers, postgraduates, graduates, and other technology enthusiasts. It presents a multidisciplinary approach to evolutionary computing trends reflecting optimization solutions through applications of evolutionary computation to design, control, and classification problems. The book aims to be the single point of reference for evolutionary computation for optimization in diverse problem-solving methodologies for Society 5.0. It explains the core theories, concepts, systems, and their practical applications of Society 5.0. It explores the aspects of evolutionary computation from research, scientific, and business perspectives for secure and scalable applications in various fields, particularly in smart-generation computing excellence. The emerging trends of the design and optimized solutions through intelligent systems are focused in light of applications pivoting around the needs of Society 5.0.

In Chapter 1, "Evolutionary Trends in Smart City Initiatives – A Consumer Perspective", the realistic concept of smart cities in Society 5.0 that can ensure the optimum availability of clean air and water, healthcare systems, waste management, adequate supply of power and energy sources, efficient transport, ample parking facilities, etc. is investigated. Precisely, IoT devices such as sensors, lights, and meters that need to be linked in a way that projected information gets automatically compiled and analyzed for further decision making to give value to the consumer's purchase are explored from the consumer's perspective.

In Chapter 2, "Human-Feedback Adaptive Learning Using Interpretable and Interactive Intelligent Systems", the state of the art in the field of interactive machine learning systems is explored. Extracting principles and guidelines for the design of interactive machine learning systems, algorithmic description and graphical representation of the workflow involved, identifying metrics for the evaluation, and proposing a human-feedback adaptive learning algorithm that adapts itself to incorporate the human expert feedback are the major highlights. The proposed approach is capable of reporting any conflict between human feedback and data along with an interpretable explanation. Moreover, it is capable of accommodating human users with a different expertise level of the domain. Establishing principles and guidelines for the design of interactive ML systems will help in standardization and building a consensus. A basic algorithm is suggested for the design of interactive machine learning systems that promises a common starting point for further improvements. Having an interface that can capture feedback from human experts with different levels of expertise will help make human experiments economical and enable leveraging of masses in improving or verifying interactive machine learning systems.

In Chapter 3, "Advertisement Detection: Image Processing and Deep Learning Approach for Effective Information Extraction from Online English Newspapers", automatic detection of advertisements in newspapers is explored for applications in Society 5.0. This chapter proposes an intelligent system to automatically detect advertisement images in online English newspapers exploiting image processing and deep learning techniques. First, a novel image processing-based technique is proposed which can successfully extract images from multiple online newspapers without making assumptions about newspaper layout or any other rule or prior knowledge. Following that, a convolutional neural network model is used which separates advertisement and non-advertisement images. An image dataset is created by gathering 11,000 images from various online newspapers in the English language. Six different convolutional neural network models are designed, trained, and evaluated using this image dataset, and a maximum accuracy of 97.82% is achieved.

In Chapter 4, "Evolutionary Computation Framework for Handling Resource and Optimization of Solar Energy Harvesting System for WSN", sun-based energy gathering for providing substitute capacity to the WSNs as it has the most powerful thickness and great productivity is suggested for varied applications of Society 5.0. Various techniques to follow the most extreme force purpose of a photovoltaic module have been proposed in this chapter to beat the restriction of effectiveness. Maximum power point tracker is utilized for removing the most extreme force from the sun-oriented photovoltaic module and moving that capacity to the heap. It is further suggested that by changing the obligation cycle, the heap impedance is coordinated with the source impedance to accomplish the most extreme force from the photovoltaic board.

In Chapter 5, "Smart Systems for Global Sustainability with Enhanced Computing", the idea of smart parking management system aiming to fit into Society 5.0 using the computational intelligence-based smart systems is explored. These can ease the situations prevailing in today's society related to the parking and other issues in context to it. Identifying the best parking slot and finding the best possible route can not only ease the situation but also save fuel and time. An energy-aware, trust-based secure routing algorithm is included in the chapter for route discovery along with the firefly algorithm (FA) with artificial neural network-based algorithms for route optimization in light of the smart parking management system.

In Chapter 6, "Intelligent Systems: Techniques for Optimized Decision Making", the need of imparting social intelligence in robotic systems for perception, navigation, and performing vital actuation for better human–robot interactions is highlighted. This chapter attempts to implement robot navigation schemes to provide better social interactions between humans and robots using the TurtleBot3 to follow human beings in social zones with intelligence applied through a Q-learning approach. The navigational schemes are implemented with the architecture developed in the Robot Operating System platform for the TurtleBot3 robot using the Q-learning approach.

In Chapter 7, "Innovations in Healthcare Using Smart Systems Equipped with Evolutionary Computation", the sequence of human body movements in which different body parts are engaged in a concurrent manner is explored. According to the computer vision perception, any kind of observation is matched with earlier defined patterns, and then it is labeled while recognizing the action. The major human activities considered in the proposed work fall under the categories such as walking, standing, upstairs, downstairs, and sitting. The evolutionary computational technique using deep learning with the use of recurrent neural network and long short-term memory model is implemented for innovations in healthcare for Society 5.0.

In Chapter 8, "Exploiting Evolutionary Computation Techniques for Service Industries", well-controlled mechanical systems that fuel the development of useful and successful automated applications to benefit the service industries in Society 5.0 are highlighted. An evolutionary machine learning framework applied to the identification and control of mechanical systems based on the widely studied cross-industry standard process for data mining (CRISP-DM) is proposed. In this framework, the actual system is excited with particular control signals to generate valuable input/output information to train and test a model. Since well-established laws from classical mechanics are available to describe mechanical systems, the system model is obtained through Lagrangian analyses. The selected model is used in the controller tuning through meta-heuristic optimization. Numerical simulations show the proposal's effectiveness for two study cases: the position regulation of a simple pendulum and the trajectory tracking with a fully actuated inverted pendulum.

In Chapter 9, "Evolutionary Computation Techniques for Strengthening Performance of Commercial MANETs in Society 5.0", the requirements for effective smart device communication are identified as reliable, cooperative, efficient, and dynamic routing protocols that meet the requirements of real-time ad hoc networking. Due to the frequent mobility of nodes in commercial mobile ad hoc networks supporting smart environments, establishing networks in the real world is difficult and demanding. In order to provide quality of service to a huge set of varied users of the Internet, there is a need to fortify the performance of such MANETs by designing intelligent routing protocols. Such intelligent algorithms are developed from evolutionary computation techniques and play a vital role in offering enhanced connectivity and data transfer capabilities to end users in a mobile ad hoc network. This chapter explores various threats to routing arising in commercial MANETs for Society 5.0 applications. It also investigates the evolutionary computation techniques in designing intelligent MANET routing protocols for strengthening the performance of commercial MANETs in Society 5.0. A comprehensive comparative analysis of emerging evolutionary computation techniques for commercial MANETs to handle challenges and threats is also presented.

In Chapter 10, "Availability Optimization of a Rice Finishing and Grading System Using Evolutionary Computation Techniques", the availability modeling and optimization of the rice finishing and grading system of a rice milling plant using particle swarm optimization and genetic algorithm evolutionary techniques is proposed. The case study presented in the chapter addresses the industrial system consisting of six subsystems, namely Abrasive Whitener, Rotary Shifter, Sizer, Polisher, Sortex, and Grader, which are subjected to failures. A mathematical model of the industrial system chosen for the study has been developed using the Markovian approach to develop various differential-difference equations that are obtained from the state transition diagram. The recursive method is applied to solve these equations and reduced to the steady state condition required for the rice finishing and grading system. The optimization is carried out in two steps to find out firstly the optimum number of generations and thereafter the optimal population size/particle size in both the approaches. The latter technique shows better results as compared to the first approach in the present case study. The particle swarm optimization gives 2.85% more availability at the 30th generation and particle size of 25 as compared to the genetic algorithm results at the 100th generation and population size of 70.

In Chapter 11, "Analysis of Sign Language Recognition System for Society 5.0 for Sensory-Impaired People", the issue of effective hand gesture recognition and its strong connection to the difficulties of interaction between mute people (unable to speak and deaf) and public are highlighted. Due to the explosive growth of software and hardware, modern strategies of human–computer interaction methods are required for providing a normal life to differently-abled person in Society 5.0. Technologies such as speech recognition and gesture recognition are gaining considerable interest in the field of human–computer interaction. The deaf and mute use sign language to interact among themselves as well as with others, which is difficult to understand for anyone who does not know the language. This chapter explores the need and related issues to build a framework that can decode gestures into text and voice. It promises to be a major step for improving the communication between the deaf community and the hearing public.

In Chapter 12, "Study and Control of Shrinkage in Gearbox Sand Casting Using Simulation and Experimental Validation", process parameters responsible for shrinkage porosity defect in green sand casting of gearbox housing are studied and controlled. With the aim of achieving various functional and fitment requirements, gearbox housings are required to be shrinkage free, to avoid any leakage or breakage under dynamic working conditions. In the existing research work, various parameters responsible for shrinkage have been studied. These parameters broadly include casting gating and feeding parameters, metal chemical composition, and raw material ingredients such as steel scrap. Vital few parameters were identified in this chapter on the basis of Pareto analysis of existing defects. Accordingly, multiple experiments were done for parameters such as ingate size, steel scrap, and profile thickness. To modify the casting gating system and material composition, commercial simulation software MAGMASoft is explored. The experimental validation is done over a batch of 2600 castings in a foundry. With the optimized parameters and tooling correction, rejection due to shrinkage porosity defect was reduced from 4.9% to 1.1%, with a saving of 32,000 USD over a period of 3 months. The implication of the current study is to identify and optimize all critical parameters of the casting gating and feeding system, the chemical composition of metal, and profiles of gearbox housings.

In Chapter 13, "An Integrated Approach Based on Structural Modeling for Development of Risk Assessment Framework for Drivers Involved in Green Supply Chain Management in India", the greening of the supply chain as the need of the hour for Society 5.0 is explored. The supply chain can be made green by considering the incoming raw material, inbound logistics, production, marketing, distribution, after-sales, and recycling of the product. For the implementation of green supply chain management, empirical research work is done and interpretive structural modeling technique has been used to understand the contextual relationship between major factors with the help of expert opinion. Cross-impact matrix multiplication applied to classification analysis is done, and it is found from the study that the top management, environmental management, material sourcing, and product design are independent enablers and are the most important factors for making green supply chain management greener in India. The analysis is also done to analyze the group of factors that are further grouped into four categories, i.e., independent, dependent, autonomous, and linkage.

In Chapter 14, "Human Resource Intelligent Systems: Rewriting the DNA of HR Function", electronic human resource management as the supplement to the traditional face-to-face direct relationships for human resource intelligent system in management applications for Society 5.0 is highlighted. This chapter includes an in-depth research study of the electronic human resource management practices and system of chosen Indian as well as multinational companies operating in India. The current research has been able to explore and identify the factors that lead to the perceptions of 'end users' (employees of selected organizations) toward their respective electronic human resource management systems and applications. The factors and the various variables studied under them identify merits, limitations, and facilitating conditions to meet business goals with competitive edge in Society 5.0 for reshaping the DNA of the human resource functions. A comparative study has been presented in the context of Indian and multinational companies regarding electronic human resource management perceptions.

In Chapter 15, "Role of Servitization in Society 5.0", the potential of a growing number of manufacturers to explore servitization as a good business model is explored. Market and customer requirements are increasingly getting complex, high-tech equipment; customers' trust on their equipment dealers for service expertise is getting demanding more than ever. Manufacturers, instead of focusing solely on selling a product, are revamping their approach to match the increasing desires of their customers. This chapter summarizes an inquiry into earlier work done in servitization and future scope of work. The shift of firms from traditional service domain to high-tech trends and impact on profitability is highlighted. The use of innovative IoT and data mining software, their amalgamation with manufacturing equipment, and customers' expectations on manufacturers to adopt newer methodologies and serve better are also investigated.

In a nutshell, this book focuses on the cutting-edge innovations in evolutionary computation. The major focus of the book includes understanding the methodologies of emerging trends of Society 5.0 and its wide application areas for developing intelligent systems. The book will help researchers and learners through its real-life applications, case studies, and examples to engage in deep conversations with their peers working in evolutionary computation and developing cutting-edge innovations, and guide them. It will cater for beginners and intermediate readers and researchers with illustrations from fundamental theories to practical and sophisticated applications of evolutionary computation in varied industries. The chapters provide insights into various platforms, paradigms, techniques, and tools used for intelligent systems in diverse fields.

Best wishes and happy reading!

Natick, MA 01760-2098 USA
Tel: 508-647-7000
Fax: 508-647-7001
E-mail: info@mathworks.com
Web: www.mathworks.com

Editors

R.S. Chauhan, PhD, is a Dean and Professor at Shri Vishwakarma Skill University, Haryana, India. He earned an Instrumentation and Control Engineering degree from the Regional Engineering College (now known as N.I.T. Jalandhar), Jalandhar, Punjab; an M.Tech. degree in Electronics and Communication Engineering from G.N.E. Ludhiana, Punjab; and PhD degrees in Electronics and Communication Engineering from Guru Jambheshwar University of Science and Technology, Hisar, Haryana. He has more than 21 years of experience in teaching and research. He has delivered keynote lectures and guest lectures and published numerous papers in journals and conferences. He is also the author of the book *Linear Control Systems*. His research areas include digital signal processing, evolutionary computation, and artificial intelligence.

Kavita Taneja, PhD, is Assistant Professor at Panjab University, Chandigarh. She earned her PhD in Computer Science and Applications from Kurukshetra University, Kurukshetra, India. She has published and presented more than 60 papers in national/international journals and conferences and has received best paper awards in many conferences including IEEE, Springer, Elsevier, and ACM. She is a reviewer on many reputed journals and has been a Technical Program Committee member of many conferences. She has also authored and edited computer books. Dr. Taneja has more than 18 years of teaching experience in various technical institutions and universities. She is also a member of BoM, Academic Council, and Board of Studies of many universities and institutions. She has guided scholars of PhD/M.Phil. and more than 100 PG students of various universities, and currently, four students are pursuing PhDs under her guidance at Panjab University. Her teaching and research activities include mobile ad hoc networks, simulation and modeling, and wireless communications.

Rajiv Khanduja, PhD, is Professor and Head of the Department of Mechanical Engineering at Jawaharlal Nehru Government Engineering College, Sundernagar, Dist. Mandi, H.P., India. He earned his B.Tech. degree from Gulbarga University, Gulbarga, India, in 1992. He earned his M.Tech. in 2002 and PhD in 2010 from the National Institute of Technology, Kurukshetra. He has 25 years of experience in teaching and research. He has published 35 research papers in reputed international and national journals. He has also presented 25 papers at international and national conferences. His research areas include industrial processes, reliability engineering, and maintainability engineering.

Vishal Kamra, PhD, earned his PhD in Business Management from LM Thapar School of Management, Thapar University, Patiala. His research interests involve marketing management, total quality management, service marketing, and healthcare. Presently, he is working as Sr. Assistant Professor at Amity School of Business,

Amity University, Noida, U.P. Dr. Kamra has 11 years of experience in teaching and 3 years in industry. He has previously served at N.I.T. Kurukshetra; Thapar University, Patiala; and GL Bajaj, Greater Noida. He has published a number of papers in reputed international and national journals. He has organized various international and national conferences. He is also on editorial boards of various reputed journals as associate editor and reviewer.

Rahul Rattan, PhD, is a Faculty and Research Investigator in the Department of Cardiology-Internal Medicine at Michigan Medicine, the University of Michigan, Ann Arbor. He earned his postdoctorate, PhD, and M.S.E. in Biomedical Engineering from the University of Michigan, Ann Arbor, USA. He is the editor for *Archives in Cancer Research* and has multiple peer-reviewed articles in *Drug Delivery and Translational Research, Nano Reviews & Experiments, Translational Biomedicine, PLOS ONE, Gene Reports*, among others. His research interests include heart failure, atherosclerosis, and glycocalyx.

Contributors

Amit Kumar Bindal
Computer Science & Engineering
Maharishi Markandeshwar Engineering College
Maharishi Markandeshwar (Deemed to be University)
Mullana, India

Pradeep Kumar Gaur
CGC Landran
India

Carlos Alberto Guerrero-León
Tecnológico Nacional de México / IT de Tlalnepantla
Mexico City, Mexico

Amit Gupta
Skill & Organization Development
Indian Oil Corporation Ltd.
India

Rahul Gupta
Amity Business School
Amity University
Uttar Pradesh, India

Axel Herroz Herrera
Tecnológico Nacional de México / IT de Tlalnepantla
Mexico City, Mexico

Pooja Jain
GGSCW
Chandigarh, India

Vishal Kamra
Amity School of Business
Amity University
Uttar Pradesh, India

Jatinder Kaur
Department of ECE
Chandigarh University
Gharuan, Mohali, India
and
CGC Jhanjeri
Mohali, India

Ramanpreet Kaur
G.G.N. Khalsa College
Ludhiana, India

Sarabpreet Kaur
Department of ECE
CGC Jhanjeri
Mohali, India

Rajiv Khanduja
Department of Mechanical Engineering
Jawaharlal Nehru Government Engineering College
Himachal Pradesh, India

Rajesh Khanna
Maharishi Markandeshawar (Deemed to be University)
Ambala, India

Rajiv Khosla
DAV Institute of Management
Chandigarh, India

Pawan Koul
Department of Management Studies
Bharati Vidyapeeth (Deemed to be University)
Off campus Navi Mumbai, India

J. Senthil Kumar
Department of Electronics & Communication Engineering
Mepco Schlenk Engineering College (Autonomous)
Sivakasi, Tamil Nadu, India

Pawan Kumar
Lovely Professional University
Phagwara, India

Shikha Mishra
Amity Business School
Amity University
Uttar Pradesh, India

Nitin Mittal
Department of ECE
Chandigarh University
Gharuan, Mohali, India

Vikas Modgil
Department of Mechanical Engineering
Deen Bandhu Chhotu Ram University of Science and Technology (DCRUST)
Murthal, Sonepat, Haryana, India

Manuel Eduardo Mora-Soto
Tecnológico Nacional de México / IT de Tlalnepantla

Miguel Ángel Paredes-Rueda
Tecnológico Nacional de México / IT de Tlalnepantla

R.S. Rai
Amity Business School
Amity University
Noida, India

Anju Rani
Computer Science & Engineering
Maharishi Markandeshwar Engineering College
Maharishi Markandeshwar (Deemed to be University)
Mullana, India

Alejandro Rodríguez-Molina
Tecnológico Nacional de México / IT de Tlalnepantla
Mexico City, Mexico

S. K. Roy
ApeeJay Stya University–Gurugram
Haryana, India

Parveen Kumar Saini
Department of Mechanical Engineering
Guru Brahma Nandji Government Polytechnic
Nilokheri, Karnal, Haryana, India

Puja Sareen
Amity Business School
Amity University
India

Kritika Raj Sharma
Department of Electronics and Communication
Chandigarh University
Gharuan, Mohali, India

Manmohan Sharma
Lovely Professional University
Phagwara, India

Neeraj Sharma
Maharishi Markandeshawar (Deemed to be University)
Ambala, India

Raman Shergill
Chandigarh University Gharuan
Gharuan, Mohali, India

Gurpreet Singh
Chandigarh University Gharuan
Gharuan, Mohali, India

Sarabjit Singh
CIHT Jalandhar
India

G. Sivasankar
Department of Electronics & Communication Engineering
Mepco Schlenk Engineering College (Autonomous)
Sivakasi, Tamil Nadu, India

José Solís-Romero
Tecnológico Nacional de México / IT de Tlalnepantla
Mexico City, Mexico

Harmunish Taneja
D.A.V. College
Chandigarh, India

Kavita Taneja
Panjab University
Chandigarh, India

Deepti Tara
Freelancer in Management and IT
Ontario, Canada

1 Evolutionary Trends in Smart City Initiatives

A Consumer Perspective

Rajiv Khosla and Deepti Tara

CONTENTS

1.1 INTRODUCTION

The United Nations World Urbanization Prospects (The 2018 Revision) report highlighted that by the year 2050, 68% of the world's population will be living in urban areas vis-à-vis 56% today. The average number of people living in rural areas worldwide in the year 1950 was 30%. Further, there is a stark dissimilarity between the urbanization trends in high-income, upper-middle-income and lower-middle-income, and low-income countries. The population of high-income countries (the USA, Australia, Belgium, UK, Italy, Germany, and Singapore) living in urban areas was 59% in 1950, which increased to 82% in 2020, and it is expected to reach 88% in 2050.

DOI: 10.1201/9781003158165-1

Upper-middle-income countries (China, Fiji, Romania, Iran, Iraq, and South Africa) witnessed an unprecedented increase in the percentage of population living in urban areas between the years 1950 (22%) and 2018 (67%), which is further anticipated to touch 83% by 2050. By contrast, the major chunk of lower-middle-income countries (India, Indonesia, Bangladesh Cambodia, Ghana, and Uzbekistan) and low-income countries (Afghanistan, Nepal, Malawi, Zimbabwe, and Ethiopia) were living in rural areas in 1950, which came down to 59% and 68%, respectively, in the year 2020. The proportion of population expected to reach urban areas by the year 2050 in lower-middle-income and low-income countries will be 59% and 50%, respectively. Table 1.1 highlights the details about how this transition is taking place every 5 years. Similarly, figures have been drawn to explicitly show the shift from rural to urban areas. Notably, the figures showcase the transformation in each region over a period of time (Figures 1.1–1.4).

Buyers in urban areas owing to increased migration of people are thus expected to face myriad challenges. Hence, there is an emergent need to ensure the optimum availability of clean air and water, healthcare systems, waste management, adequate supply of power and energy sources, efficient transport, ample

TABLE 1.1
Population at Midyear Residing in Urban Areas in Different Regions: 1950–2050 (in Percent)

Year	World	High-Income Countries	Middle-Income Countries	Upper-Middle-Income Countries	Lower-Middle-Income Countries	Low-Income Countries
1950	29.61	58.51	19.89	22.08	17.21	9.32
1955	31.64	61.14	22.17	25.08	18.56	10.39
1960	33.75	63.76	24.55	28.40	19.85	11.90
1965	35.59	66.25	26.68	31.26	21.17	13.47
1970	36.59	68.68	27.83	32.19	22.59	15.66
1975	37.72	70.45	29.36	33.59	24.33	17.47
1980	39.35	71.85	31.66	36.28	26.35	19.11
1985	41.20	73.11	34.27	39.79	28.16	20.89
1990	42.96	74.43	36.71	42.91	30.02	22.76
1995	44.78	75.67	39.14	46.44	31.59	24.35
2000	46.68	76.80	41.64	50.26	33.14	25.68
2005	49.16	78.59	44.72	55.02	35.02	27.16
2010	51.66	80.05	47.86	59.76	37.12	28.91
2015	53.93	80.95	50.83	64.12	39.25	30.93
2020	56.17	81.85	53.74	68.20	41.60	33.17
2025	58.33	82.84	56.46	71.75	44.18	35.65
2030	60.43	83.92	59.05	74.76	46.95	38.34
2035	62.48	85.05	61.51	77.28	49.89	41.20
2040	64.47	86.21	63.86	79.34	52.92	44.16
2045	66.42	87.34	66.11	81.02	55.99	47.17
2050	68.36	88.39	68.32	82.56	59.03	50.21

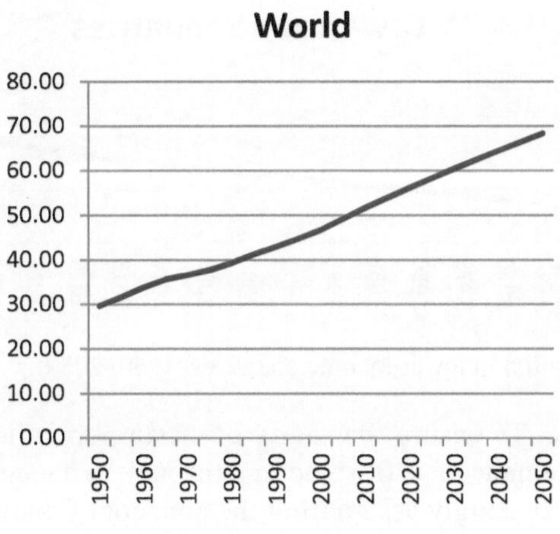

FIGURE 1.1 Worldwide urbanization 1950–2050.

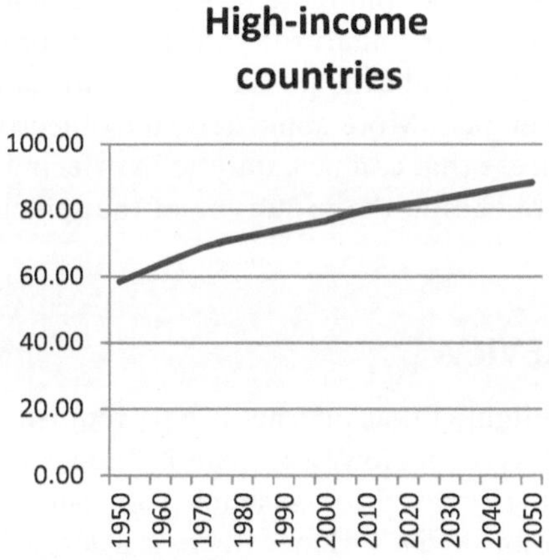

FIGURE 1.2 Urbanization in high-income countries (1950–2050).

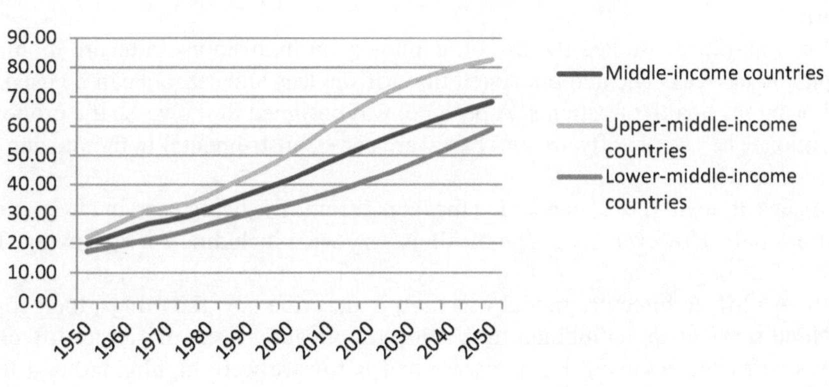

FIGURE 1.3 Urbanization in middle-income, upper-middle-income, and lower-middle-income countries (1950–2050).

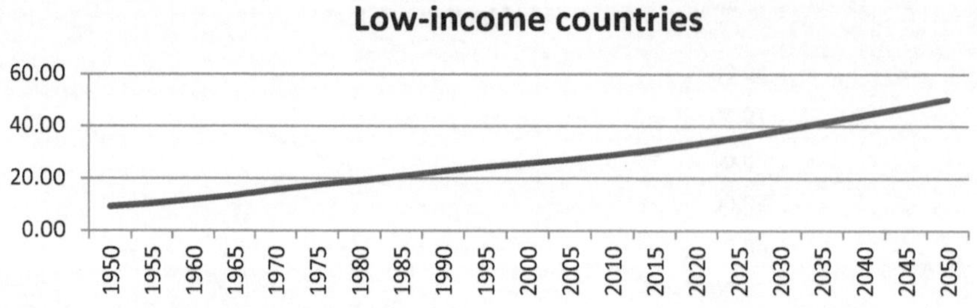

FIGURE 1.4 Urbanization in lower-income countries (1950–2050).

parking facilities, etc. To ensure the safety of urban population and to ascertain the optimum management of infrastructure through reduced costs and energy, policymakers are increasingly associating the concept of disruptive technologies or Internet of things (IoT) with smart cities. In simple terms, smart cities are expected to use Internet-enabled electronic gadgets that are made up of components such as sensors, lights, and meters, which are mutually connected, and the data so generated roll out the analyzed information for decision making. The present study intends to highlight the possible alternatives through which future smart cities may be developed. More appositely, the objectives of this study were to assess the strategic areas that can be catalytic in offering value to the consumers in smart cities and to exhibit the prototype of the selected strategic areas for creating smart cities.

1.2 LITERATURE REVIEW

Heeret et al. (2011) highlighted that Internet is a prerequisite for all the communication devices in an IoT system. However, there is a security concern with respect to the Internet protocol (IP)-based IoT systems. Conventional end-to-end IP may not support communication devices. Hence, there is a desperate need to design new Internet protocols that can ensure high security. Further, all layers responsible for communication have their security concerns that are also required to be plugged in amicably.

Jara et al. (2013) studied the use of technological innovations in health monitors and patient devices. The authors stated that IoT devices and sensors can be used for monitoring the health of patients. A protocol was designed to showcase the proposed innovation. The use of software and hardware can be instrumental in monitoring the health of patients.

Khajenasiri et al. (2017) concluded that at present IoT is being used in few application areas only. However, the scope of IoT is very wide including energy saving. The use of IoT through smart energy control systems in energy saving is a service to the society. An IoT architecture model was used to describe how the things, the people, and cloud services are combined to facilitate application tasks. Different IoT challenges confronted by using IoT software and hardware were highlighted and their resolution discussed.

Alavi et al. (2018) in their study stated that the spread of urbanization in modern cities causes smart solutions to address the vital issues, viz. health care and energy. IoT was seen as a tool for building smart cities through an extensive use of communication technologies. The study investigated several challenging issues such as traffic management, smart parking, smart traffic management, and smart waste collection and tried to address them by using IoT. The study concluded that IoT is a future weapon to address the challenging issues.

Eunil et al. (2018) in their study discussed the different aspects of IoT technologies. The study recommended that it is imperative to promote government-led demonstrations for the endorsement of smart city industry and markets, which will facilitate the participants to evaluate the efficiency, feasibility, and effects of IoT technologies. Besides, a data-oriented smart city infrastructure should also be established. Further, the authors opined that professional panels of experts from diverse research fields including urban development, information and communication technologies, transportation, and environmental policies should sit together while formulating policies about the creation of smart cities.

Kumar et al. (2019) lamented that IoT developers and researchers are working together to provide the benefits of technology to the society. Several issues and challenges that IoT developer should consider while developing an improved model were considered in the study. Important application areas of IoT, viz. environment, healthcare smart city, transport, and vehicles, were discussed. The authors also highlighted the importance of big data analytics as a tool to unfold IoT applications.

Behrendt (2019) in her study found that policy documents on the European Commission website that are relevant to transport and mobility and IoT in smart cities have discussed the smart mobility with a strong focus on cars and ignoring cycles. The study while comparing cars to cycles challenged the continuation of cars in automobiles. The inclusion of cycling in the policy documents is strongly recommended. Further, it is stated that although the study is restricted to Western and European viewpoint, it carries an international perspective.

Similar studies are conducted (Bharadwaj et al., 2016; Vadillo et al., 2017; Aleyadeh and Taha, 2018; Murad and Hidayanto, 2018; Paula et al., 2019), and it has been identified that strategic areas can be expedited to transform toward the smart cities without many hassles.

1.3 METHODOLOGY

A review of the above studies clearly demonstrates that the authors have hinged around three main areas that can help create smart cities: smart housing, smart transport, and smart waste management. However, the majority of the studies have not collectively discussed the three issues. Each study focused on either one or two of these issues. These three broad areas further contain a gamut of subareas that can add value proposition to consumers' purchase. All the three areas are discussed in detail in the following sections. However, for the sake of simplicity, we are using a figure that entails the identified areas and subareas (Figure 1.5).

FIGURE 1.5 IoT-enabled services in different areas of smart cities in Society 5.0.

1.4 SMART HOUSING AND SMART INFRASTRUCTURE

This section primarily emphasizes on the projected and proposed applications of smart devices that facilitate the technologically enabled living in terms of homes, buildings, and other such areas. These facilities are becoming the major factor for the movement toward the urbanization. A few of the IT-enabled services are discussed in the following paragraphs:

1.4.1 Smart Housing

Thanks to the advancements in artificial intelligence, smart homes will be able to anticipate the consumers' needs.

- Smart lighting and air conditioning systems will detect occupants in the room and start functioning. Also, it will adjust lighting as required, i.e., lesser lights during daytime and vice versa. In addition, smart thermostats will monitor the outside temperature and regulate the home temperature accordingly.
- Smart TVs connected with Internet, on the basis of historical data, will automatically display news, sports, videos, or music channels at fixed time periods (Figure 1.6).
- Smart locked doors and windows will detect the arrival or departure of residents and when they are near get opened up or locked accordingly. Forced breaking of door or windows will flash message to the police control room and to the owner (who can access on his mobile camera, if there is any event of trespassing). Even when an online parcel comes, the delivery boy will be permitted to place it in the house as the owner will open the lock through mobile phone and at the same time witness the activities of the delivery boy.

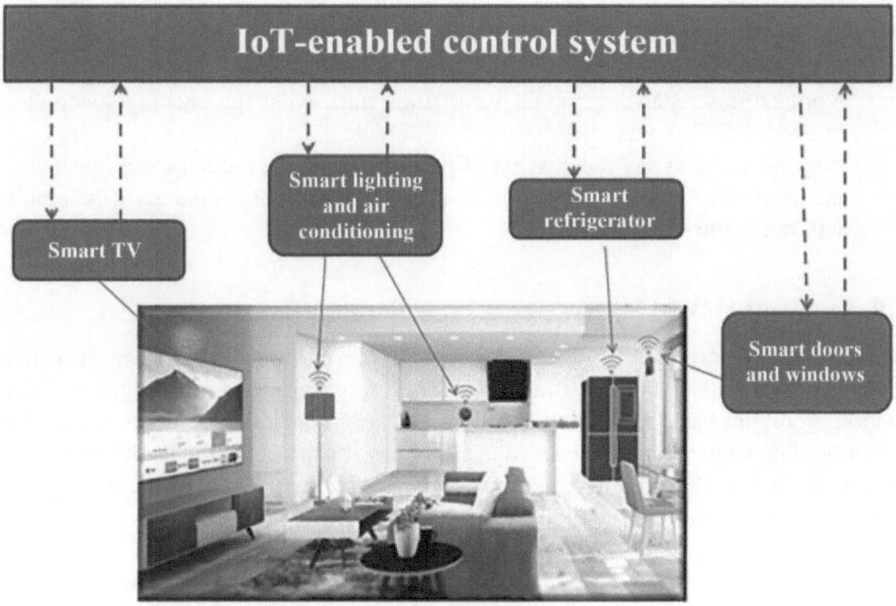

FIGURE 1.6 IoT-enabled smart housing services in Society 5.0.

One may also see their calendar on the door, with a nice real-time reminder. They can check their to-do list before leaving the house. Also, one may see some suggestions based on the weather outside, such as "It is rainy today, take your umbrella!". The door should not open automatically for stringers! So, user expects the door to recognize them, their relevant, and the housemaid. It should also send building owner a report about the strangers who tried to use it!

- Smart refrigerators will automatically send signals to the preselected online stores for the replenishment of food items such as fruits, vegetables, cheese, and milk.

1.4.2 Smart Buildings and Infrastructure

- Smart lights will brighten up when there are passersby and will get dim when no one is on the road.
- The smart parking system will help locate the vacant parking space in the public places. The technology-enabled sensors will also communicate the information about the time and duration of parking the vehicle in the public places such as shopping mall, office spaces, and airports to the smartphone of the driver and will also have portal to pay for the parking fee where applicable.
- The smart buildings will also have provision for charging stations where the electronic vehicles will be charged.
- Here, sensor will reflect signals to the mobile phone of a person looking for a parking lot for his vehicle in a particular area. Regular updation about the

free parking space and parking map will flash on the mobile phone of the needy. Thus, the driver will find and reach the ideal parking spot without wasting much time.

- Notifications regarding the elapse of time until what the parking space is booked will also reach the driver of the vehicle. Further, in case, some shopping store has to pay the charges of parking, the same message will flash at the designated store for payment, and when the payment is made the receipt will reach the vehicle owner too.

1.4.3 Smart Health Sector

A digital watch worn by its owner will assemble information about his health such as blood pressure, pulse, and heart rate and store it too. In case of any exigency, information on urgent basis will be shared with a nearby hospital. The doctors from the data stored in the user's smart watch can easily have an access to the health records of the patient without letting the patient undergo lengthy tests. Thus, medical treatment of the patient gets initiated without wasting much time (Figure 1.7).

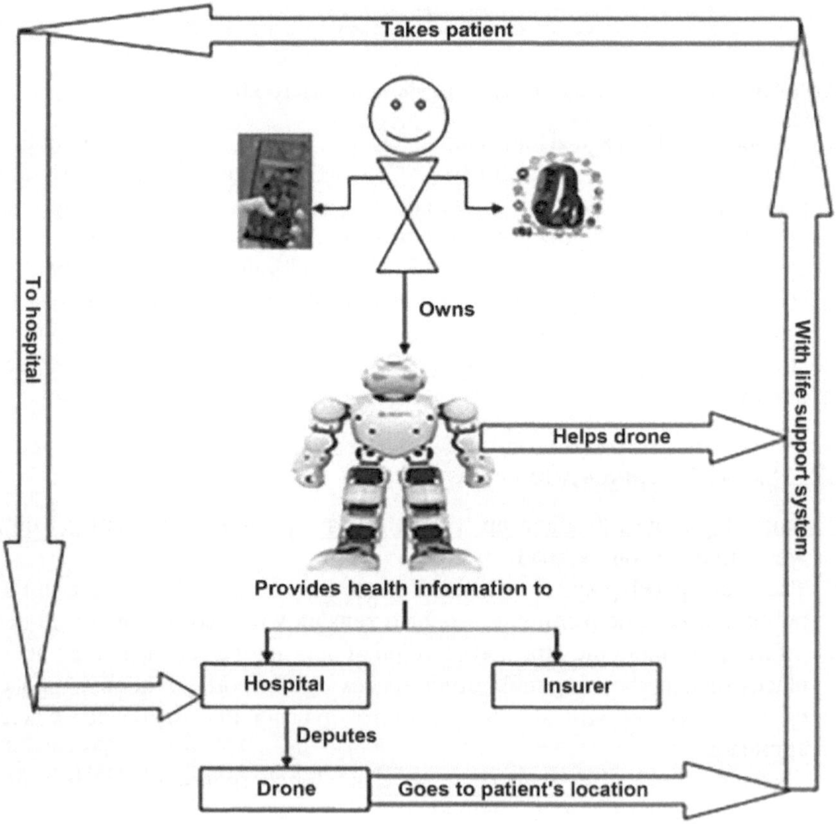

FIGURE 1.7 IoT helping in health services sector in Society 5.0.

- In yet another case, information about the patient can also be remitted to the insurance company from where the patient had purchased insurance. The insurance company then catches up the hospital records related to the tests of the patient. Finally, insurer processes this information and verifies it once again from the concerned hospital records. By the time, the patient recovers all his medical expenses get cleared by the insurance company and he reaches back to his place happily.

1.5 SMART TRANSPORT AND SMART VEHICLES

The advanced applications of information technology in Society 5.0 could prove to be the boon to the transportation sector be it private vehicle or the public one. The advent of IoT and advanced sensors integrated into vehicles will not only save time and effort but also be helpful in dealing the emergency situations, which in turn can save the human life. Some of the developments in this direction are discussed as follows:

1.5.1 Smart Transport

- Sensors embedded in vehicles in Society 5.0 will remit signal to the driver at designated time asking him if to switch on the ignition and let the vehicle warm up before driving.
- Such sensors will automatically detect the real-time updates of traffic flow and select the next best alternative route to reach the destination.

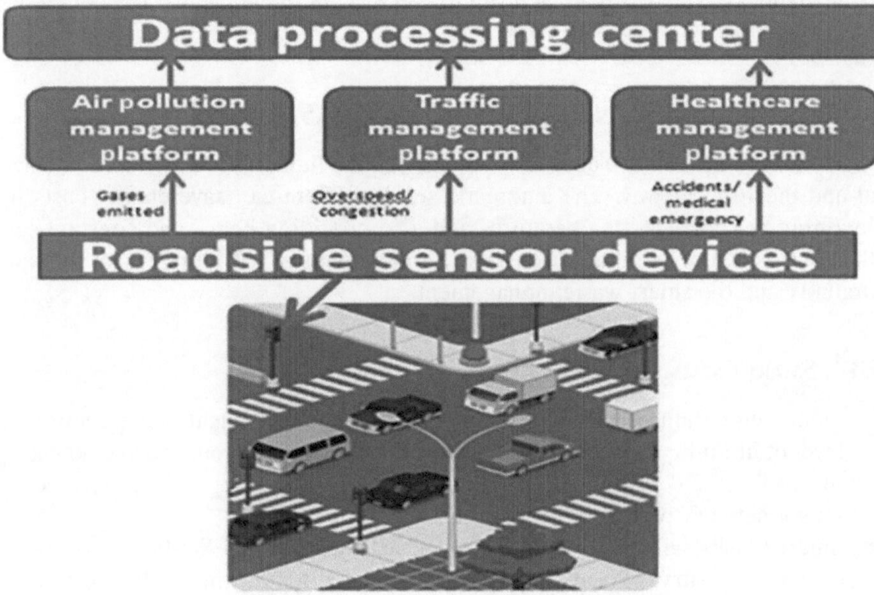

FIGURE 1.8 IoT-enabled smart transport management in Society 5.0.

- Vehicle-to-vehicle sensors will mechanically determine the distance to be maintained between two vehicles. Even the speed limit at different roads will be governed, and in case the vehicle is seen going beyond the permitted speed, automatic brakes will apply or slow it down. Thus, it will help in preventing accidents and promoting safety for the drivers.
- Smart traffic signals would not let everyone wait equally for the green signal. If there is less rush of vehicles, red signal will hold the vehicles for lesser duration of time.
- IoT-enabled devices will detect the misuse of lights, thereby ensuring lesser wastage of resources (Figure 1.8).

1.5.2 Public Vehicles

The technology-enabled public buses will have a unique ticketing system, by which there would be an alert system in the vehicle that would alert the passenger when the requisite stop comes. This would help illiterate persons and persons with reading problems or the passengers who may fall asleep to deboard the bus at their destination. The passenger seats will have the vibration system that would be used as an alert mechanism. The destination station can be sensed from the ticket or the pass that the person would be having.

- The automatic blood pressure machine and pulse detector installed with the seat of passenger would also keep a track of persons' health, and if anything goes beyond the prescribed limit, an alert will be triggered and the driver will be made aware of the medical emergency. This will also navigate through the nearest hospital and reroute the bus toward the medical facility.

1.6 SMART COMMUNITY AND SMART WASTE MANAGEMENT

The IoT-enabled devices are constantly impacting the human society in a more powerful and meaningful way. The automatic sensor system can save energy costs by maintaining the balanced temperatures and also providing better and constant surveillance. In this unit, we shall discuss the developments of IoT with respect to smart community and the smart waste management.

1.6.1 Smart Community Services

- Smart temperature control systems will monitor the outside temperature, level of humidity, and pressure of air in the environment outside and adjust the cooling and heating needs of the property accordingly. The wireless sensor network will ensure the proper ventilation as well.
- Smart kiosks will provide a variety of services such as Wi-Fi access, IP cameras for surveillance purposes, and some announcements. It will also keep the record of visitors. These will provide the information about the nearby public facilities such as clinics, restaurants, and grocery stores.

- The smart parking systems installed in societies would notify the residents if some guests come to visit them. Moreover, the record of visit would be kept and the visitor would be provided parking space in parking lot. This will help in implementing more security and safety of the residents.

1.6.2 Smart Water Management

- IoT-enabled water management systems will be helpful in monitoring the quality of water in the city. This system would also be responsible for keeping track of leakages in water pipes.
- IoT and connected devices will help in identifying the waste of factories in large water bodies such as rivers and seas. A separate mechanism will also be capable of identifying flood conditions in the river if the water level goes beyond the limit and an alert will be generated.
- The temperature of water will be controlled according to the weather. The cold water and heated water will be supplied in the swimming pool automatically by monitoring the outside temperature.

1.6.3 Smart Waste Management

- In the context of waste collection, the households will have to dispose off the waste at a nearby big bin kept in the pit. The big bins will be attached with the sensors. When a certain amount of waste is emptied into them, waste disposal companies will get a message and they will take it to a place where waste will be processed or so on. Thus, overflowing of bins thereby creating stinking conditions will get avoided. It will help to keep the city clean and air healthy. Apart from these evolutionary trends in smart housing, smart transport, and smart waste management, other associated trends that may facilitate the creation of smart cities contain the following.
- Regular monitoring of air quality will help to provide information about the air we breathe, and in case quality dips below a certain point, the source of pollution is also indicated. It will help to keep the environment clean and citizens healthy.
- The use of IoT in industry will help industry create intelligent equipment. Artificial intelligence will enable companies to ship the products to its consumers anticipating the need for replacement based on their past purchase record. The present "shopping-to-shipping" model of ordering goods and then shipping of the product by the concerned company will be replaced by the "shipping-to-shopping" model.
- Despite this, IoT-enabled devices can be helpful in maintaining the water level of soil so as to ensure the fulfillment of agricultural needs of the crops. The automatic water sprinklers can be programmed to judge the consistency of water into soil, and the sprinklers can automatically be turned on or off as and when the level of water goes below or above the designated limit for the crop. The user will only need to set the limit, and the rest of the thing will be done automatically.

- The geographical areas that experience dense snowfall during winters need to employ manpower to clear the roads and walkways. The IoT-enabled heating stations and the snow blowers can make the life easy by automatically performing this operation.
- In addition to this, the automatic lawn mower drones can be used to cut the grass in the residential and commercial areas during summers. These would also have the capability of clearing out the weeds from the grass, thereby ensuring the avoidance of allergies caused due to weed.
- The automatic garage would be able to open up when the car comes near to it. It would also have the capability to determine whether there is enough space to accommodate new car and will open accordingly. This can also be linked with car washers, which will be able to wash the vehicles.
- In businesses, asset management can be facilitated. If a business has reasonably high assets, inventory, machinery, shipments, fleet, etc., then keeping track of everything cannot be a small task. A better insight into these assets promotes efficiency, reduces expenses, and improves operations. IoT sensors can measure a variety of metrics in real time, supplying the businessman minute details so that he may make better business decisions, which further help to save on costs, and drive operational efficiency. Supply chain visibility lets the businessman know if assets are missing, deviated from a route, or they have arrived in time. Knowing the condition of the assets whether perishables are being stored at the right temperature or fragile cargo has been damaged in transit can save business money and keep the consumers happy. Monitoring asset usage helps increase efficiency by assessing whether there are too little or too many equipments or goods deployed in a specific location. Thus, the use of IoT in businesses helps improve efficiency and savings by maximizing utilization of assets and resources, and streamline operations and use predictive maintenance to reduce equipment failure.

Other vital areas of businesses where IoT can help include the following:

- Timely and quick delivery helps to serve the consumers and their retention also. It is done by the way of tracking and monitoring the shipments, optimizing the routes, identifying the possible choke points, and verifying delivery and cold chain integrity.
- Similarly, IoT helps to reduce the risk of theft and improve the security of goods and assets, and get real-time alerts for tampering, route deviations, cold chain breaches, or unnecessary/unexpected stops. The same also assists in tracking systems for cold chain, heat, and humidity monitoring, especially in food and pharmaceutical perishables to help prevent spoilage or contamination with sensors calibrated to the National Standards. It also encourages the efficient use of key tools and equipment in yards and warehouses. IoT supports anticipating the availability, needs, and potential equipment shortages based on usage trends, so that businessman may ensure that equipment operates at peak capacity, performance, and efficiency.

- If the business is using vehicles and the business owners want to know their condition and where they are, how they are being driven, and whether they are early or late, on time, or idle, IoT may be roped in that will supply incessant information on the same that can help business take significant decisions and thus facilitate lower costs. IoT will also make possible the improved safety and driver performance. Immediate in-vehicle feedback remittance means driver will remain careful of the over speeding, braking hard, or driving in other ways that increase risk and cost. By monitoring vehicle traffic, road surfaces, and weather, delays and hazards can be condoned as real-time information that will help in revising the routing or instructions.

1.7 CONCLUSION

From the above discussion, it may be inferred that smart infrastructure will allow more efficient management of the overall environment of Society 5.0 and buildings. More electric energy can be saved. The applications such as smart transportation and smart health sector will help in integration of various departments through gadgets in Society 5.0. The transportation, hospitals, and insurance companies will be able to work in close proximity to each other. The overall health of the society will also be managed by smart waste management system and water management system. All the applications discussed in the chapter have a direct or indirect link to the betterment of the society in general. The smart housing applications will be helpful in ensuring the safety and security of the citizens by keeping an eye on the unauthorized access to smart homes in Society 5.0. This may be helpful in reducing crimes and frauds. Moreover, these technological gadgets will also make the life of human being easy, thereby enhancing the lifestyle of humans.

REFERENCES

Alavi, A.H., Jiao, P., Buttlar, W.G. and Lajnef, N. (2018), "Internet of Things-enabled smart cities: state-of-the-art and future trends", *Measurement*, Vol. 129, pp. 589–606

Aleyadeh, S. and Taha, A.M. (2018), "An IoT-based architecture for waste management", *Proceedings of the IEEE International Conference on Communications Workshops*, Kansas City, MO, 20–24 May.

Behrendt, F. (2019), "Cycling the smart and sustainable city: analyzing EC policy documents on Internet of Things, mobility and transport, and smart cities", *Sustainability*, Vol. 11, No. 3, p. 763.

Bharadwaj, A.S., Rego, R. and Chowdhury, A. (2016), "IoT based solid waste management system: A conceptual approach with an architectural solution as a smart city application", *Proceedings of the IEEE Annual India Conference*, Bangalore, India, 16–18 December.

Eunil, P., Del Pobil, A.P. and Kwon, S.J. (2018), "The role of Internet of Things (IoT) in smart cities: technology roadmap-oriented approaches", *Sustainability*, Vol. 10, No. 5, p. 1388.

Heer, T., Garcia-Morchon, O., Hummen, R., Keoh, S.L., Kumar, S.S. and Wehrle, K. (2011), "Security challenges in the IP-based internet of things", *Wireless Personal Communication*, Vol. 61, pp. 527–542.

Jara, A.J., Zamora-Izquierdo, M.A. and Skarmeta, A.F. (2013, September), "Interconnection framework for mHealth and remote monitoring based on the Internet of Things", *IEEE Journal on Selected Areas in Communications*, Vol. 31, No. 9, pp. 47–65.

Khajenasiri, I., Estebsari, A., Verhelst, M. and Gielen, G. (2011), "A review on Internet of Things solutions for intelligent energy control in buildings for smart city applications", *Energy Procedia*, Vol. 111, pp. 770–779.

Kumar, S., Tiwari, P. and Zymbler, M. (2019), "Internet of Things is a revolutionary approach for future technology enhancement: a review", *Journal of Big Data,* Vol. 6, No. 1, pp. 1–21.

Murad, D.F. and Hidayanto, A.N. (2018), "IoT for development of smart public transportation system: a systematic literature review", *International Journal of Pure and Applied Mathematics*, Vol. 118, No. 18, pp. 3591–3604.

Paula, H.T.L., Gomes, J.B.A., Affonso, L.F.T., Rabêlo, R.A.L. and Rodrigues, J.J.P.C. (2019), "An IoT-based water monitoring system for smart buildings", *Proceedings of the SECSDN 2019 in Conjunction with IEEE ICC*, Shanghai, China, 20–24 May.

Vadillo, L., Martín-Ruiz, M.L., Pau, I., Conde, R. and Valero, M.Á (2017), "A smart telecare system at digital home: perceived usefulness, satisfaction, and expectations for healthcare professionals", *Journal of Sensors*, Vol. 2017, pp. 1–12.

2 Human-Feedback Adaptive Learning Using Interpretable and Interactive Intelligent Systems

Pawan Kumar and Manmohan Sharma

CONTENTS

DOI: 10.1201/9781003158165-2

2.1 INTRODUCTION

Humans and machines have their unique strengths, and collaboration between these two has the potential to improve machine learning (ML) systems further. For any such collaboration, human users must be able to interpret the behavior of ML systems. Moreover, human users should have a provision to give feedback to ML systems based on their domain knowledge. Consequently, human interpretability and the ability to interact with human experts are becoming crucial parameters, in addition to accuracy, for intelligent systems.

The solution of an ML problem involves the optimization of one or more internal metrics. The most popular among these internal metrics include accuracy, precision, recall, AUC values, and F1 score. Solution space of an ML problem may have multiple solutions that are equally good in terms of these internal optimization metrics. However, these solutions may differ in terms of their alignment with the user's perspective of the problem domain. Including human experts in this exploration of solution space of a problem has the potential to help to search a solution that not only satisfies threshold for internal metrics but also has improved agreement with the user's perspective. Moreover, the involvement of human experts has the potential to accelerate this exploration of solution space, therefore helping in terms of algorithmic complexity.

The goal of this chapter is to propose a framework for designing interpretable and interactive intelligent systems capable of human-feedback adaptive learning. The objective is to make underlying ML model adapt to human expert feedback in case there is a lack of agreement with the human domain experts regarding the perception of the problem domain. Additional goals of this chapter are as follows: (i) extracting principles and guidelines for the design of interpretable and interactive intelligent systems, (ii) identifying metrics for the evaluation of interpretable and interactive ML systems, and (iii) enabling human users with different expertise levels of the domain to interact with and provide feedback to ML systems. Establishing principles and guidelines for the design of interactive ML systems will help in standardization and building a consensus. A set of commonly agreed-upon metrics will help the evaluation of interactive ML systems. Having an interface that can capture feedback from human experts with different levels of expertise will help make human experiments economical and enable leveraging of masses in improving or verifying interactive ML systems.

The rest of the chapter consists of the following sections: Section 2.2 presents the research gaps in the existing work. Section 2.3 gives an algorithmic description of an interactive ML system along with its flowchart. Section 2.4 describes the proposed framework.

Section 2.5 gives details of the experimental setup. Section 2.6 compiles the results and observations from the experimental work. Section 2.7 proposes metrics that can be used for the evaluation of interactive ML systems. Section 2.8 describes novel improvements to improve the working of interactive ML systems in form of human-feedback adaptive learning. Section 2.9 summarizes conclusions and possible lines for future work.

2.2 RELATED WORK

ML has been a consistently talked about field during the recent years owing to scope of its application in complex problem domains. A few examples of its recent applications include advertisement classification of online newspapers using convolutional neural networks (Jain, Taneja & Taneja, 2021), recommender systems for life insurance (Rani et al., 2021a, b), and commercial mobile ad hoc networks (Taneja, Taneja & Kaur, 2021).

To take ML to the masses, end users must trust ML-based solutions. To facilitate this trust building, these users should be enabled to interpret the decision-making process of an ML model. During recent years, a lot of work has been done in making ML models interpretable. Interpretable ML is gaining attention due to its potential advantages such as facilitating winning trust, debugging model, and ensuring fairness and RTE (right-to-explanation) obligation (Ribeiro, Singh & Guestrin, 2016; Lipton, 2018; Doshi-Velez & Kim, 2017). The existing approaches for conferring interpretability to ML models can be categorized as model-specific and model-agnostic (Molnar, 2020). Model-agnostic approaches apply to any ML model and thus offer advantages such as flexibility in the choice of the underlying ML model. A few of the most prominent model-agnostic work in conferring interpretability to ML models include partial dependence plots (Friedman, 2001), individual conditional expectation plots (Goldstein, Kapelner, Bleich & Pitkin, 2015), feature interaction plots (Friedman & Popescu, 2008), feature importance plot (Breiman, 2001), global surrogate models (Molnar, 2020), local interpretable model-agnostic explanations (LIME) (Ribeiro et al., 2016), and Shapley explanations (Shapley, 1953).

Interactive ML: Humans and machines have a distinct set of capabilities that can complement each other via collaboration between the two. Humans are good at making decisions in never-seen-before situations based on their prior experiences. Machines are good at processing a large volume of data without getting tired. A collaboration of humans and machine has the potential to affect the learning process positively in terms of accelerating exploration of solution space. The idea is to have this search space exploration process guided via human interaction. Such collaboration has other potential advantages such as reducing engineering efforts, learning with a lesser number of human experts, or training data. Moreover, a collaboration of humans and machines is useful in reaching out to ML solutions that have better user agreement. It involves iterative interaction cycles between the two sides. As human users who are not ML experts are getting access to ML tools, it becomes imperative that these users understand the

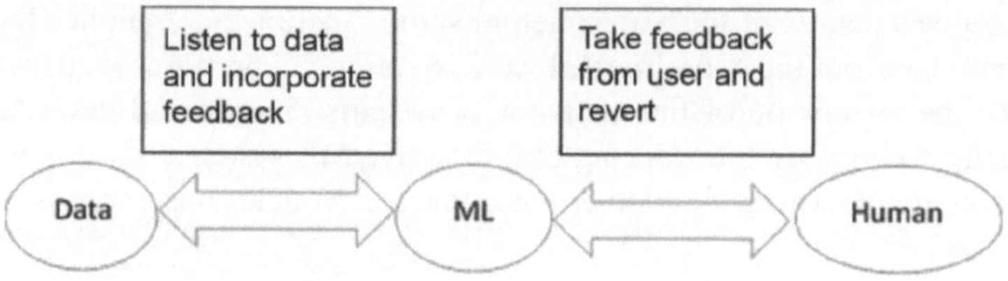

FIGURE 2.1 Human–machine interaction.

decision-making process of the model. Also, they should be enabled to provide feedback to the ML model based on their domain expertise. So, the ML community needs to continue improving ML systems in terms of interpretability and the ability to interact with human users (Figure 2.1).

A hybrid human–machine intelligence approach has been proposed (Yang, Kandogan, Li, Sen & Lasecki, 2019) where the model is refined to incorporate the feedback given by human experts. A user interface called "Ruleslearner" has been used to present the learned model as a set of rules in disjunctive normal form and collect feedback from users in the form of addition, deletion, modification, ranking, or filtering of rules.

The involvement of humans in the exploration of solution space can positively affect computationally hard problems (Holzinger, Plass, Katharina, Gloria & Cris, 2019). The ant colony optimization (ACO) algorithm was used for solving a traveling salesman problem (TSP). The key idea has been to increase the probability of selection of a path that is traversed by a human being, by artificial agents also.

An interactive ML model "iBCM" that enables a two-way communication (Kim, 2015) with human experts has been proposed for clustering. Human-to-ML model communication is in terms of prototypes and subspaces. ML model reverts through explanations after incorporating user feedback.

The democratization of ML by providing a bigger role to end users in the design of ML systems has been advocated (Amershi, Cakmak, Knox & Kulesza, 2014). Involving users in the learning process results in rapid and specific incremental updates in the ML model. This involvement brings the challenge of understanding capabilities, behavior, and needs of the end user.

Interdisciplinary expertise (Rosé, Mclaughlin, Liu & Koedinger, 2019) helps to develop explanatory learning models capable of providing interpretable and achievable insights. The interdisciplinary expertise spans AI/ML engineers, and cognitive, education, and UI/UX designers.

Research Gaps: The potential advantages associated with a human–machine collaboration include accelerating exploration of solution space in computationally hard problems, making ML economic via reducing the number of human users required, reducing engineering efforts, and facilitating trust in ML-based systems. The existing work has focused on providing a bigger role to human users in the ML-based systems by establishing bidirectional communication between the ML model and humans.

The associated future lines of work in the design of interactive and interpretable ML-based systems include (i) understanding capabilities and needs of each user, (ii) developing a common language across different domains, (iii) establishing principles and guidelines for the design of interactive ML systems, (iv) developing metrics for the evaluation of interactive ML systems, (v) leveraging human expertise in reducing computation cost of problems that involve the exploration of a large solution space, (vi) utilizing established practices in UI/UX design, and (vii) developing innovative and intuitive ideas to accommodate human expert feedback.

Using feature importance is an intuitive idea to collect feedback from human experts regarding prevailing domain knowledge. Existing work has made use of prototypes and subspaces as medium of taking feedback in a clustering problem. The proposed approach aim to utilize human perception to evaluate learning acquired by an ML model and make it adapt itself to incorporate human expert feedback.

2.3 ALGORITHMIC DESCRIPTION OF AN INTERACTIVE INTELLIGENT SYSTEM

This section describes workflow involved in an interpretable and interactive ML system in the form of an algorithm and flowchart. The workflow in an interactive ML system has been presented using Algorithm 1.

Algorithm 2.1: Interpretable and Interactive ML System with the Capability to Interact with Human Expert

Step 1. Build an accurate ML model learning from the provided dataset
Step 2. Present the ML model to the human expert for interpretation
Step 3. Collect the agreement between user's perspective and ML model
Step 4. Prompt the human user to provide feedback to the ML model
Step 5. IF (User feedback conflict with data)
    Report conflict in an interpretable manner
    ELSE
    Incorporate feedback and revert with a revised model
Step 6. IF (User is done with giving feedback)
    Go to Step 7
    ELSE go to Step 4
Step 7. Collect agreement between user and model
Step 8. Measure the statistical significance of the change in user agreement
    before (Step 3) and after (Step 6) interaction with human
Step 9. If (Change in the agreement is significant)
    Human interaction is fruitful
    ELSE
    Interaction is not fruitful in terms of increasing user agreement

First, an accurate ML model is developed using the provided dataset and an appropriate ML algorithm. The developed model may be a black-box model and a white-box model. The behavior of this ML model is explained to the human domain expert using appropriate interpretability techniques such as feature importance plot. After presenting the ML model to the humans, the system collects the level of agreement between the ML model and user's perspective of the problem domain. The Likert scale is one commonly used technique to collect user agreement. The choice of the scale can be kept flexible and be decided as per the requirement of the problem domain. A commonly used scale is 1–5, where the range starts from strongly disagree (represented by 1) and goes up to strongly agree (represented by 5). After collecting user agreement, the system prompts the human user to provide feedback, if any, to the ML model in an intuitive manner. A dedicated user interface is used to collect this feedback. This interface provides users with an option to play with existing model behavior such as modifying feature importance and modifying rules. As a next step, the system goes back to incorporate the feedback provided and revert with a refined model to the user. In case the user feedback is conflicting with what data are saying, the system reverts with a conflict and its explanation. Again, the system prompts the human user to provide feedback on the revised version of the model or resubmit feedback in case a conflict is reported. This iterative cycle of collecting and incorporating feedback from a human user goes until the human user is done with providing feedback. After the human user has no more feedback for the ML model, the system collects agreement between the human user and the ML model again. The difference in user agreement before and after human interaction is measured and checked for statistical significance using statistical tests such as two-sided Wilcoxon signed-rank test. If the interaction has resulted in an improvement in the user agreement, it is termed as a fruitful interaction else a non-fruitful interaction. Figure 2.2 shows the graphical representation of Algorithm 2.1 using a flowchart.

2.4 PROPOSED FRAMEWORK

The objective is to develop a framework that can present a ML model to a human expert in an interpretable manner. Additionally, it is able to collect feedback from human experts regarding their perception of the problem domain. The proposed framework must be capable of measuring agreement with the human user and validating the statistical significance of this agreement. Moreover, in case the agreement between ML model and human expert is lacking, it should be able to adapt itself to incorporate the human feedback. While attempting to incorporate the human feedback, the ML model should remain within acceptable limits of accuracy. Whenever there is a conflict between data and what the human says, the same should be reported in a human-interpretable manner. The acronyms used in the proposed framework are listed in Table 2.1.

The basic idea is to listen to the dataset, the model, and the human domain experts separately in terms of ranking of features. A higher degree of agreement between

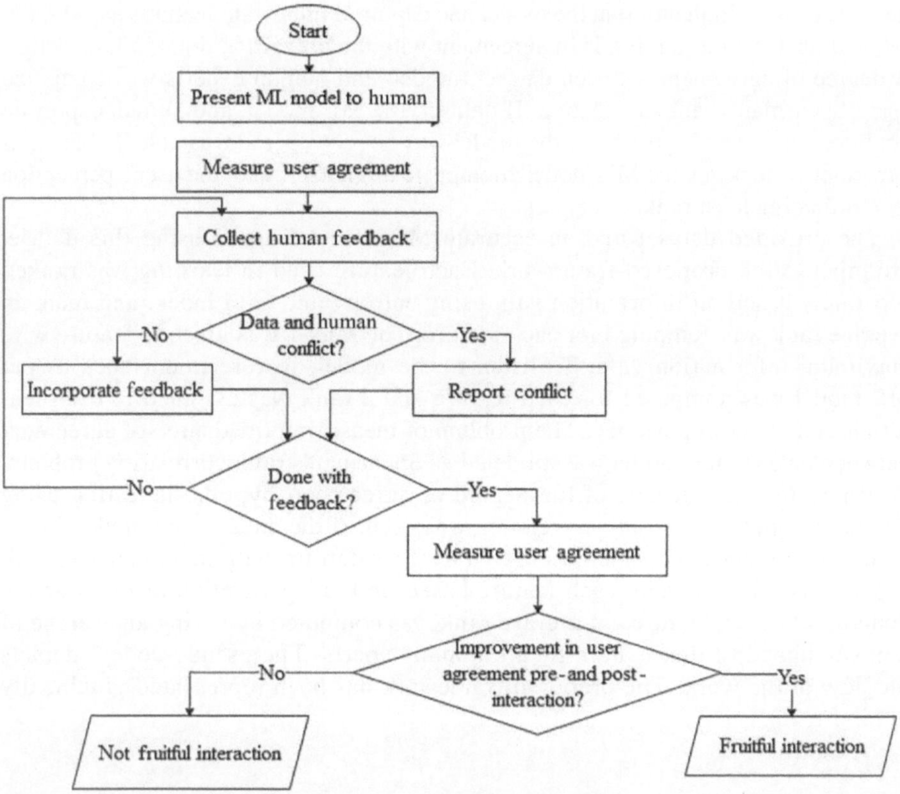

FIGURE 2.2 Graphical representation of an interactive ML system.

TABLE 2.1
Acronyms Used in the Proposed Framework

Acronym	Description
IG	Information gain
$IG_{\{entropy, f\}}$	Information gain for feature "f" using entropy
$IG_{\{gini, f\}}$	Information gain for feature "f" using Gini index
$R_{\{entropy, f\}}$	Rank assigned to the feature "f" using $IG_{\{entropy, f\}}$
$R_{\{gini, f\}}$	Rank assigned to the feature "f" using $IG_{\{gini, f\}}$
$R_{\{dataset, f\}}$	Average rank to a feature "f" using $IG_{\{entropy, f\}}$ and $IG_{\{gini, f\}}$
Imp_f	Importance measure for feature "f" as per model
$R_{\{Model, f\}}$	The rank assigned to a feature as per model
$R_{\{k, f\}}$	The rank assigned to a feature "f" by the *k*th human expert
$R\{_{Human, f\}}$	The average rank assigned to a feature by human experts

these three is an indicator that the model had captured important features as per dataset, and its decision-making is in agreement with the prevailing domain knowledge. A degree of agreement between dataset, model, and human experts was computed using Spearman's rank correlation. If dataset, the ML model, and human expert do not have a degree of agreement, the model was not considered trustable. In case this agreement is lacking, the ML model attempts to align itself with the user's perception by eliminating least ranked features.

The provided dataset and an accurate ML model learned using this dataset are input to the proposed framework. Each feature used in learning was ranked separately based on information gain using entropy and Gini index, and then, an average rank was computed for each feature. The rank 1 was given to feature with maximum information gain. To listen to the model, feature importance as per ML model was computed for each feature and a rank was assigned to each feature based on its importance. The problem of measuring the degree of agreement between dataset and model was modeled as Spearman's rank correlation problem. To verify the significance of the degree of agreement, hypothesis testing using one-tailed Spearman's rank correlation was used. If the dataset and model do not agree, the model is considered unreliable. To listen to the human experts, each expert was asked to rank each feature based on their perception of the problem domain. For each feature, an average rank was computed by taking an average of ranks assigned to that feature by all human experts. The pseudo-code 1 depicts the flow of the work. The proposed framework has been represented graphically in Figure 2.3.

Pseudo Code 1: Human-Feedback Adaptive Learning

FEATURES = The pool of features employed for learning
HUMANS = The pool of human experts
Input: The dataset used for learning; an ML model extracted using the features set FEATURES
Output: Interpreting model decision-making, verifying learning, facilitating agreement with user
Procedure

1. [Listening to what data says?]
   for each "f" in FEATURES
   Calculate $IG_{\{entropy, f\}}$, $IG_{\{gini, f\}}$
   end for
2. [Ranking of features as per data]
   for each "f" in FEATURES
      Assign $R_{\{entropy, f\}}$, using $IG_{\{entropy, f\}}$,
      Assign $R_{\{gini, f\}}$ using $IG_{\{gini, f\}}$,
   end for
3. Calculate $R_{\{dataset, f\}}$ as average of $R_{\{entropy, f\}}$ and $R_{\{gini, f\}}$
4. [Listening to what ML Model says?]
   for each "f" in FEATURES
      Calculate Imp_f
   end for

5. [Ranking of features as per ML Model]
   for each "f" in FEATURES
      Assign $Rank_{\{Model, f\}}$ using Imp_f
   end for
6. Measure and analyze the agreement between what dataset and ML Model
7. IF Dataset and Model do not agree
      Model is unreliable. Exit.
8. [Listening to Human expert]
   for each "k" in HUMANS and "f" in FEATURES
      Get $Rank_{(k, f)}$
   end for
9. Compute $Rank_{\{human, f\}}$ as average rank assigned
10. IF Dataset, Model and Human agree
      Model is trustworthy
   else
      Rebuild Model using HARF algorithm
11. IF Dataset, Model and Human agree
      IF Model is still accurate enough
         Model is trustworthy
      else
         Report as a conflict

end procedure

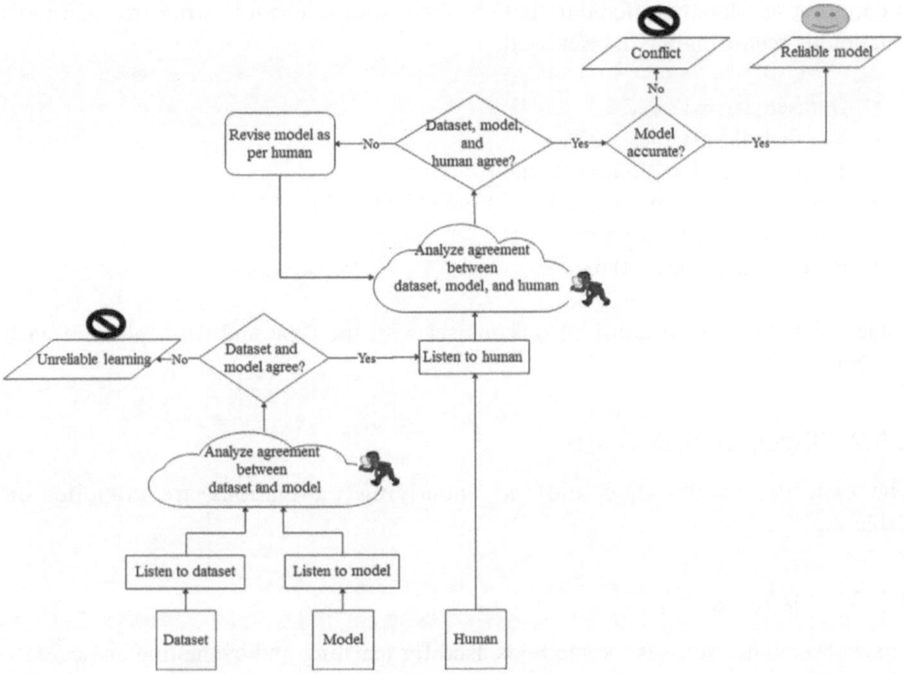

FIGURE 2.3 Human-feedback adaptive learning framework.

2.5 EXPERIMENTATION

This work is based on an intuitive idea that the underlying dataset, ML model learned using that dataset, and feedback from human domain experts are important pillars of an ML-based solution. Intuitively, in any ML-based solution, there should be an agreement between dataset, ML model, and human experts. The objective of this research work is to demonstrate a framework capable of verifying agreement between the dataset, the ML model, and human experts regarding the problem domain. The proposed framework aims to listen to each of these three pillars (dataset, model, and human expert) and measure the degree of agreement between these. By "listening to", the meaning is to get ranking for features in terms of importance as per the dataset, model, and human expert. In case of lack of agreement, the framework makes the ML model revise itself to align with the user's perspective.

2.5.1 Material (Dataset)

The dataset consisted of students enrolled at an institute of higher education in North India during admission year 2018 (Kumar, Kumar & Sobti, 2020). The objective was to use ML to foresee which of the enrolled students were actually likely to join the institute. The dataset consisted of information of enrolled students such as demographic details, academic details, status of availing different facilities of the institute, and kind of scholarship offered to them by the institute. The following are the important details regarding the dataset used:

- Dataset size = 13,125
- Number of variables = 25
- Target variable = "JoiningStatus"
- Classes of the target variable = {"Joined", "Lost"}
- Joined = 8374; Lost = 4751
- Baseline accuracy = 0.638

Table 2.2 shows the structure of our dataset with the type and description of each attribute.

2.5.2 Experiments Designed

The experiments designed and the underlying motivations are compiled in Table 2.3.

2.5.3 Methods

This subsection discusses the methods used for learning and evaluating an accurate ML-based model. It also discusses methods used for identifying dataset characteristics, interpreting the behavior of the learned model, and taking feedback from human experts.

TABLE 2.2
Demography of the Subjects

Attribute	Data Type	Description
RegistrationNumber	Int	8-digit unique ID for each student
AdmissionMonth	Int	Month of admission, e.g., 5 = May and 6 = June
Gender	Factor	Student gender: F – female; and M – male
State	Factor	Home state of student
HomeTownType	Factor	Rural, urban (metropolitan), urban (town)
BatchYear	Factor	Admission year, i.e., 2017 or 2018
ProgramName	Factor	BBA, MBA, B. Arch., etc.
Discipline	Factor	Agriculture, management, etc.
QualifyingExam	Factor	Eligibility qualification, e.g., 10, +2, and graduation
MarksPercent	Factor	Marks in the qualifying examination
CategoryCode	Factor	General, SC, ST, etc.
TransportAvailed	Factor	Transport facility of university availed? [Yes/No]
LoanLetter	Factor	Education loan availed? [Yes/No]
PreviouslyStudied	Factor	Whether the student studied earlier? [Yes/No]
HostelAvailed	Factor	Hostel availed or not? [Yes/No]
MessAvailed	Factor	Mess availed or not? [Yes/No]
ScholarshipPercentage	Numeric	Scholarship amount as a percentage of tuition fee
ScholarshipBracket	Factor	High, low, medium
EconomicCondition	Factor	Above average, average, below average, good
FeePaidPercentage	Numeric	Percentage of the fee paid so far by the student
MediumOfStudy	Factor	English, non-English
FeePaidCategorized	Factor	High, low, medium
MarksCategory	Factor	Excellent, fail, first division, second division, third division
HostelOrTransport	Factor	Hostel or transport availed? [Yes/No]
StudentStatus	Factor	Joined, Lost (did not join)

TABLE 2.3
Experiments Designed and the Underlying Objective

S. No.	Experiment Designed	Objective of the Experiment
1	Learning an accurate ML model	Exploring multiple ML algorithms to build an accurate model
2	Listening to the provided dataset, learned ML model, and human domain experts	Identifying important features separately as per the dataset, ML model, and human domain experts
3	Analyzing the agreement between dataset, model, and human expert	Analyzing the degree of agreement between dataset, model, and human expert in terms of ranking of features
4	Revising ML model to align with user's feedback (if required)	To align with the user's perception of the domain knowledge

2.5.4 Learning Algorithm and Evaluation Metrics

Logistic regression (LR), naive Bayes (NB), classification and regression trees (CART), and random forests (RF) were explored for learning. LR uses the logistic function to estimate the probabilities of each class of the target variable. NB uses the famous Bayes theorem to calculate the probability that a new instance belongs to a particular class. CART is a decision tree algorithm that can perform classification (Breiman, Friedman, Stone & Olshen, 1984). RF is an ensemble approach based on bagging technique and improves the accuracy of CART. Classification accuracy (acc), sensitivity (Se) or recall (R), specificity (Sp), precision (p), F1 score, and area under the ROC curve (AUC) values were used as performance metrics.

2.5.5 Listening to the Dataset by Identifying Important Features

For listening to data, information gain was used as measure of importance of a feature. Entropy and Gini index were used to compute information gain for each feature. Entropy is a measure of disorder or impurity and is used to measure information gain if a particular feature is selected for splitting while constructing trees (Hausser & Strimmer, 2014). Gini index is a measure of inequality in distribution (Handcock & Morris, 2006) and is always in the range of 0–1.

$$\text{Entropy}(E) = -\sum \big(p(x)\big)\log 2\big(p(x)\big) \tag{2.1}$$

$$\text{Gini index}=1-\sum \big(p(x) * p(x)\big) \tag{2.2}$$

where $p(x)$ represents probability of a class "x", $x = \{0, 1\}$

2.5.6 Listening to the ML Model by Interpreting Its Behavior

Listening to the model referred to identifying features important to the decision-making behavior of the ML model. Feature importance was used to interpret ML model behavior. It is a measure of the increase in the model's error rate when the values of a feature are permuted. A higher increase in error rate indicates that higher is the importance of the corresponding feature. The increase in error rate was calculated as 1-AUC, using "iml" package in *R* (Molnar, 2018). The variable importance measures were computed using the "randomForestExplainer" package in R (Zhao, Williams & Huang, 2017). The average rank assigned to a feature "f" as per model is computed by finding the average of ranks assigned to a feature using mmd, acc_d, gini_d, times_root, and f_{imp}.

2.5.7 Listening to Human Domain Experts

Listening to human experts referred to taking feedback from them as per their perspective of the problem domain. While taking feedback from human experts, outcomes of listening to data and interpreting ML model behavior were not shared with them to get unbiased feedback. No knowledge of ML was assumed on their part.

Human Domain Experts in This Problem Domain: Admission counselors working in the Division of Admissions were picked as human domain experts. Their job profile was handling queries of aspiring students in-person or telephonically. Based on their conversation with prospective students over the years, they were having a perception developed in their mind regarding the factors that affect the joining behavior of freshmen students. The objective of taking feedback from these human experts was to extract inputs from their experience-based perception.

Eligibility Condition: Minimum 5 years of work experience and recommendation from their project head were set as eligibility criteria for selection as a human expert. The project head was the reporting manager of these admission counselors.

The Process Followed for Taking Feedback: To ensure that their feedback was based on their experience-based perception only, the outcomes from dataset measures and understanding of the model behavior were not shared with the identified human experts. Each expert was asked to give feedback on each of the features used for developing ML-based model. For each feature, an expert had to answer whether that feature is important, to mention the level of importance, and to assign a relative rank to each feature.

The Orientation of Human Experts: To ensure that the feedback from a human expert is based on their experience only, a training session was organized before taking feedback. Before taking feedback, it was ensured that all experts understand the feedback format and questions.

The average rank assigned to a feature "f" by "*n*" human experts is given by

$$R_{\{\text{Human, f}\}} = \left(\sum_{k=1}^{n} \text{R}(k, \text{f})\right)\Big/ n \tag{2.3}$$

2.5.8 Statistical Tests and Hypothesis Testing

The problem of measuring agreement between the dataset, ML model, and human expert was modeled as Spearman's rank correlation problem. In a rank correlation problem, there are two judges and "n" participants. Each of the judges is asked to rank each participant. The objective of the test is to check whether the two judges are in sync in terms of their rating of the participants. The value of the correlation coefficient (ρ) varies from −1 to +1. Each feature was considered as equivalent to a participant. Dataset, model, and human expert were considered as judges. A higher value of correlation coefficient indicates higher agreement between dataset, ML model, and human perception. Spearman's rank correlation coefficient is computed using the equation

$$\rho = 1 - \frac{6\sum d^2}{n(n^2 - 1)} \tag{2.4}$$

where "n" = number of features used for learning and d = paired differences in ranks.

Hypothesis testing using one-tailed test was used to verify the statistical significance of the degree of agreement. The following three hypotheses were framed:

Hypothesis Test 1: Agreement between Dataset and Model

Null Hypothesis (H_0): Ranking of features by dataset and model is independent ($\rho = 0$)
Alternate Hypothesis (H_1): Ranking of features by dataset and model has a positive association ($\rho > 0$)

Hypothesis Test 2: Agreement between Model and Human Experts

Null Hypothesis (H_0): Ranking of features by model and human experts is independent ($\rho = 0$)
Alternate Hypothesis (H_1): Ranking of features by model and human experts has a positive association ($\rho > 0$)

Hypothesis Test 3: Agreement between Dataset and Human Experts

Null Hypothesis (H_0): Ranking of features by data and human experts is independent ($\rho = 0$)
Alternate Hypothesis (H_1): Ranking of features by data and human experts has a positive association ($\rho > 0$)

2.6 RESULTS AND DISCUSSION

This section presents the observations from the experimental work and the related discussion.

2.6.1 Evaluating ML Model Performance

Table 2.4 mentions the performance evaluation metrics computed from the confusion matrix.

It was observed that all the learning algorithms gave around 78%–80% accuracy, a considerably good improvement from baseline accuracy of 63.8%. All algorithms competed well in terms of sensitivity, specificity, precision, F1 score, and AUC values. RF model has given the best classification accuracy. All models generalized well from training to test data. RF model was taken as input for the next stages of the experimental work.

TABLE 2.4
Performance Evaluation Metrics of ML Models

Model	Acc		Se		Sp		Precision		F1 Score		AUC	
	Trng	Test	Trng	Test	Trng	Test	Trng	Test	Trng	Test	Trng	Test
LR	0.818	0.81	0.654	0.633	0.911	0.91	0.806	0.8	0.722	0.707	0.862	0.861
NB	0.796	0.79	0.672	0.657	0.867	0.866	0.742	0.735	0.705	0.694	0.849	0.845
CART	0.802	0.789	0.66	0.627	0.883	0.88	0.762	0.748	0.707	0.682	0.793	0.78
RF	0.832	0.816	0.653	0.625	0.933	0.925	0.848	0.825	0.738	0.711	0.834	0.835

TABLE 2.5
Ranking of Features as Per Data

Feature	$IG_{\{entropy, f\}}$	$Rank_{\{entropy, f\}}$	$IG_{\{gini, f\}}$	$Rank_{\{gini, f\}}$	$Rank_{\{dataset, f\}}$
ScholarshipBracket	0.197	1	0.125	1	1
MarksCategory	0.116	2	0.067	2	2
HostelorTransport	0.1	3	0.062	3	3
FeePaidCategorized	0.091	4	0.052	4	4
LoanLetter	0.027	5	0.015	5	5
HomeTownType	0.022	6	0.014	6	6
PreviouslyStudied	0.005	7	0.003	7	7
QualifyingExam	0.002	8	0.001	8	8
AdmissionMonth	0.001	9	0.001	9	9

2.6.2 Listening to the Dataset in Terms of Dataset Characteristics

Table 2.5 presents the ranking of features using entropy and Gini index. The features are listed in descending order of $IG_{entropy}$. Moreover, each feature has been assigned a rank in descending order of $IG_{entropy}$ and IG_{gini}. The last column "$Rank_{dataset}$" is the average of the above two ranks. It is observed that the features giving maximum information gain are "ScholarshipBracket", "MarksCategory", "HostelorTransport", "FeePaidCategorized", and "LoanLetter". The rank assigned to each feature is the same using entropy and Gini index.

2.6.3 Listening to the Model through a Diagnosis of Its Decision-Making Behavior

Table 2.6 presents the importance given to a feature by the learned ML model. Top features to which model outcome is sensitive to include "ScholarshipBracket", "FeePaidCategorized", "HostelorTransport", and "MarksCategory".

TABLE 2.6
Feature Importance as Per ML Model

Feature	Imp_f	$R_{\{Model, f\}}$
ScholarshipBracket	1.634198	1
MarksCategory	1.104211	4
HostelorTransport	1.111017	3
FeePaidCategorized	1.135687	2
LoanLetter	1.029349	8
HomeTownType	1.033603	7
PreviouslyStudied	1.04211	6
QualifyingExam	1.093152	5
AdmissionMonth	1.634198	1

2.6.4 Feedback from Human Domain Experts

Table 2.7 compiles the feedback from each human expert against each feature. Each human expert (E) assigned a unique rank from 1 to 9 to the nine features used in learning. The feature that is perceived as affecting joining behavior the most was assigned rank 1. Successive ranks were assigned in the same order. The average rank assigned to each feature was computed.

2.6.5 Agreement between Dataset, Model, and Human Experts in Terms of Ranking of Features

After collecting rankings of features as per the three judges (dataset, model, and human experts), the degree of agreement between these three in terms of ranking of features was computed and verified for statistical significance. Table 2.8 presents ranking of features in terms of their importance as per data, ML model, and human expert feedback.

TABLE 2.7
Ranking of Features as per Human Experts

Attribute	E1	E2	E3	E4	E5	E6	E7	E8	E9	E10	Average Rank ($R_{\{Human, f\}}$)
Scholarship bracket	1	1	1	1	1	1	1	1	1	1	1.0
Marks category	2	3	2	2	2	4	2	6	2	2	2.7
Hostel or transport	6	6	7	8	6	9	3	7	4	4	6.0
Fee paid categorized	7	5	9	4	9	2	9	5	5	8	6.3
Loan letter	8	7	5	3	4	5	7	9	3	3	5.4
Home town type	9	9	8	6	7	3	8	3	9	6	6.8
Previously studied	5	8	6	5	5	8	4	4	6	7	5.8
Qualifying exam	3	2	3	7	8	7	5	8	7	9	5.9
Admission month	4	4	4	9	3	6	6	2	8	5	5.1

TABLE 2.8
Ranking of Features as Per the Dataset, Model, and Human Experts

Feature	$R_{\{dataset, f\}}$	$R_{\{model, f\}}$	$R_{\{human, f\}}$
ScholarshipBracket	1	1	1
MarksCategory	2	6	2.7
HostelorTransport	3	3	6
FeePaidCategorized	4	2	6.3
LoanLetter	5	9	5.4
HomeTownType	6	4	6.8
PreviouslyStudied	7	8	5.8
QualifyingExam	8	7	5.9
AdmissionMonth	9	5	5.1

TABLE 2.9
Degree of Agreement and Hypothesis Testing

Pair of Judges	Hypothesis	Test Statistic (TS)	Critical Value (CV)	Statistical Test Outcome
Dataset and model	Hypothesis Test 1	0.516667	0.6	H_0 is accepted
Model and human expert	Hypothesis Test 2	0.456333	0.6	H_0 is accepted
Dataset and human expert	Hypothesis Test 2	0.694667	0.6	H_0 is rejected

Table 2.9 compiles the value of Spearman's rank correlation coefficient to measure agreement between data, model, and human feedback, taking two at a time. Also, it compiles the outcomes of three hypothesis tests formulated to verify the significance of the observed agreement. Referring to the table of critical values for one-tailed Spearman's ranked correlation coefficient, the critical value for $n = 9$ (number of features used for learning) was 0.6 at 5% level of significance.

2.6.6 Observations

i. **Dataset and Model**: A positive value of correlation coefficient (test statistic) was observed between the ranking of features as per dataset and model; however, the magnitude of correlation was "moderate" only. As the test statistic was smaller than the critical value, the null hypothesis was accepted. It indicates that rank orders of features as per dataset and model are independent.
ii. **Model and Human Expert**: A positive value of correlation coefficient (test statistic) was observed between the ranking of features as per ML model and human experts. However, the magnitude of the correlation was "moderate" only. As the test statistic was smaller than the critical value, the null

hypothesis was accepted. It indicates that rank orders of features as per ML model and human experts were independent.

iii. **Dataset and Human Expert**: A strong positive value of correlation coefficient (test statistic) was observed between the ranking of features as per dataset and human experts. As the test statistic was greater than the critical value, the null hypothesis was rejected. It indicates that there was a statistically significant positive association between rankings of features by dataset and human experts.

iv. As two of three hypothesis tests resulted in the rejection of the alternate hypothesis, it was concluded that the dataset, model, and human experts were not in sync in terms of rank orders assigned to features.

2.6.7 Revising ML Model to Adapt to Human Expert Feedback

To achieve a positive degree of association between rank orders by dataset, model, and human experts, "HomeTownType", the least average ranked feature by human experts, was removed as a predictor. The ML model was rebuilt using all earlier features except "HomeTownType". Table 2.10 compiles the revised ranks assigned to features as per dataset, model, and human experts.

Table 2.11 compiles the value of Spearman's rank correlation coefficient taking two judges at a time. Also, it compiles the outcomes of three hypothesis tests

TABLE 2.10
Revised Ranks as Per Dataset, Model and Human Experts

Feature	$R_{\{dataset, f\}}$	$R_{\{model, f\}}$	$R_{\{human, f\}}$
ScholarshipBracket	1	1	1
MarksCategory	2	4	2.7
HostelorTransport	3	3	6
FeePaidCategorized	4	2	6.3
LoanLetter	5	8	5.4
PreviouslyStudied	6	7	5.8
QualifyingExam	7	6	5.9
AdmissionMonth	8	5	5.1

TABLE 2.11
Degree of Agreement and Hypothesis Testing after Adapting to Human Feedback

Pair of Judges	Hypothesis	Test Statistic (TS)	Critical Value (CV)	Statistical Test Outcome
Dataset and model	Hypothesis Test 1	0.766667	0.643	H_0 is rejected
Model and human expert	Hypothesis Test 2	0.688333	0.643	H_0 is rejected
Dataset and human expert	Hypothesis Test 3	0.795	0.643	H_0 is rejected

formulated in Section 2.6 to verify the statistical significance of the agreement observed. The value of "n" was revised to 8 as now total of eight features were used for learning revised ML model. Referring to the table of critical values for one-tailed Spearman's ranked correlation coefficient, the critical value for $n=8$ was 0.643 at 5% level of significance.

2.6.8 Observation(s)

i. A positive value of the correlation coefficient (test statistic) was observed for each pair of judges. The positive values indicated agreement between dataset measures, model behavior, and human perception.
ii. As the test statistic was greater than the critical value, the null hypothesis was rejected for each pair of judges. So, a statistically significant positive association between rankings of features by dataset, model, and human experts was observed. It was concluded that the model has been able to extract important features of the dataset into its learning and this learning is also in sync with the prevailing domain knowledge as per human domain experts.
iii. The accuracy of the revised model dropped from 0.832 to 0.82, which is a marginal decrease only. Any gain in facilitating trust in the model at such a minimal cost in accuracy is worth trading-off in many problem domains.

2.7 METRICS FOR EVALUATION OF INTERACTIVE ML SYSTEMS

Future ML systems shall be expected to be accurate, interpretable, and interactive. The field of evaluating ML models in terms of prediction accuracy is quite established. Accuracy, precision, recall, F1 score, and AUC values have established themselves as performance evaluation metrics of ML models. The evaluation of ML models in terms of interpretability is still evolving. Evaluation metrics for human interpretability that has been proposed include (i) size of the explanation, e.g., number of nodes in the tree, and (ii) rating of explanations by human users. The evaluation of ML models in terms of interact ability is still evolving, and there is a lack of established metrics. The following metrics are proposed for evaluating interactive ML systems:

a. **Improvement in User Agreement**: A crucial expectation from an interactive ML system is its capability to align itself with the user's perspective of the problem domain. So, an ML system should be capable of collecting agreement between ML model and human user in terms of their perception of the problem domain. Also, it should be capable of verifying whether the interaction with the human expert has been fruitful or not in terms of improvement in user agreement.
   i. **Collecting User Agreement**: Agreement of the ML model with the user's perspective of the domain knowledge can be collected using the following techniques:

Likert Scale: Each human expert is asked to rate the ML model in terms of its alignment with the perspective of that human expert about the problem domain using the Likert scale. The most commonly used value of the Likert scale is 1–5 representing range from "Strongly disagree" to "Strongly agree". Agreement of the ML model with the user's perspective is collected before and after human interaction.

Spearman's Rank Correlation Coefficient: In this approach, the problem of measuring agreement between the ML model and human expert is modeled as Spearman's rank correlation problem. The features used for learning are ranked in terms of importance both by the ML model and by the human expert. Spearman's rank correlation is computed between these two vectors of feature ranks. The value of this coefficient indicates the magnitude and type of agreement between ML model and human expert.

ii. **Measuring Improvement in User Agreement**: The improvement in user's agreement can be measured using the following alternatives:

Hypothesis Testing: To verify the statistical significance of the change in user agreement before and after interaction of the ML model with human users, hypothesis testing is used.

An example of the hypothesis has been framed below:

Null Hypothesis (H_0): There is no change in the user agreement level before and after interaction.

Alternate Hypothesis (H_1): There is a positive change in user agreement level after interaction.

The above hypothesis can be tested using statistical significance tests such as Wilcoxon's signed-rank test or Spearman's rank correlation.

Human Support (HS): It refers to the number of human users feeling that ML model agrees with their perspective. The number is counted before and after the interaction. A positive change can be taken as an improvement in user agreement after the ML model refined itself to accommodate feedback from human experts. The threshold for happiness can be kept at >4 or can be made available for fine-tuning depending on the requirement of the problem domain.

Happiness Average (HA): To keep HS independent of the number of human experts used, change in the average user agreement before and after the interaction is computed. The average is taken of Likert scale values collected.

b. **Accelerated Exploration**: Gain in terms of decrease in computational cost can be measured to check how much acceleration has been possible as a result of involving a human expert. Comparison is made with the computation cost without involving a human expert.

c. **Economical Learning**: In problem domains, having less training data or expensive availability of human experts for evaluating ML systems, iterative human interaction has the potential to make the learning process economical. The economic gain can be measured in terms of the decrease in the number of human experts required or the associated cost.

d. **UI/UX**: An interactive ML system must provide an interface through which human users can provide feedback to the system. The desired characteristics of this interface include (i) user-friendliness, (ii) provision of "Playground" or "What-if" analysis to foresee impact before submitting feedback to the system, and (iii) intuitiveness of the medium for taking feedback from human experts.
e. **Conflict Reporting**: Whenever human user feedback is different from what the underlying dataset says, the ML system must be capable of detecting and reporting this conflict. An intuitive rule to detect a conflict can be fixing a threshold in terms of prediction accuracy. If incorporating feedback from human user results in a decrease in accuracy below the agreed-upon predecided threshold value, it is reported as a conflict. This threshold can be decided after consultation with the human domain experts. Additionally, the system should be able to explain this conflict to the human expert so that human expert can interpret the conflict and resubmit the feedback.

2.8 OTHER NOVEL OPPORTUNITIES

This section describes the novel improvements having the potential to improve the state of the art in the field of interpretable and interactive ML systems. The idea is to improve the agreement of the ML system with the user's perspective by making the ML model adapt to human feedback without losing much on accuracy.

2.8.1 Human-Feedback Adaptive Random Forest (HARF) – Weighted Selection of Features

The selection of a feature subspace for growing decision trees is a key step in building RF models. The most commonly used approach for the selection of feature subspaces has been to randomly select a subset of features. However, this random selection can be replaced by weighted subspace selection where features are selected as per feature weightage provided as an input vector to the RF algorithm. This approach has been demonstrated to improve the classification performance of RF models in case of high-dimensional data (Xu, Huang, Williams, Wang & Ye, 2012; Zhao et al., 2017). In such a scenario, random selection may not be the appropriate choice as many features may not be having even decent predictive power.

On the same lines, this approach can be applied to the design of interpretable and interactive ML systems where feature importance is being used as an intuitive medium of collecting feedback from human experts. Using a dedicated user interface, the feature importance as per the user's perspective of the domain knowledge is collected as feedback to the ML model. The ML model then adapts itself to incorporate the feedback provided by the human expert. The key idea is to use feature importance as per user feedback while selecting a feature subspace instead of going for random selection. As a result, the refined version of the ML model will aim to align itself more with the user in terms of importance assigned to feature.

2.8.2 User Interface That Can Adapt to Human Capability

As different users from the same domain may not have the same level of expertise, there is a need to have a flexible way of collecting feedback from the user. For example, in case of feature importance being used as a medium for collecting user feedback, the following flexibility can help address the issue of different users possessing a different expertise in terms of knowledge of the domain under study:

i. Allowing the human user to play with (increasing or decreasing) the size (width) of the bars showing the importance of each feature.
ii. Showing feature importance as high, medium, or low and allowing the user to play with this grouping in terms of changing labels.
iii. Showing feature importance as binary values such as {Useful, Not Useful} and allowing the user to change this labeling as per his or her perception of the problem domain.

This approach attempts to match experts with varying domain expertise. It allows users to submit detailed or subjective feedback. Users can choose between giving feedback as a continuous value interval, discrete {High, Medium, Low}, or even binary {Useful, Not Useful}.

2.9 CONCLUSIONS AND FUTURE WORK

Accuracy, interpretability, and ability to interact with human users are going to be important characteristics and evaluation criteria for future ML systems. ML community needs to come up with variants of existing ML algorithms, which have established supremacy in terms of accuracy. These variants should have the ability to explain their decision-making process to their users and adapt themselves as per feedback from domain experts. There is a need to bring a consensus on principles, guidelines, and formal evaluation metrics regarding interpretable and interactive ML systems. These systems must take care of the fact that the human users whom these are going to interact with are not ML engineers and do not have the same expertise level about the problem domain.

Solution space of an ML problem generally has multiple solutions that have accuracy within acceptable limits but differ in terms of agreement with the human user's perception of the domain. Interaction of an ML system with human domain experts has the potential to help an ML system reach out solutions with the improved user agreement and acceptable accuracy. Any gain in terms of agreement with a human user, if feasible without resulting in an inaccurate model, is a good trade-off in many of the problem domains. Improved user agreement facilitates developing trust of human users in the ML system.

The provided dataset, the learned ML model, and human domain experts can be considered as the three pillars that can collaborate for the development of interpretable and interactive ML systems. The problem of measuring agreement between them can be modeled as a rank correlation problem. One-tailed hypothesis testing can be used to verify the significance of the improvement in the user agreement.

Future lines of work include (i) evaluating the proposed framework for human-feedback adaptive learning in other problem domains, (ii) experimental work to evaluate the other variants of the human-feedback adaptive algorithms, (iii) evaluating the efficacy of metrics for evaluation of interactive and interpretable systems, and (iv) designing user interface to enable interaction of human users with the ML model.

REFERENCES

Amershi, S., Cakmak, M., Knox, W. B., & Kulesza, T. (2014). Power to the People : Interactive Machine Learning. *Ai Magazine*, 35(4), 105–120.

Breiman, L. (2001). Random forests. *Machine Learning*, *45*(1), 5–32. Doi: 10.1023/A:1010933404324.

Breiman, L., Friedman, J., Stone, C. J., & Olshen, R. A. (1984). *Classification and Regression Trees*. Wadsworth Inc.

Doshi-Velez, F., & Kim, B. (2017). *Towards A Rigorous Science of Interpretable Machine Learning*. (Ml), 1–13. Retrieved from http://arxiv.org/abs/1702.08608.

Friedman, J. H. (2001). Greedy function approximation: a gradient boosting machine. *Annals of Statistics*, *29*(5), 1189–1232. Doi: 10.1214/aos/1013203451.

Friedman, J. H., & Popescu, B. E. (2008). Predictive learning via rule ensembles. *Annals of Applied Statistics*, *2*(3), 916–954. Doi: 10.1214/07-AOAS148.

Goldstein, A., Kapelner, A., Bleich, J., & Pitkin, E. (2015). Peeking inside the black box: visualizing statistical learning with plots of individual conditional expectation. *Journal of Computational and Graphical Statistics*, *24*(1), 44–65. Doi: 10.1080/10618600.2014.907095.

Handcock, M. S., & Morris, M. (2006). *Relative Distribution Methods in the Social Sciences*. Springer Science & Business Media.

Hausser, J., & Strimmer, K. (2014). Entropy: Estimation of Entropy, Mutual Information and Related Quantities. *Cran R*.

Holzinger, A., Plass, M., Katharina, M. K., Gloria, H., & Cris, C. (2019). Interactive machine learning : experimental evidence for the human in the algorithmic loop a case study on ant colony optimization. *Applied Intelligence*, *49*(7), 2401–2414.

Jain P., Taneja K., & Taneja H. (2021). Convolutional neural network based advertisement classification models for online English newspapers. *Turkish Journal of Computer and Mathematics Education (TURCOMAT)*, *12*(2), 1687–1698. Doi: 10.17762/turcomat.v12i2.1505.

Kim, B. (2015). *Interactive and Interpretable Machine Learning Models for Human Machine Collaboration*. Doctoral dissertation, Massachusetts Institute of Technology.

Kumar, P., Kumar, V., & Sobti, R. (2020). Predicting joining behavior of freshmen students using machine learning-a case study. *2020 International Conference on Computational Performance Evaluation, ComPE 2020*, 141–145. Doi: 10.1109/ComPE49325.2020.9200167.

Lipton, Z. C. (2018). The mythos of model interpretability. *Communications of the ACM*, *61*(10), 35–43. Doi: 10.1145/3233231.

Molnar, C. (2018). iml: an R package for interpretable machine learning. *Journal of Open Source Software*, *3*(26), 786. Doi: 10.21105/joss.00786.

Rani, A., Taneja, K., & Taneja, H. (2021a). Life insurance-based recommendation system for effective information computing. *International Journal of Information Retrieval Research (IJIRR)*, *11*(2), 1–14.

Rani, A., Taneja, K., & Taneja, H (2021b). Multi Criteria Decision Making (MCDM) based preference elicitation framework for life insurance recommendation system. *Turkish Journal of Computer and Mathematics Education (TURCOMAT)*, *12*(2), 1848–1858. Doi: 10.17762/turcomat.v12i2.1523.

Ribeiro, M. T., Singh, S., & Guestrin, C. (2016). "Why should I trust you?" Explaining the predictions of any classifier. *Proceedings of the ACM SIGKDD International Conference on Knowledge Discovery and Data Mining*, 13–17-August, 1135–1144. Doi: 10.1145/2939672.2939778.

Rosé, C. P., Mclaughlin, E. A., Liu, R., & Koedinger, K. R. (2019). Explanatory learner models: Why machine learning (alone) is not the answer. *British Journal of Educational Technology, 50*(6), 2943–2958. Doi: 10.1111/bjet.12858.

Taneja, K., Taneja, H., & Kaur, R. (2021). Evolutionary computation techniques for intelligent computing in commercial mobile adhoc networks. *International Journal of Next-Generation Computing, 12*(2), 209–217.

Xu, B., Huang, J. Z., Williams, G., Wang, Q., & Ye, Y. (2012). Classifying very high-dimensional data with random forests built from small subspaces. *International Journal of Data Warehousing and Mining, 8*(2), 44–63. Doi: 10.4018/jdwm.2012040103.

Yang, Y., Kandogan, E., Li, Y., Sen, P., & Lasecki, W. S. (2019). A study on interaction in human-in-the-loop machine learning for text analytics. *CEUR Workshop Proceedings, 2327.*

Zhao, H., Williams, G. J., & Huang, J. Z. (2017). Wsrf: an R package for classification with scalable weighted subspace random forests. *Journal of Statistical Software, 77*(1). Doi: 10.18637/jss.v077.i03.

3 Advertisement Detection

Image Processing and Deep Learning Approach for Effective Information Extraction from Online English Newspapers

Pooja Jain, Kavita Taneja, and Harmunish Taneja

CONTENTS

DOI: 10.1201/9781003158165-3

3.1 INTRODUCTION

Advertisements play a major role in our lives. Newspapers, with their massive reach, are one of the biggest platforms for advertisements. Many job aspirants rely on newspaper advertisements for their job search. Students may look for admission notices or information about coaching classes, etc., in newspaper advertisements. Interested contractors look for various tenders, bids, auctions, etc., in newspapers. Product sales and promotional advertisements attract many buyers. Apart from these advertisements, newspapers have political advertisements, public notices, lost and found, matrimonial ads, remembrance messages, etc., which are of general interest. For a timely action, one needs quick access to desired advertisements. Online newspapers have made it possible to access multiple newspapers anytime and anywhere on our mobile phones, laptops, and desktops without waiting for printed copies of different newspapers. Home-locked situations in pandemics have further boosted the trend of online newspapers many folds. But even online newspapers do not give automatic advertisement search options. Also, no search portal facilitates the searching of advertisements through various online newspapers based on keyword matching. As a result, one has to sequentially manually search multiple newspapers page by page to find a relevant advertisement. To mitigate this problem, an intelligent system is required, which can automatically retrieve advertisement images across a range of newspapers making them instantly accessible for a timely action and hence adding value toward Society 5.0.

Automatic advertisement detection from multiple newspapers is not only crucial for personalized advertisement search across a range of newspapers but it can also help companies and individuals to ensure that their advertisements appear in the newspapers for which they were charged. It is also helpful in removing advertisements before tracking articles. Online newspapers are mostly available in .pdf format, and detecting advertisements in these files is not straightforward. We humans can easily detect an advertisement image by just looking at it, but automatic advertisement detection using computer program is a challenging task. It is a typical image recognition and classification problem, which can be solved by first extracting all the images in the newspapers and then classifying those images into advertisements or non-advertisements. In this way, all the advertisement images can be extracted from the newspapers and can be utilized for different purposes. When clubbed with optical character recognition (OCR) technologies (Jain et al., 2021a), the text from these advertisement images can be retrieved and keyword-based advertisement search can be facilitated.

Advertisement detection problem can be broadly subdivided into (i) newspaper layout segmentation for extracting images and (ii) image classification into advertisements and non-advertisements.

3.1.1 Newspaper Layout Segmentation

To facilitate automatic advertisement detection in online newspapers, the first step is to separate all the images from the articles in the newspaper .pdf files. Advertisements in newspapers are usually well-bounded so that they stand out and are not missed by readers. To separate advertisement images, well-bounded regions are identified and finally these images are extracted as separate files. Using this method, not only all the advertisement images are extracted but some article-related images are also extracted. Article-related images (non-advertisements) are filtered out in the next step.

3.1.2 Image Classification into Advertisements and Non-advertisements

After the images from the newspapers have been extracted, these images are classified as advertisements or non-advertisements so that non-advertisements can be filtered out. By simply looking at images, humans can differentiate between an advertisement image and non-advertisement image. To do the same, a computer program needs to understand the visual features of both advertisements and non-advertisements and then classify the images based on the visual features present in them. There are several classification techniques available including K-nearest neighbor (K-NN) (Cover & Hart, 1967), decision trees (Murthy, 1998), naïve Bayes classifiers (Lewis, 1998), and support vector machines (SVM) (Cortes & Vapnik, 1995), but convolutional neural network (CNN) (LeCun et al., 2010; Krizhevsky et al., 2012; Gu et al., 2018) has contributed to several image classification breakthroughs (He et al., 2016) and is considered as the most popular choice for image classification and visual recognition tasks (Razavian et al., 2014).

This study presents a novel image processing-based technique to separate out well-bounded images from online English newspapers' .pdf files. These images are input to trained CNN model to separate advertisement and non-advertisement images. After filtering out non-advertisements, only advertisement images are left, which can further serve several different purposes. The major contributions of the proposed research include (i) novel image processing technique for extracting images from different online English newspapers, (ii) image dataset compiled from different online English newspapers using the proposed technique, and (iii) evaluation of six different advertisement image classification models to classify advertisements and non-advertisements in online English newspapers.

The rest of the chapter is structured as follows: Literature review of "newspaper layout segmentation" and "advertisement image classification" is presented in Section 3.2. Section 3.3 introduces the image processing technique for separating the images from newspapers' .pdf files along with the extraction results obtained. Section 3.4 presents and compares six classification models for separating advertisement and non-advertisement images. This study is concluded in the conclusion section, and future work is also discussed in this section.

3.2 RELATED LITERATURE

3.2.1 Review of Newspaper Layout Segmentation

Layout segmentation is used to divide a document image into various regions such as headings, columns, paragraphs, drawings, images, and tables. Kaur and Jindal (2017) presented a survey of various newspaper layout segmentation techniques.

Gatos et al. (1999) used a rule-based approach to identify the image components from the newspaper image including lines, title blocks, drawing, and images and identified articles in the newspaper using rules.

Liu et al. (2001) proposed a bottom-up approach for newspaper layout segmentation. First, connected components are identified, and then, line, text, or graph components are separated as basic components. The heuristic rule is applied to merge these basic components based on the attributes of the components.

Mitchell and Yan proposed a number of bottom-up approaches (Mitchell & Yan, 2001, 2004a, b) for segmentation of newspaper images into components. Adjacent rectangular regions formed the patterns of these components.

Furmaniak (2007) proposed a text similarity-based newspaper page segmentation approach to segment out complete articles by finding out the blocks belonging to the same article using OCR and geometric layout rules.

Chaudhury et al. (2009) presented a rule-based approach for article segmentation used in "Google Newspaper Search program" using block segmentation, headline detection, and rule-based classifier.

A logical segmentation method was presented by Palfray et al. (2012) using a conditional random field (CRF) model and a set of grouping rules to segment articles from digitized old newspapers. Here, a generic layout model is assumed and a set of simple rules is used to segment out newspaper articles.

Antonacopoulos et al. (2013) evaluated different methods for layout analysis of scanned historical newspapers. The ability of the methods to correctly segment regions is evaluated along with the whole pipeline of segmentation with a purpose of evaluating region classification for text extraction. All the evaluated methods used bottom-up approach and gave a similar performance.

To learn the structure and layout of newspaper documents, a machine learning framework was proposed by Bansal et al. (2014), which could extract the complete articles from the newspaper images by learning relationship between neighboring blocks using the fixed-point model proposed in Ref. (Li et al., 2013).

Chu and Chang (2016) presented a segmentation approach for detecting the regions containing advertisements in the front pages of "China Times" (scanned images). First, connected components are detected, and then, rules such as placement and/or size of image in the newspaper are used to identify advertisements.

Meier et al. (2017) presented a fully convolutional neural network (FCN)-based approach to separate out articles from newspapers. Their approach segments the newspaper page into different regions but does not identify the type of element in each segment.

Almutairi and Almashan (2019) proposed Mask R-CNN-based approach for segmenting the newspaper page into its main elements. The results show that articles and newspaper headers of different newspapers in different languages are successfully segmented. However, the results do not take into account the small textual advertisements. Also, once a region is identified as advertisement region, the proposed technique does not separate different advertisements, which are very close to each other.

This review of newspaper layout segmentation techniques reveals that the complete article extraction is the focus of the research and mostly rule-based approaches are used to segment out the main elements of the newspaper page. But rule-based approaches do not work well for extracting advertisements from different newspapers because the rules for placement and permissible sizes of advertisements may differ. Moreover, the rules may change with time and according to the need of customers. Hence, using predefined rules does not serve the purpose of advertisement extraction across a range of newspapers. Also, most of the approaches worked with one type of newspapers and assumed a generic layout of the newspaper page. But different newspapers have different layouts, and previous approaches did not account for the diverse styles and layouts of various newspapers. This chapter presents an image processing-based technique to extract images from newspapers' .pdf files of various different newspapers without using any rules and without making any assumptions about newspapers' layout. These images are later classified as advertisements and non-advertisement images.

3.2.2 Review of Advertisement Image Classification

Peleato et al. (2000) used textual content of newspaper advertisements to classify them into various categories (vehicles, employment, real estate, and others) using naïve Bayes classifiers.

Duan et al. (2006) used multimodal analysis for segmenting, categorizing, and identifying commercials in TV streams. Spatial–temporal features are used for detecting the boundaries of commercials, and speech, text, edge, and color features are used to classify commercials into multiple categories with SVM classifier. Another work on commercial detection in digital videos is presented by Zhang et al. (2007). Audio and video cuts are used to segment out the video shots, and these video shots are classified as commercial shots or program shots using SVM classifier. This system uses spatial–temporal content in commercial shots for accurate classification. But newspaper advertisements do not have spatial–temporal content in them.

Li et al. (2007) used text, link, visual content, and visual layout of Web pages to detect the advertisement images on the Web pages using AdaBoost (Freund & Schapire, 1997) and SVM classifiers. Gong and Zhu (2010) proposed a neural network-based classifier to classify contextual ads on the Web pages from non-contextual ones using features such as text, link (hyperlink and URL), layout, and style (cascading style sheets, etc.). Features such as link, layout, and style are specific to Web pages and are not relevant to newspaper domain.

Ouji et al. (2011) presented a color segmentation approach with K-NN and AdaBoost classifiers for detecting advertisements in the digitized press documents including variety of magazines and newspaper pages. The approach suffers from very average precision values and high confusion rate between advertisements and Text_and_Graphic classes.

Jung et al. (2012) presented an advertisement classification technique that effectively expands its vocabulary by learning ad-specific features, and useful semantic associations in the "terms" (words) present in the textual ads. Again, the limitation of this work is that it relies on the advertisement's textual content for classification. Also, graphical advertisements, which usually have very less text, are not considered here.

Chu and Chang (2016) classified advertisements present in website snapshots and newspaper images using both semantic (text) and visual features. VIREO374 package is used to extract semantic features used for classification. Visual features are extracted using CNN, and finally, SVM is used for the classification of images into various advertisement classes including education, real estate, wholesale, and manufacturing.

Banerjee (2017) used a machine learning approach to classify political advertisements from a dataset of online advertisements. Here, textual features and non-textual features including color and facial features are used to identify political advertisements.

Vo et al. (2017) presented an image classification model (nLmF-CNN) to identify online advertisements in website snapshots. This CNN-based model is used to classify advertisements as displayed clearly or not. Trained on online website snapshot dataset, the model finds the suitable value of hyper-parameters including number of convolutional layers (n) and number of filters in each convolutional layer (m) through experiments. Finding suitable CNN architecture and hyper-parameters for classifying advertisements and non-advertisement images in online English newspapers remains an open challenge.

Almgren et al. (2018) used CNN-based approach to classify advertisements from articles from a dataset of images collected from various magazines on the basis of visual features only. This work is limited to the classification of advertisement blocks from the article blocks. Separation of individual advertisements (in advertisement block) is not considered here.

From the review of advertisement image classification, it is evident that most of the classification tasks use textual features for classification, but online newspaper advertisements are available in image format and not in text format. Hence, their textual content is not directly available. Moreover, newspapers also contain graphical advertisements, which usually have very less text in them making visual features more important here rather than textual features. This study presents a novel CNN-based approach to classify advertisement and non-advertisement images using visual features only.

3.3 IMAGE EXTRACTION FROM ONLINE ENGLISH NEWSPAPERS

3.3.1 Image Extraction Technique

This study uses a novel image extraction technique based on "adaptive thresholding" and finding "connected components" for separating out images from the newspapers' .pdf files. This technique does not make any assumption on newspaper layout and does not make use of any rule on advertisement size or placement. The independence from newspaper layout and advertisement size and placement rules makes this technique capable of extracting images from various different newspapers without any difficulty. This technique first identifies the well-bounded regions. Then, bounding rectangles are drawn around these regions, and finally, images are extracted by saving these regions as separate image files.

Figure 3.1 presents the flowchart for adaptive thresholding and connected component-based image processing technique. The main steps of the technique are explained as follows:

(i) **Conversion of All the Pages of a Newspaper .pdf File into Separate .jpeg Files**: Newspapers have multiple pages, and their .pdf file is also multipage with one page in the pdf file representing one newspaper page. Every page of the newspaper .pdf file is converted into a separate .jpeg file.
(ii) **Conversion of Each page of a .jpeg File into Grayscale:** Each page of the .jpeg file is converted into grayscale irrespective of whether it is black and white or colored or a mixture of both.
(iii) **Find Edges by Applying "Adaptive Thresholding"**: To find well-bounded regions, edges are identified in the image by applying thresholding. Different thresholding techniques can be applied here, but the best results are obtained by using "adaptive thresholding" (Roy et al., 2014) technique.
(iv) **Find "Connected Components"**: Similar pixel values form a "connected component" in an image. All such components are identified, and each such reasonably big region is given different intensity (color /hue) values to separate these regions from each other. Black color is given to image pixels, which are not a part of any connected component.
(v) **Apply Image Blurring and Image Erosion**: After obtaining the connected components, the next step is to blur the page image using Gaussian blur (Basu, 2002). Image erosion (Haralick et al., 1987) is performed thereafter. These two techniques help in obtaining better results when contours are extracted.
(vi) **Draw Rectangular Boundaries by Finding Contours**: Once the page image is blurred and erosion is performed, contours of each of the connected components are extracted. Contours are the boundary lines of an image that have the same intensity pixels. The shape of the newspaper advertisements is rectangular or square most of the time. Hence, rectangular boundaries are obtained from each contour.
(vii) **Saving Each Bounded Region as a Separate Image**: Each bounded region is saved as a separate image file (.png). Very small areas are discarded here as advertisements are always assumed to be big enough to be clearly visible.

3.3.2 Image Extraction Results and Discussion

Implemented in Python 3.7, the proposed image extraction technique uses OpenCV (https://pypi.org/project/opencv-python/) image processing library and it is tested on various different online English newspapers. Images from all the pages of newspaper .pdf files of different newspapers are successfully extracted. Figures 3.2–3.4 present some of the image extraction results obtained.

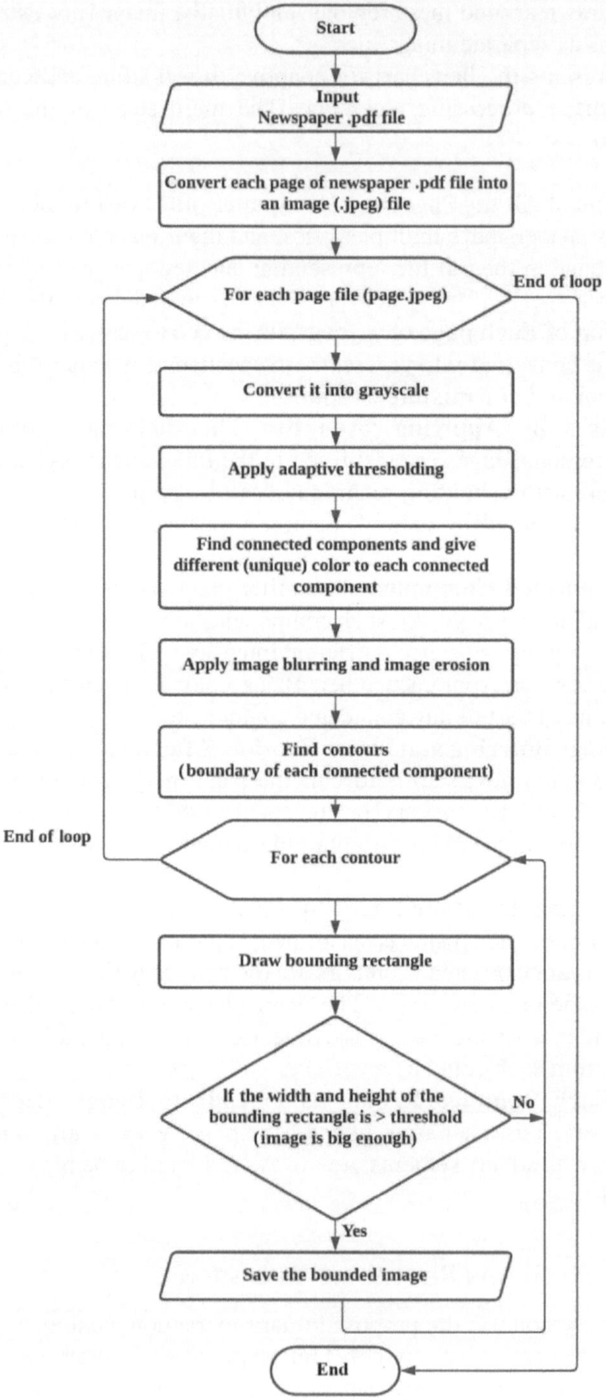

FIGURE 3.1 Image extraction technique to separate images from online newspapers.

Figure 3.2c shows that both advertisement and non-advertisement images are successfully extracted. Very small images whose size is below the threshold size are not extracted as they are not the probable candidate for advertisements.

In Figure 3.3, a half-page advertisement and other article-related images are extracted. It is clearly evident that the proposed technique has no difficulty in extracting images even as big as half-page.

Figure 3.4 shows that the proposed technique can easily extract even very close advertisement images. Here, five closely placed advertisements are successfully extracted along with other non-advertisement images.

3.4 ADVERTISEMENT IMAGE CLASSIFICATION

3.4.1 Using CNN for Advertisement Image Classification

For image classification, many supervised learning approaches such as K-NN, naïve Bayes, decision trees, and SVM are available. These classification techniques work well when the data are available in the form of features. But data may not be available in the form of features. Instead, data may be present in the form of videos, images, text, and speech. In such cases, another algorithm is required to first extract the useful features from the data and the classification technique is applied thereafter. Using two algorithms in this way leads to an optimization problem as instead of optimizing one algorithm; we need to optimize two totally unrelated algorithms. But CNN works in a different manner. It automatically extracts features from a large amount of data without requiring any other feature extraction technique or manual feature selection and extraction (Shaheen et al., 2016). CNN has an input layer, many hidden layers (convolutional and pooling layers), and an output layer. Hidden layers perform feature selection, and the output layer does the classification task on these features. In this way, when CNN is trained end-to-end, it performs both the tasks of finding good features and classification. This capability of CNN has revolutionized many fields including image classification (Krizhevsky et al., 2012). When it comes to the classification of images from online English newspapers into advertisements and non-advertisements, data are available in the form of images only without any knowledge of useful features for classification, making CNN the best choice for classifying images into advertisements and non-advertisements (Jain et al., 2021b).

3.4.1.1 Working of CNN

As shown in Figure 3.5, CNN is made up of many hidden layers, which include convolutional layers, pooling layers, flatten layer, and fully connected layers. Here, convolution layers are the most important layers responsible for feature selection. In convolutional layers, important low-level and high-level features are detected by applying various filters to the sample image. A new matrix called feature map is obtained as a result of filtering. This feature extraction process can be repeated several times based on the number of features that need to be extracted. Usually, the extracted features are subjected to rectified linear units (ReLU) (Nair & Hinton, 2010), which is a nonlinear activation function. Convolutional + ReLU layer is followed by a pooling layer. These convolutions + pooling blocks can be

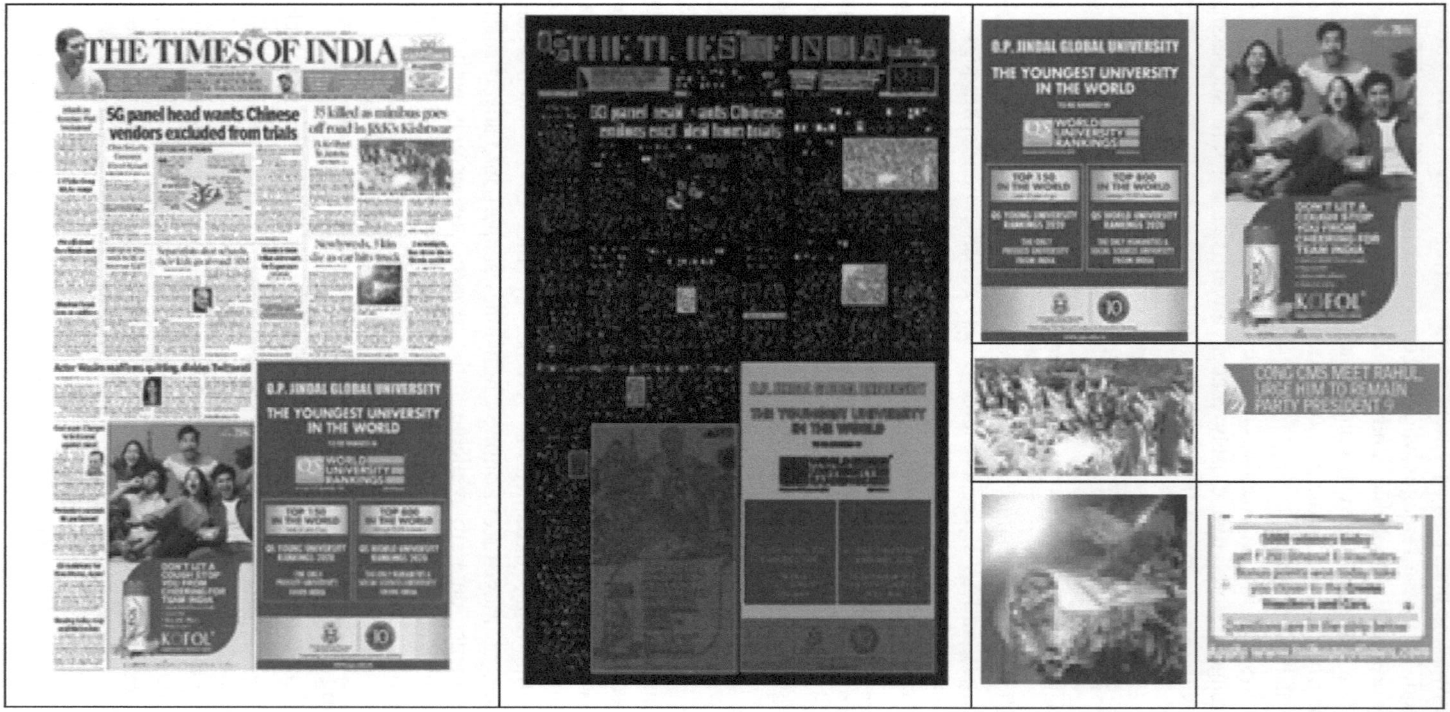

(a) (b) (c)

FIGURE 3.2 (a) Newspaper page (original .jpeg file). (b) Connected components are labeled with different colors (hue values). (c) Images (advertisements and non-advertisements) are extracted.

(a) (b) (c)

FIGURE 3.3 (a) Newspaper page (original .jpeg file). (b) Connected components labeled with different colors (hue values). (c) Advertisement image as big as half-page is extracted along with other images.

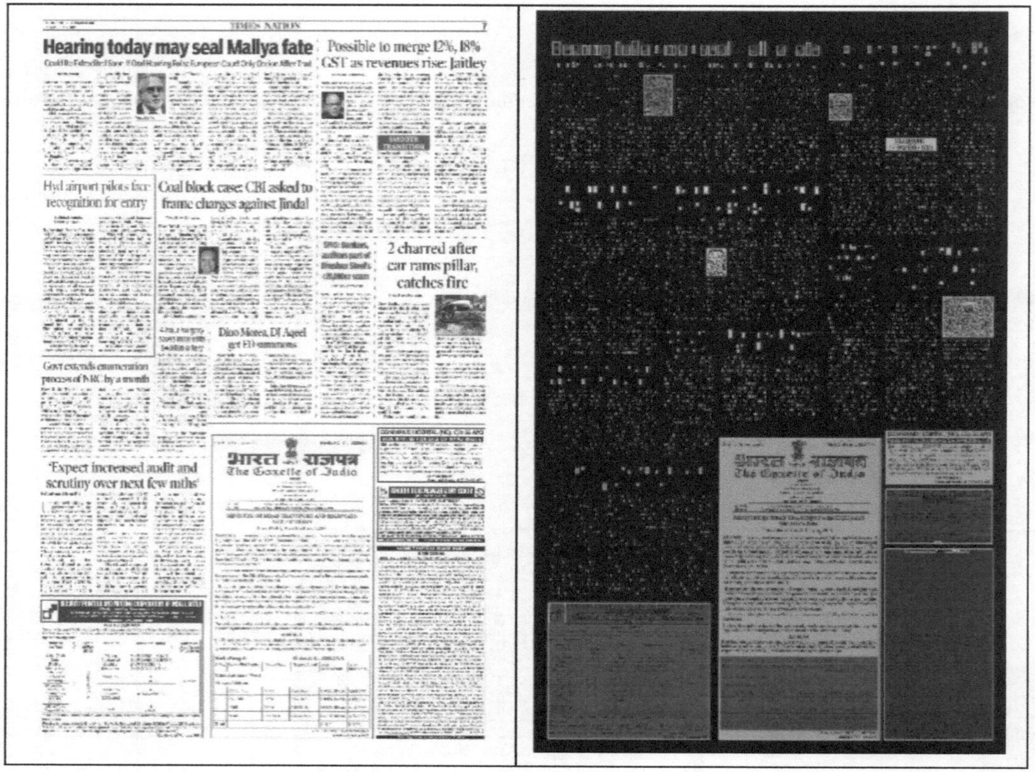

(a) (b)

FIGURE 3.4 (a) Newspaper page (original .jpeg file). (b) Connected components are labeled with different colors (hue values). (c) Image extraction results of (a) (job advertisement, government notification, tenders, and other non-advertisement images are extracted).

(Continued)

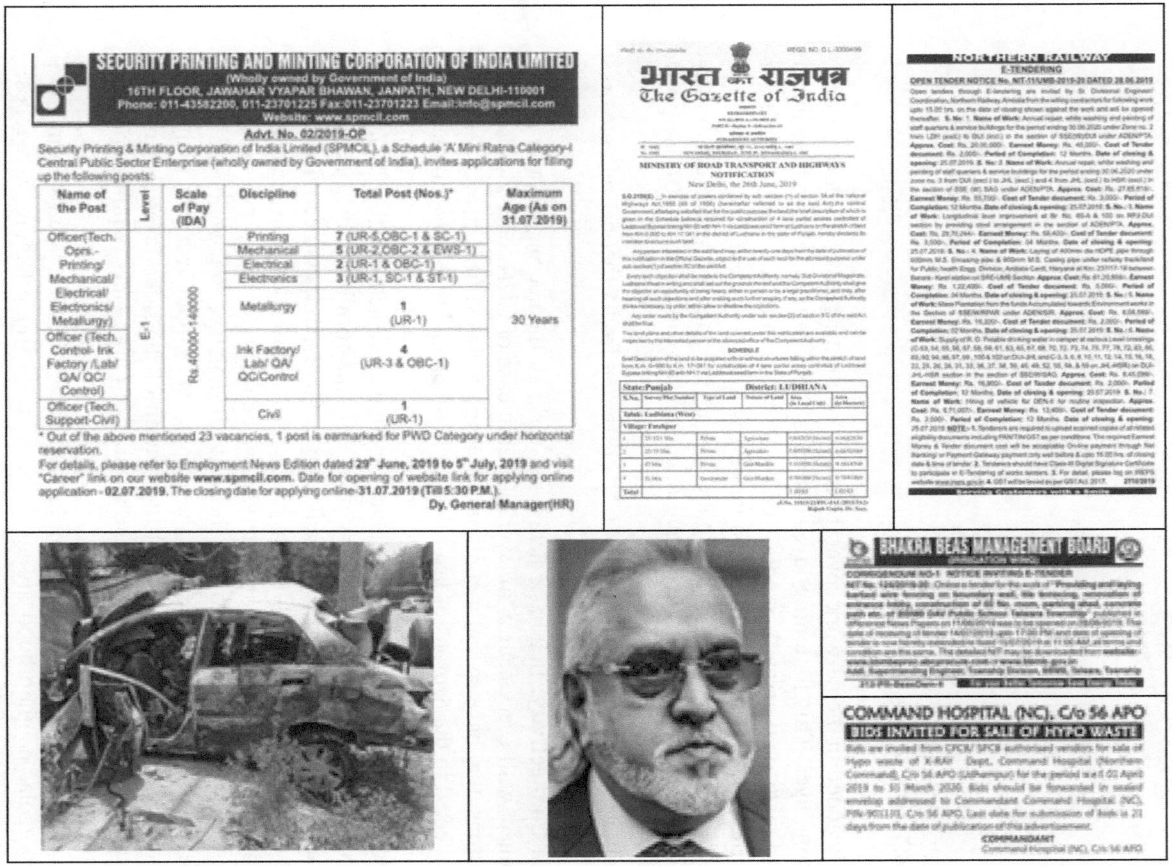

(c)

FIGURE 3.4 ***(CONTINUED)*** (a) Newspaper page (original .jpeg file). (b) Connected components are labeled with different colors (hue values). (c) Image extraction results of (a) (job advertisement, government notification, tenders, and other non-advertisement images are extracted).

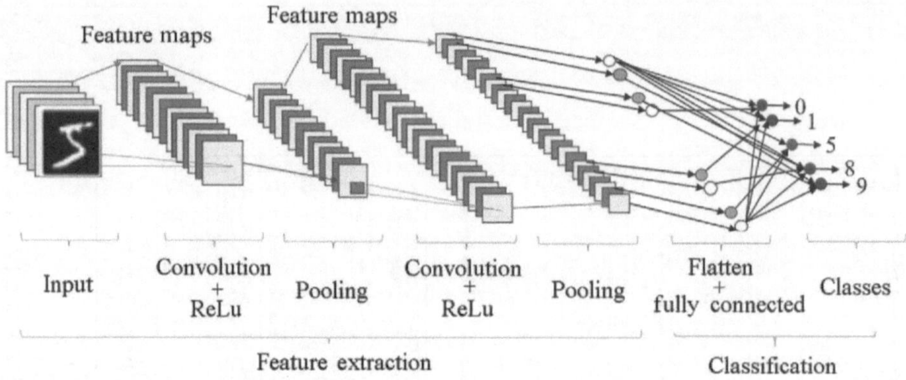

FIGURE 3.5 CNN block diagram.

repeatedly added in the network several times. A flattening layer is added after the last pooling layer. This flattening layer converts the resulting feature map matrix to a one-dimensional array. It is followed by a fully connected layer, which connects all the inputs of this one-dimensional array to neurons creating a multilayer perceptron (MLP) structure. Fully connected layers may be repeated multiple times. Finally, a classification technique (e.g., Softmax (Bridle, 1990)) is applied to obtain the output.

The following three properties make CNN different from other neural networks:

Sparse Connectivity: In a basic neural network, all the neurons in layer $m-1$ connect with every neuron in layer m. But in case of CNN, only a subset of neurons (that have spatially contiguous receptive fields) from layer $m-1$ are connected to the neurons of layer m. The logic behind the sparse connectivity is that there are strong correlations between local pixels as compared to long-distance pixels that have weak correlations, and hence, rather than preserving the relation between every pixel only the correlation between local pixels needs to be preserved. The advantage of this sparse connectivity is that the number of parameters for neurons decreases dramatically in each layer.

Weight Sharing: Inside neural networks, neurons of the previous layer $(m - 1)$ act as input to neurons in the next adjacent layer (m), which are considered as outputs. To determine the output, every input neuron is assigned its own weight. But in CNN, all the neuron of the same layer shares the same bias and weight reducing the number of parameters required for the whole network.

Pooling: In pooling (a down sampling technique), all the features from the rectangular neighborhood are aggregated into a single feature. It reduces the dimension of the network and avoids overfitting.

Aiming at dimensionality reduction, these three characteristics significantly reduce the computation time of training a CNN (Hasan et al., 2019). To further reduce the training time of deep neural networks, graphical processing units (GPUs) can be used.

3.4.2 English Newspaper Image Dataset

A standard dataset of advertisements from the English newspapers is not available, and therefore, the authors have compiled their own dataset of images (Figure 3.6) from a variety of online English newspapers that were available for free download at the time of data collection (May 2019 to September 2020) including "Hindustan Times", "The Tribune", "Times of India", and "Indian Express". Adaptive thresholding and connected component-based image extraction technique (presented in Section 3.3.1) are used to extract images from different online English newspapers. Non-advertisement images are manually separated from advertisements, and a balanced dataset of 5500 advertisements and 5500 non-advertisement images is created.

3.4.3 Transfer Learning

To automatically learn features from data, CNN usually requires a huge dataset for training, especially with high-dimensional input samples such as images. Training a CNN model using a small dataset may lead to overfitting where the model simply learns the training samples but does not classify the new samples correctly and gives low classification accuracies. Transfer learning (Pan & Yang, 2010; Torrey & Shavlik, 2010; Yosinski et al., 2014) is a simple technique for fitting a CNN model with a limited dataset and still achieves higher accuracies. In transfer learning, a pre-trained model (which is trained on a huge dataset) is used with the new dataset. The weights learned by the pre-trained model in performing "task A" are transferred to a new "task B". In a way, the knowledge gained by the pre-trained model in doing one task is used in another task as a starting point resulting in short computation and training time and better performance without needing a lot of data.

There are different approaches to use transfer learning including (i) fixed feature extraction, (ii) fine-tuning, and (iii) re-training the whole model (Yosinski et al., 2014).

Fixed Feature Extraction: In this approach, pre-trained model's classifier is removed, and a new classifier with the number of classes corresponding to the new task is added at the end of the model. The resulting model is re-trained on the new dataset while keeping weights of all the convolutional

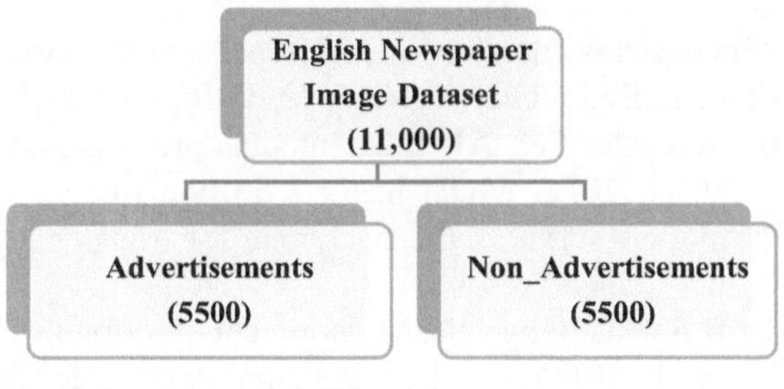

FIGURE 3.6 English newspaper image dataset.

layers as fixed and only learning the news weights for the classifier. By freezing the convolutional base, the pre-trained model acts as fixed feature extractor for the new dataset. This approach does not overfit the model even if the new dataset is small as weights of all convolutional layers are kept fixed and only the classifier is re-trained using the new dataset.

Fine-Tuning: In this approach, again pre-trained model's classifier is replaced by a new classifier suitable for the new task. Here, not only the new classifier is re-trained but also the few top layers are re-trained with the new dataset. In a neural network, initial layers extract general and simple features and the top layers extract complicated features more specific to dataset. Re-training the top layers along with the classifier while keeping the initial layers frozen allows higher-order feature extraction more relevant to the new dataset and may lead to improved accuracy. The n number of layers from the top of the pre-trained model should be re-trained for maximum accuracy using experimentation only. If the dataset is small, then training more layers along with the classifier may lead to overfitting.

Re-training the Whole Model: If the dataset is large enough, whole model (all the layers of the pre-trained model along with the classifier) can be re-trained on the new dataset. But this approach is very time-consuming and resource-consuming as all the weights in the network need to be fine-tuned in each epoch. Instead of training the model from the scratch (random weights), pre-trained weights of the model can be used for initialization. Due to this initialization, the model can be trained in lesser number of epochs as compared to training the model from the scratch.

Transfer learning techniques pose certain restrictions:

1. The data sample size of the new dataset must match with the input size on which the pre-trained model was initially trained. If not, the data sample size of the new dataset should be tailored to match with the same.
2. While fine-tuning the models, lower learning rates should be used in comparison with the learning rates used when training the model from scratch. The reason for this is that pre-trained weights are expected to be relatively good and too much and too quick distortion in them should be avoided.

3.4.4 Pre-Trained ResNet50 Model for Transfer Learning

When it comes to image classification, a lot of pre-trained models are available for transfer learning including VGG (VGG16, VGG19), GoogLeNet (Inceptionv3), ResNet (ResNet-50), and AlexNet. All these models are trained on the ImageNet dataset (Krizhevsky et al., 2012) which has 1.2 million images in 1000 different classes. These CNN models serve as the backbone for transfer learning for visual recognition tasks.

After winning the ILSVRC image classification competition in 2015, residual networks (ResNets) (He et al., 2016) have become quite popular for image recognition and classification tasks. ResNets differ from conventional CNNs and use a technique known as "shortcut" or "**skip connections**", which connects the input from the previous layers directly to the output skipping training from few layers in between. The logic behind "skip connections" is that if any layer hurts the network performance, it will be skipped by regularization. This allows the training of deep neural networks without the issue of vanishing/exploding gradients.

There are many variants of ResNet such as 18-layer ResNet (ResNet-18 model), 34-layer ResNet (ResNet-34 model), 50-layer ResNet (ResNet-50 model), 101-layer ResNet (ResNet-101 model), and even deeper 152-layer ResNet (ResNet-152 model). ResNet-50 (48 convolutional layers + 1 max pooling + 1 average pooling layer) is one of the most popular CNN model used for transfer learning. It has five convolutional blocks (Convl–Conv5), which are stacked on top of one another. Figure 3.7 presents a basic architecture of ResNet-50 model (Mahmood et al., 2020). Each convolutional block (except Convl) consists of "bottleneck" units, and every "bottleneck" unit is made up of three convolutional layers. These "bottleneck" units are repeated several times in each block. In Figure 3.7, the number above the convolutional blocks shows the repetition of "bottleneck" units in them. As shown in Figure 3.7, ResNet-50 is made up of $3 + 4 + 6 + 3 = 16$ "bottleneck" units and hence $16 \times 3 = 48$ convolutional layers. The notation ($k \times k$, n) denotes the filter size ($k \times k$) and the number of output channels (n) for each layer at different stages of ResNet-50 model. The five convolutional blocks are followed by an average pooling layer. After average pooling, a fully connected layer (fc) with "Softmax" activation function (Nwankpa et al., 2018; Sharma et al., 2020) is added. "fc" layer has 1000 output features corresponding to 1000 classes in ImageNet dataset. The convolutional blocks extract the useful features in the input data, whereas "fc" layer with "Softmax" activation function classifies the input in one of the 1000 classes.

Figure 3.8 depicts the working of one basic "bottleneck" unit. In a "bottleneck" unit, the first convolutional layer has a kernel of 1×1 and it reduces the input dimensions to ¼th. As shown in Figure 3.8, the first layer reduces dimensions from 256 to 64. The second convolutional layer has a kernel of size 3×3, and it does not change the dimensions. The third convolutional layer with 1×1 kernel size increases the dimensions to four times, restoring the input dimensions. In Figure 3.8, the third layer restores the dimensions from 64 to 256. This trick greatly reduces the number of parameters, and as a result, the amount of calculations needed is reduced to a great

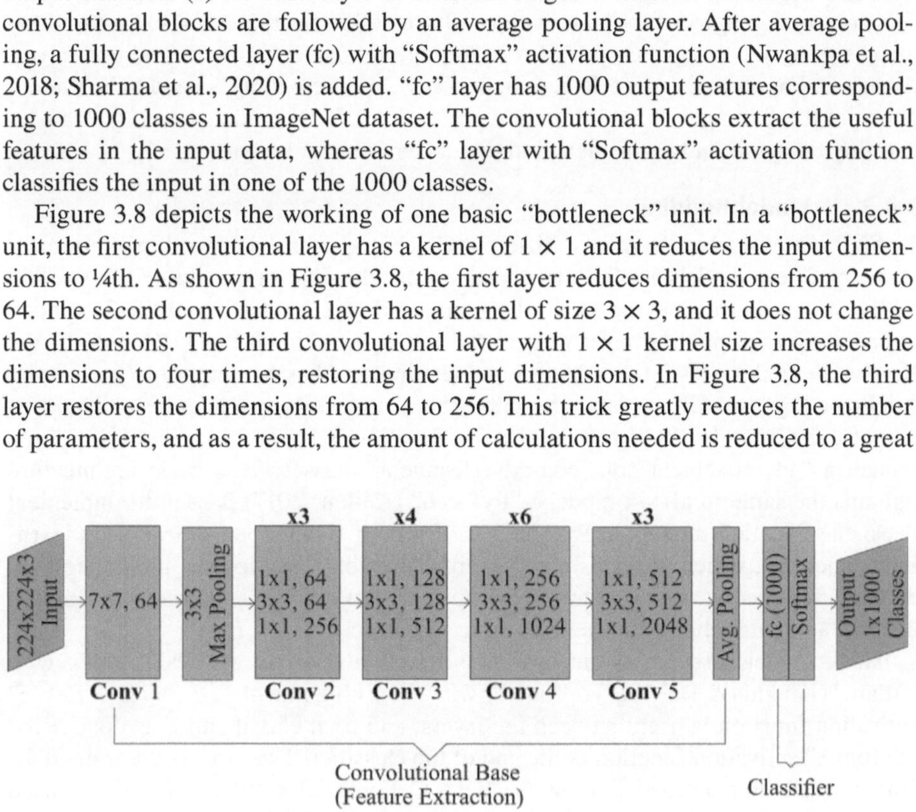

FIGURE 3.7 Basic architecture of ResNet-50 model.

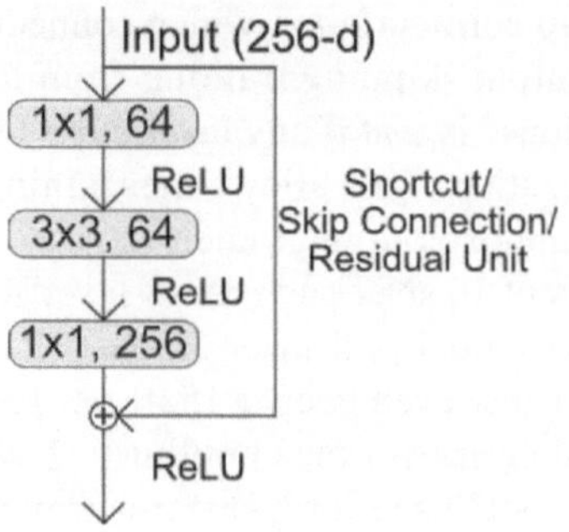

FIGURE 3.8 A sample bottleneck unit of ResNet-50.

extent (Hong et al., 2019). Despite being deeper, ResNet-50 has around 23 million trainable parameters in comparison with 60 million parameters of AlexNet and 138 million parameters of VGG16.

3.4.5 CNN Models for Advertisement Classification

The proposed image extraction technique (discussed in Section 3.3.1) extracts all the images from the newspaper .pdf files including both advertisements and non-advertisements. For advertisement detection, non-advertisement images need to be filtered out from these images. This subsection presents six CNN models based on transfer learning to classify advertisement and non-advertisement images.

3.4.5.1 Model Building

To filter out non-advertisements, different CNN models are designed using transfer learning. These models are built by replacing the classifier of pre-trained ResNet-50 with a new classifier having two fully connected layers ("fc-1" and "fc-2") added after the average pooling layer next to the last convolutional block (Conv5) of the ResNet-50 model. "fc-1" takes 2048 input features and has 64 output features, and it is followed by "ReLU" activation function. "fc-2" takes 64 input features and has two output units. Finally, "Softmax" activation function is applied for classifying input image in "advertisement" or "non-advertisement" classes. This basic architecture remains the same in all six models. "PyTorch" (Ketkar, 2017) is used to implement these classification models in Python 3.8. "PyTorch" is an open-source deep learning framework, which provides both usability (flexible, easy to use, and debug) and computation speed (support automatic differentiation and GPU acceleration) and has become a popular choice for deep learning research.

These models use "cross-entropy" loss function (Nasr et al., 2002) along with "Adam" (Kingma & Ba, 2017) optimizer with a learning rate of 1e-3 (0.001). "ReLU" activation function is used between the layers, and final classification is done using "Softmax" activation function at the end of the classifier. The models are trained for 100 epochs with a batch size of 32. NVIDIA GeForce GTX 1650 4 GB GPU is used to train Model 1, Model 2, Model 3, and Model 4. Model 5 and Model 6 are computationally extensive (number of trainable parameters are much higher than Model

1–Model 4), and these models are trained on Tesla V100 SXM2 16 GB GPU available on "Google Colab" (https://colab.research.google.com).

The proposed CNN models are trained and tested using the English newspaper image dataset (Figure 3.6). This dataset has 11,000 images including 5500 advertisements and 5500 non-advertisements. Table 3.1 shows the classification of image dataset. 80% (8800) of the total images are kept for training, while 20% (2200) are used for testing the models. Training data are split into two categories: data for training use and data for validation use in the ratio of 80:20, i.e., 7040 images for training use and 1760 images for validation use.

3.4.5.2 Model Evaluation

The proposed CNN models are evaluated using different accuracy measures such as confusion matrix and accuracy curves. One-dimensional indicators including precision, recall, and F1 score (Goutte & Gaussier, 2005) are also used to measure the performance of the models. Finally, the accuracy of each model is calculated on the test set. Table 3.2 illustrates the formulae used for computing precision, recall, F1 score, and accuracy.

3.4.5.3 Model 1 (Feature Extractor + Classifier)

In this "transfer learning"-based model, ResNet-50 (pre-trained on ImageNet dataset) is used for feature extraction. All the convolutional blocks (Conv 1–Conv 5) of ResNet-50 model are frozen, and only the classifier is trained on the English newspaper image dataset. Figure 3.9 shows the basic architecture of Model 1. The test results of Model 1 in terms of precision, recall, and F1 score for both advertisement and non-advertisement classes are displayed in Table 3.3. The test set includes 1105 advertisements and 1095 non-advertisements (support) making a total of 2200 images in the test set. The confusion matrix (Figure 3.10a) shows that out of 1105 advertisements

TABLE 3.1
Classification of Image Dataset

Dataset	No. of Images
Data for training use	7040
Data for validation use	1760
Data for testing	2200
Total	11,000

TABLE 3.2
Formulae for Precision, Recall, F1 Score, and Accuracy

Precision = Number of True Positives / (Number of True Positives + Number of False Positives)

Recall = Number of True Positives / (Number of True Positives + Number of False Negatives)

F1-score = 2 × (Recall x Precision) / (Recall + Precision)

Accuracy = Total no. of correctly classified images / Total no. of Test samples.

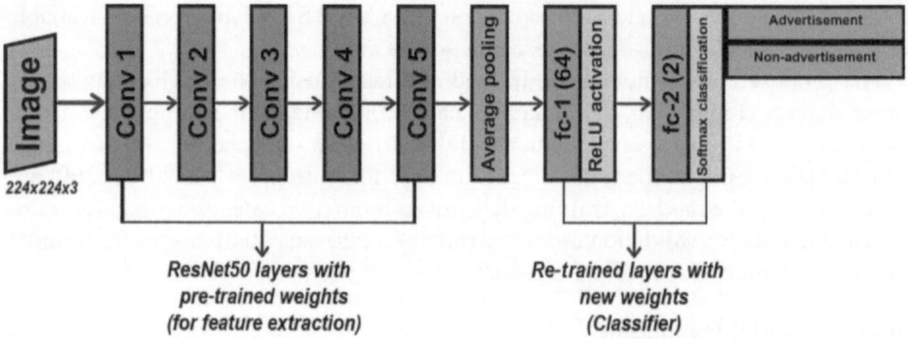

FIGURE 3.9 Model 1 (feature extractor + classifier) architecture.

in the test set, 1061 are correctly classified, whereas 44 are misclassified as non-advertisements. Similarly, out of 1095 non-advertisements, 1061 non-advertisements are correctly classified and 34 are misclassified. Hence, a test accuracy of (1061 + 1061 = 2122)/2200 = 96.45% is achieved. Training and validation accuracy curves are shown in Figure 3.10b. Training accuracy of nearly 1.00 (100%) is achieved in 100 epochs, whereas the maximum validation accuracy remains below 0.97 (97%).

3.4.5.4 Model 2 (Fine-Tuning Top Most Conv Block)

In this model, the first four convolutional blocks (Conv 1–Conv 4) of ResNet-50 are frozen, and the top most block (Conv 5) is trained (fine-tuned) along with the classifier using the English newspaper image dataset. Figure 3.11 shows the basic architecture of Model 2. Table 3.4 displays precision, recall, and F1 score values of advertisement and non-advertisement classes for Model 2. Figure 3.12a illustrates the confusion matrix of Model 2. 1066 out of 1084 advertisements and 1086 out of 1116 non-advertisements are correctly classified giving the test accuracy of (1066 + 1086 = 2152)/2200 = 97.82%. The training and validation curves of Model 2 for 100 epochs are shown in Figure 3.12b. A training accuracy of 1.00 (100%) is achieved here, whereas the validation accuracy remains below 0.98 (98%) with maximum at 97.73%.

3.4.5.5 Model 3 (Fine-Tuning Top Two Conv Blocks)

In Model 3, the top two convolutional blocks (Conv4 and Conv5) of ResNet-50 model and the classifier are re-trained on the English newspaper image dataset while keeping the weights of initial blocks (Conv 1, Conv2, and Conv 3) unchanged. Figure 3.13 shows the architecture of Model 3. The evaluation results of Model 3 are displayed in Table 3.5.

TABLE 3.3
Test Results from Model 1 (Feature Extractor + Classifier)

	Precision	Recall	F1 Score	Support
Advertisements	0.97	0.96	0.96	1105
Non-Advertisements	0.96	0.97	0.96	1095

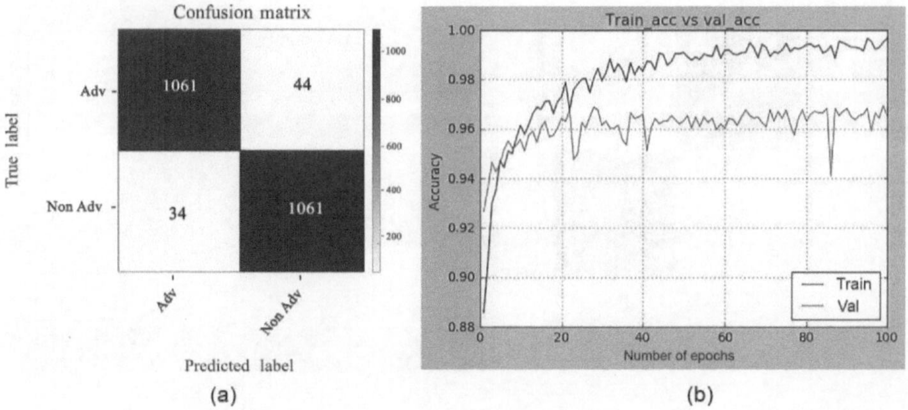

FIGURE 3.10 (a) Confusion matrix for Model 1. (b) Training and validation accuracy (Model 1).

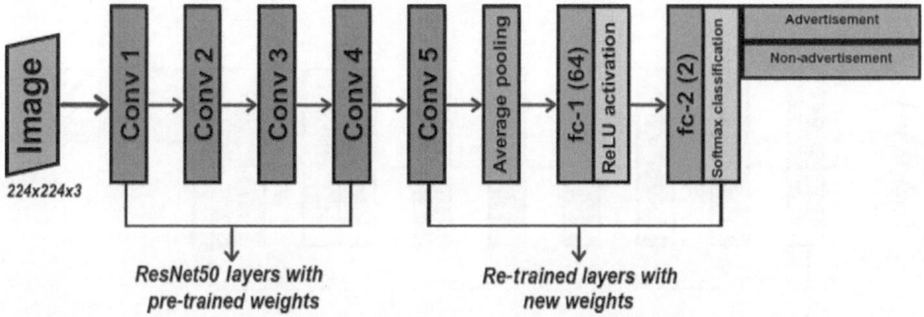

FIGURE 3.11 Model 2 (fine-tuning top most Conv block) architecture.

TABLE 3.4
Test Results from Model 2 (Fine-Tuning Top Most Conv Block)

	Precision	Recall	F1 Score	Support
Advertisements	0.97	0.98	0.98	1084
Non_Advertisements	0.98	0.97	0.98	1116

The test set includes 1078 advertisements and 1122 non-advertisements making a total of 2200 images. The confusion matrix of Model 3 (Figure 3.14a) shows that 1041/1078 advertisements and 1088/1122 non-advertisements are correctly classified, and a test accuracy of (1041 + 1088 = 2129)/2200 = 96.77% is achieved. Figure 3.14b shows that the training accuracy of 1.00 (100%) is achieved, whereas the maximum validation accuracy (97.56%) remains below 98% in 100 epochs.

3.4.5.6 Model 4(Fine-Tuning Top Three Conv Blocks)

As shown in Figure 3.15, Model 4 trains the top three convolutional blocks (Conv3, Conv4, and Conv5) of ResNet-50 model along with the classifier. Test results of Model 4

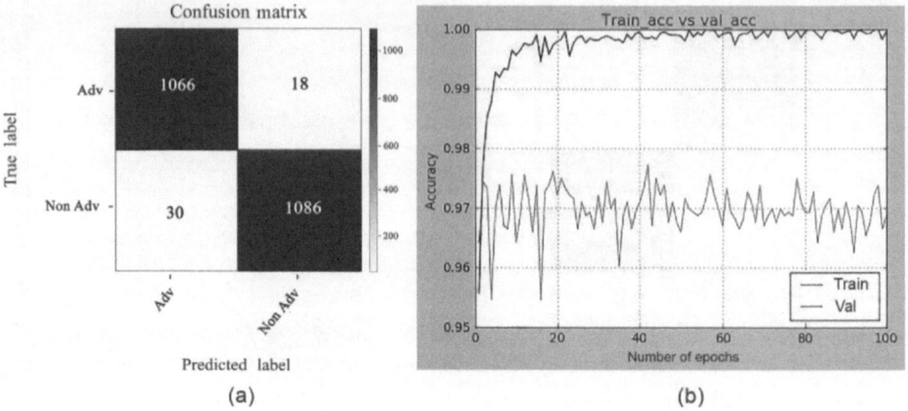

FIGURE 3.12 (a) Confusion matrix for Model 2. (b) Training and validation accuracy (Model 2).

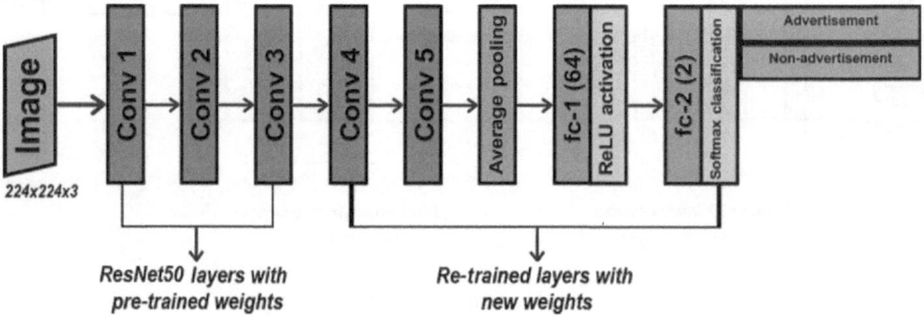

FIGURE 3.13 Model 3 (fine-tuning top two Conv blocks) architecture.

TABLE 3.5
Test Results from Model 3 (Fine-Tuning Top Two Conv Blocks)

	Precision	Recall	F1 Score	Support
Advertisements	0.97	0.97	0.97	1078
Non_Advertisements	0.97	0.97	0.97	1122

including precision, recall, and F1 score for advertisements and non-advertisement categories are shown in Table 3.6. Figure 3.16a shows the confusion matrix of Model 4. There are 1088 advertisements and 1112 non-advertisements (support) in the test set out of which 1044 advertisements and 1082 non-advertisements are correctly classified. Test accuracy of (1044 + 1082 = 2126/2200) 96.63% is achieved. Training and validation accuracies of Model 4 for 100 epochs are shown in Figure 3.16b.

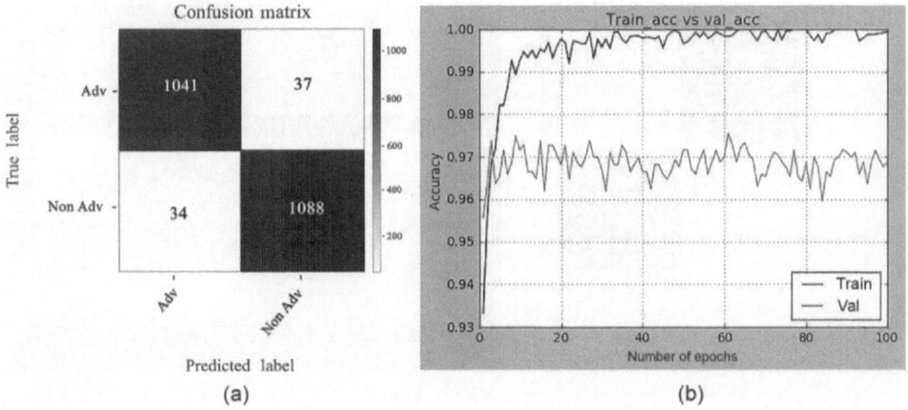

FIGURE 3.14 (a) Confusion matrix for Model 3. (b) Training and validation accuracy (Model 3)

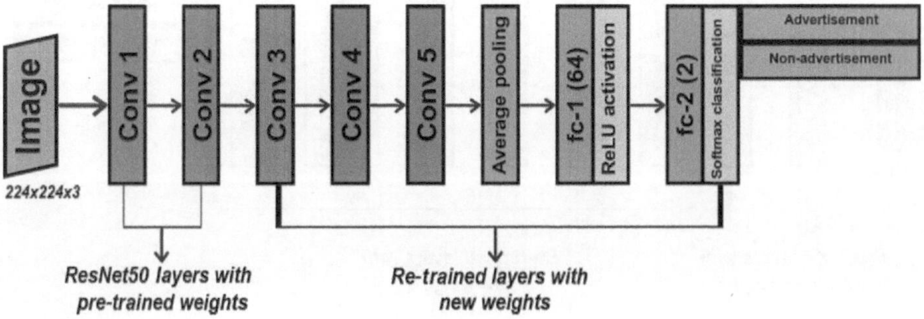

FIGURE 3.15 Model 4 (fine-tuning top three Conv blocks) architecture.

TABLE 3.6
Test Results from Model 4 (Fine-Tuning Top Three Conv Blocks)

	Precision	Recall	F1 Score	Support
Advertisements	0.97	0.96	0.97	1088
Non_Advertisements	0.96	0.97	0.97	1112

The training accuracy curve shows that 1.00 (100%) training accuracy is achieved, whereas the validation accuracy curve shows that the validation accuracy remains below 97% (maximum at 96.93%).

3.4.5.7 Model 5 (Fine-Tuning Top Four Conv Blocks)

In Model 5 (Figure 3.17), the top four convolutional blocks (Conv2 to Conv5) of ResNet-50 model are retrained along with the classifier. Evaluation results of Model 5 are illustrated in Table 3.7, which displays precision, recall, and F1 score for both

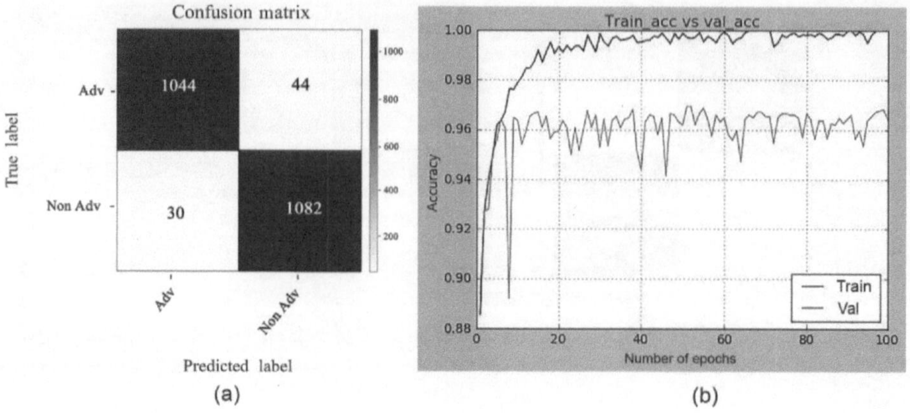

FIGURE 3.16 (a) Confusion matrix for Model 4. (b) Training and validation accuracy (Model 4).

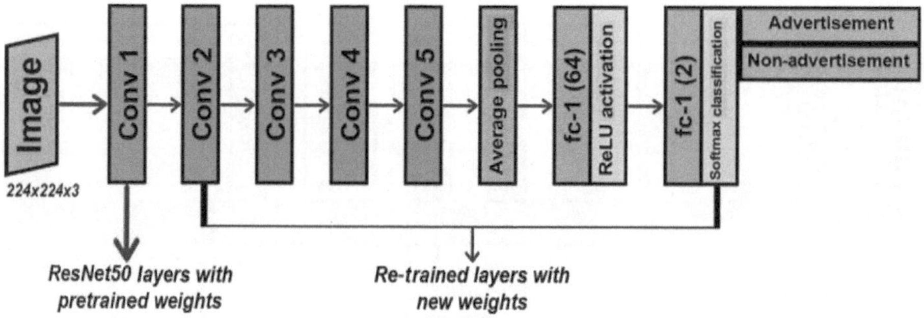

FIGURE 3.17 Model 5 (fine-tuning top four Conv blocks) architecture.

TABLE 3.7
Test Results from Model 5 (Fine-Tuning Top Four Conv Blocks)

	Precision	Recall	F1 Score	Support
Advertisements	0.96	0.96	0.96	1102
Non_Advertisements	0.96	0.96	0.96	1098

advertisement and non-advertisement categories. The confusion matrix (Figure 3.18a) of Model 5 depicts that out of 1–102 advertisements in the test set, 1061 advertisements are correctly classified, whereas 41 advertisements are misclassified as non-advertisements. Also, out of 1098 non-advertisements, 1059 non-advertisements are correctly classified, whereas 39 non-advertisements are misclassified as advertisements. Hence, Model 5 achieves a test accuracy of (1061 + 1059 = 2120)/2200 = 96.36%. Figure 3.18b shows the 100% training accuracy is achieved in 100 epochs, whereas the maximum validation accuracy remains below 97% with maximum at 96.98%.

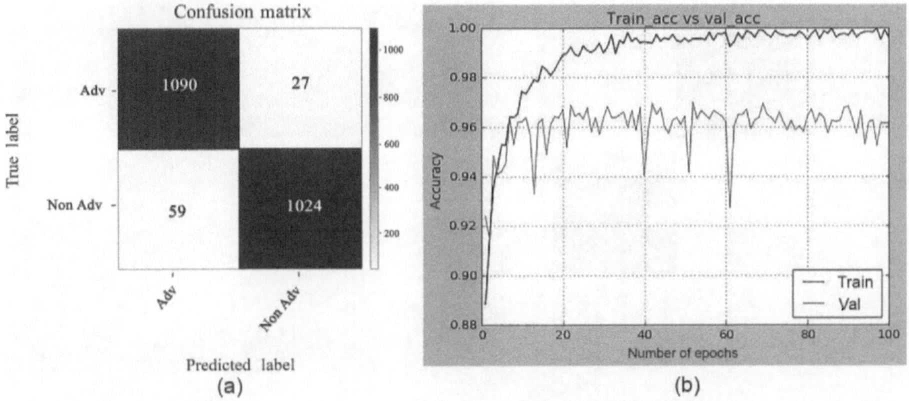

FIGURE 3.18 (a) Confusion matrix for Model 5. (b) Training and validation accuracy (Model 5).

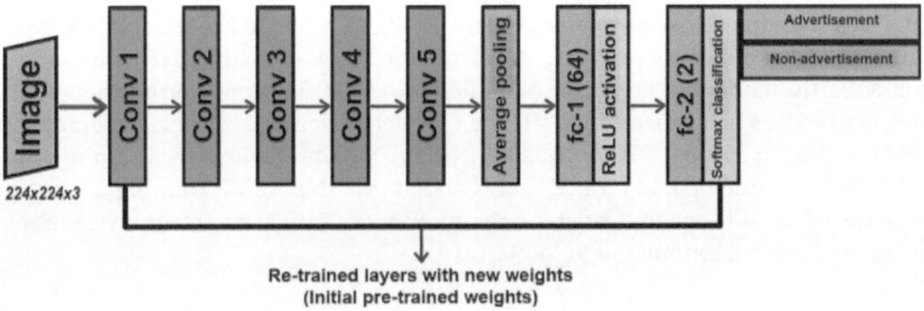

FIGURE 3.19 Model 6 (Re-training whole model and classifier) architecture.

3.4.5.8 Model 6 (Re-training Whole Model + Classifier)

In Model 6 (Figure 3.19), whole ResNet-50 model (Conv 1–Conv 5) along with the classifier is retrained on the English newspaper image dataset using pre-trained weights for initialization. Table 3.8 displays the classification report (test results) of Model 6 including precision, recall, and F1 score for both advertisement and non-advertisement categories. Figure 3.20a shows the confusion matrix of Model 6, and it illustrates that out of 1106 advertisements in the test set,

TABLE 3.8
Test Results from Model 6 (Re-training Whole Model and Classifier)

	Precision	Recall	F1 Score	Support
Advertisements	0.97	0.95	0.96	1106
Non_Advertisements	0.95	0.97	0.96	1094

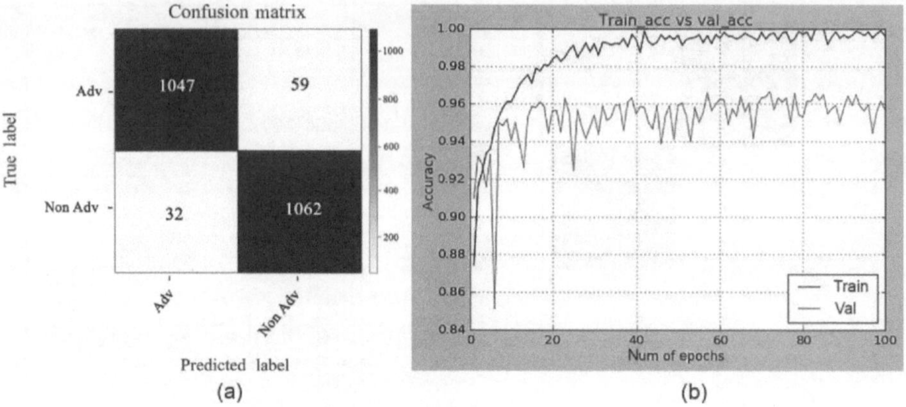

FIGURE 3.20 (a) Confusion matrix for Model 6. (b) Training and validation accuracy (Model 6).

1047 advertisements are correctly classified, whereas 59 advertisements are misclassified as non-advertisements. Also, out of 1094 non-advertisements, 1062 non-advertisements are correctly classified, whereas 32 non-advertisements are misclassified as advertisements. Hence, Model 6 achieves a test accuracy of (1047 + 1062 = 2109) / 2200 = 95.86%. Figure 3.20b displays the training and validation accuracy curves for Model 6 and shows that 100% training accuracy is achieved in 100 epochs, whereas the maximum validation accuracy remains below 97% with maximum at 96.64%.

3.4.6 Classification Results and Analysis

The test accuracies of all the six advertisement classification models evaluated on the English newspaper image dataset are summarized in Table 3.9. It is found that training the classifier alone (Model 1) did not give the best results as in this model, only the features learned from ImageNet dataset are used, and the features specific to the English newspaper image dataset are not considered. Top convolutional blocks extract the complex features specific to dataset, and training the top Conv blocks may help in increasing the accuracy of the model by extracting useful features for classification. Model 2 fine-tunes the top convolutional block (Conv 5) of ResNet-50 along with the classifier and gives the highest accuracy of 97.82% among all the six transfer learning models evaluated here, and it is discovered to be the most effective model for classifying advertisement and non-advertisement images from online English newspapers. Increasing the number of convolutional blocks for re-training (Model 3, Model 4, Model 5, and Model 6) on the English newspaper dataset has not helped either. As the numbers of Conv blocks for re-training (fine-tuning) are increased, model accuracies rather decrease. The reason for the same is that the English newspaper image dataset is limited to 11,000 images only. When Model 3, Model 4, Model 5, and Model 6 retrain more Conv blocks on this small dataset,

TABLE 3.9
Comparison of Test Accuracies of Evaluated Models

S. No.	Model	Accuracy (%)
1.	Model 1 (feature extractor + classifier)	96.45
2.	**Model 2 (fine-tuned top 1 layer + classifier)**	**97.82**
3.	Model 3 (fine-tuned top 2 layers + classifier)	96.77
4.	Model 4 (fine-tuned top 3 layers + classifier)	96.63
5.	Model 5 (fine-tuned top 4 layers + classifier)	96.36
6.	Model 6 (training whole model + classifier)	95.86

test accuracies decrease due to overfitting where models start learning the training samples but could not classify the test samples with higher accuracies. If the size of the dataset is increased (around 50,000 image samples), we may expect better classification accuracies on re-training more Conv blocks.

3.5 CONCLUSION

Automatic advertisement detection is very useful for performing personalized advertisement search in online newspapers. This study presents an intelligent system, which uses image processing and deep learning techniques to automatically detect advertisements across a range of online English newspapers without any human intervention or prior knowledge or any other aid. First images are extracted from online newspaper .pdf files using a novel adaptive thresholding and connected component-based image processing technique, and then, these images are classified as advertisements and non-advertisements using transfer learning-based CNN models. Six different transfer learning-based CNN models are designed, trained, and tested for classifying advertisement and non-advertisement images, and the model that fine-tunes the top convolutional block of ResNet-50 along with classifier was discovered to be the best model for online English newspaper image classification task giving 97.82% test accuracy. In this way, non-advertisements can be filtered out and only advertisements are left, which can be utilized for different purposes including advertisement search and advertisement studies and analysis helping us steer toward Society 5.0. The future work may include adding more images to the English newspaper image dataset so that more convolutional blocks can be re-trained without overfitting and better classification accuracies may be achieved. The future extension may also include the classification of advertisement images into various advertisement categories including education, job ads, public notices, tenders, sales and promotions, remembrance messages, and matrimonial advertisements This classification would allow the user to perform advertisement search in the category of own choice. Using OCR techniques, the text of the advertisement images can be obtained making it possible to perform keyword-based advertisement searches across several advertisement categories.

REFERENCES

Almgren, K., Krishnan, M., Aljanobi, F., & Lee, J. (2018). AD or Non-AD: a deep learning approach to detect advertisements from magazines. *Entropy*, 20(12), 982. Doi: 10.3390/e20120982.

Almutairi, A., & Almashan, M. (2019). Instance segmentation of newspaper elements using Mask R-CNN. *2019 18th IEEE International Conference on Machine Learning and Applications (ICMLA)*, 1371–1375. Doi: 10.1109/ICMLA.2019.00223.

Antonacopoulos, A., Clausner, C., Papadopoulos, C., & Pletschacher, S. (2013). ICDAR 2013 competition on historical newspaper layout analysis (HNLA 2013). *2013 12th International Conference on Document Analysis and Recognition*, 1454–1458. Doi: 10.1109/ICDAR.2013.293.

Banerjee, B. (2017). *Machine Learning Models for Political Video Advertisement Classification* [Capstones]. Iowa StateUniversity.https://lib.dr.iastate.edu/creativecomponents/365.

Bansal, A., Chaudhury, S., Roy, S. D., & Srivastava, J. B. (2014). Newspaper article extraction using hierarchical fixed point model. *2014 11th IAPR International Workshop on Document Analysis Systems*, 257–261. Doi: 10.1109/DAS.2014.42.

Basu, M. (2002). Gaussian-based edge-detection methods-a survey. *IEEE Transactions on Systems, Man and Cybernetics, Part C (Applications and Reviews)*, *32*(3), 252–260. Doi: 10.1109/TSMCC.2002.804448.

Bridle, J. S. (1990). Probabilistic interpretation of feedforward classification network outputs, with relationships to statistical pattern recognition. *Neurocomputing*, 227–236. Doi: 10.1007/978-3-642-76153-9_28

Chaudhury, K., Jain, A., Thirthala, S., Sahasranaman, V., Saxena, S., & Mahalingam, S. (2009). Google newspaper search-image processing and analysis pipeline. *2009 10th International Conference on Document Analysis and Recognition*, 621–625. Doi: 10.1109/ICDAR.2009.272.

Chu, W.-T., & Chang, H.-Y. (2016). Advertisement detection, segmentation, and classification for newspaper images and website snapshots. *2016 International Computer Symposium (ICS)*, 396–401. Doi: 10.1109/ICS.2016.0086.

Cortes, C., & Vapnik, V. (1995). Support-vector networks. *Machine Learning*, *20*(3), 273–297.

Cover, T., & Hart, P. (1967). Nearest neighbor pattern classification. *IEEE Transactions on Information Theory*, *13*(1), 21–27. Doi: 10.1109/TIT.1967.1053964.

Duan, L. Y., Wang, J., Zheng, Y., Jin, J. S., Lu, H., & Xu, C. (2006). Segmentation, categorization, and identification of commercial clips from TV streams using multimodal analysis. *Proceedings of the 14th Annual ACM International Conference on Multimedia - MULTIMEDIA '06*, 201–210. Doi: 10.1145/1180639.1180697.

Freund, Y., & Schapire, R. E. (1997). A decision-theoretic generalization of on-line learning and an application to boosting. *Journal of Computer and System Sciences*, *55*(1), 119–139. Doi: 10.1006/jcss.1997.1504.

Furmaniak, R. (2007). Unsupervised newspaper segmentation using language context. *Ninth International Conference on Document Analysis and Recognition (ICDAR 2007) Vol 2*, 1263–1267. Doi: 10.1109/ICDAR.2007.4377118.

Gatos, B., Mantzaris, S. L., Chandrinos, K. V., Tsigris, A., & Perantonis, S. J. (1999). Integrated algorithms for newspaper page decomposition and article tracking. *Proceedings of the Fifth International Conference on Document Analysis and Recognition. ICDAR '99 (Cat. No.PR00318)*, 559–562. Doi: 10.1109/ICDAR.1999.791849.

Gong, C., & Zhu, F. (2010). On detection of contextual advertisements. *2010 2nd International Asia Conference on Informatics in Control, Automation and Robotics (CAR 2010)*, 29–32. Doi: 10.1109/CAR.2010.5456544.

Goutte, C., & Gaussier, E. (2005). A probabilistic interpretation of precision, recall and F-score, with implication for evaluation. In D. E. Losada & J. M. Fernández-Luna (Eds.), *Advances in Information Retrieval*, 345–359. Springer. Doi: 10.1007/978-3-540-31865-1_25.

Gu, J., Wang, Z., Kuen, J., Ma, L., Shahroudy, A., Shuai, B., Liu, T., Wang, X., Wang, G., Cai, J., & Chen, T. (2018). Recent advances in convolutional neural networks. *Pattern Recognition*, *77*, 354–377. Doi: 10.1016/j.patcog.2017.10.013.

Haralick, R. M., Sternberg, S. R., & Zhuang, X. (1987). Image analysis using mathematical morphology. *IEEE Transactions on Pattern Analysis and Machine Intelligence*, *PAMI-9*(4), 532–550. Doi: 10.1109/TPAMI.1987.4767941.

Hasan, M., Ullah, S., Khan, M. J., & Khurshid, K. (2019). Comparative analysis of SVM, ANN and CNN for classifying vegetation species using hyperspectral thermal infrared data. *ISPRS - International Archives of the Photogrammetry, Remote Sensing and Spatial Information Sciences*, *XLII-2/W13*, 1861–1868. Doi: 10.5194/isprs-archives-XLII-2-W13-1861–2019.

He, K., Zhang, X., Ren, S., & Sun, J. (2016). Deep residual learning for image recognition. *2016 IEEE Conference on Computer Vision and Pattern Recognition (CVPR)*, 770–778. Doi: 10.1109/CVPR.2016.90.

Hong, J., Cheng, H., Zhang, Y.-D., & Liu, J. (2019). Detecting cerebral microbleeds with transfer learning. *Machine Vision and Applications*, *30*(7–8), 1123–1133. Doi: 10.1007/s00138-019-01029-5.

Jain, P., Taneja, K., & Taneja, H. (2021b). Convolutional neural network based advertisement classification models for online English newspapers. *Turkish Journal of Computer and Mathematics Education (TURCOMAT)*, *12*(2), 1687–1698. Doi: 10.17762/turcomat.v12i2.1505.

Jain, P., Taneja, K., & Taneja, H. (2021a). Which OCR toolset is good and why? A comparative study. *Kuwait Journal of Science*, *48*(2). Doi: 10.48129/kjs.v48i2.9589.

Jung, J.-Y. (2012). Vocabulary expansion technique for advertisement classification. *KSII Transactions on Internet and Information Systems (TIIS)*, *6*(5), 1373–1387. Doi: 10.3837/tiis.2012.05.007.

Kaur, R. P., & Jindal, M. K. (2017). A survey on newspaper image segmentation techniques. *International Journal of Advanced Research in Science, Engineering*, *6*(10), 1789–1797. http://ijarse.com/images/fullpdf/1509353899_GNC890_ijarse.pdf

Ketkar, N. (2017). Introduction to PyTorch. *Deep Learning with Python*, 195–208. Apress. Doi: 10.1007/978-1-4842-2766-4_12.

Kingma, D. P., & Ba, J. (2017). Adam: a method for stochastic optimization. *ArXiv:1412.6980 [Cs]*. http://arxiv.org/abs/1412.6980.

Krizhevsky, A., Sutskever, I., & Hinton, G. E. (2012). ImageNet classification with deep convolutional neural networks. *Advances in Neural Information Processing Systems*, *25*, 1097–1105. Doi: 10.1145/3065386.

LeCun, Y., Kavukcuoglu, K., & Farabet, C. (2010). Convolutional networks and applications in vision. *Proceedings of 2010 IEEE International Symposium on Circuits and Systems*, 253–256. Doi: 10.1109/ISCAS.2010.5537907.

Lewis, D. D. (1998). Naive (Bayes) at forty: the independence assumption in information retrieval. *European Conference on Machine Learning: ECML-98*, 4–15. Springer. Doi: 10.1007/BFb0026666.

Li, D., Wang, B., Li, Z., Yu, N., & Li, M. (2007). On detection of advertising images. *2007 IEEE International Conference on Multimedia and Expo*, 1758–1761. Doi: 10.1109/ICME.2007.4285011.

Li, Q., Wang, J., Wipf, D., & Tu, Z. (2013). Fixed-point model for structured labeling. *International Conference on Machine Learning*, 214–221. Doi: 10.5555/3042817.3042843.

Liu, F., Luo, Y., Yoshikawa, M., & Hu, D. (2001). A new component based algorithm for newspaper layout analysis. *Proceedings of Sixth International Conference on Document Analysis and Recognition*, 1176–1180. Doi: 10.1109/ICDAR.2001.953970.

Mahmood, A., Ospina, A. G., Bennamoun, M., An, S., Sohel, F., Boussaid, F., Hovey, R., Fisher, R. B., & Kendrick, G. A. (2020). Automatic hierarchical classification of kelps using deep residual features. *Sensors*, *20*(2), 447. Doi: 10.3390/s20020447.

Meier, B., Stadelmann, T., Stampfli, J., Arnold, M., & Cieliebak, M. (2017). Fully convolutional neural networks for newspaper article segmentation. *2017 14th IAPR International Conference on Document Analysis and Recognition (ICDAR)*, 414–419. Doi: 10.1109/ICDAR.2017.75.

Mitchell, P. E. & Yan, H. (2001). Newspaper document analysis featuring connected line segmentation. *Proceedings of Sixth International Conference on Document Analysis and Recognition*, 1181–1185. Doi: 10.1109/ICDAR.2001.953971.

Mitchell, P. E., & Yan, H. (2004a). Connected pattern segmentation and title grouping in newspaper images. *Proceedings of the 17th International Conference on Pattern Recognition, 2004. ICPR 2004*, *1*, 397–400. Doi: 10.1109/ICPR.2004.1334135.

Mitchell, P. E, & Yan, H. (2004b). Newspaper layout analysis incorporating connected component separation. *Image and Vision Computing*, *22*(4), 307–317. Doi: 10.1016/j.imavis.2003.11.001.

Murthy, S. K. (1998). Automatic construction of decision trees from data: a multi-disciplinary survey. *Data Mining and Knowledge Discovery*, *2*(4), 345–389. Doi: 10.1023/A:1009744630224.

Nair, V., & Hinton, G. E. (2010). Rectified linear units improve restricted Boltzmann machines. *International Conference on Machine Learning*, ICML, 807–814. Doi: 10.5555/3104322.3104425.

Nasr, G. E., Badr, E. A., & Joun, C. (2002). Cross entropy error function in neural networks: forecasting gasoline demand. *FLAIRS Conference*, 381–384. Doi: 10.5555/646815.708603.

Nwankpa, C., Ijomah, W., Gachagan, A., & Marshall, S. (2018). Activation functions: comparison of trends in practice and research for deep learning. *ArXiv:1811.03378 [Cs]*. http://arxiv.org/abs/1811.03378.

Ouji, A., Leydier, Y., & Lebourgeois, F. (2011). Advertisement detection in digitized press images. *2011 IEEE International Conference on Multimedia and Expo*, 1–6. Doi: 10.1109/ICME.2011.6011890.

Palfray, T., Hebert, D., Nicolas, S., Tranouez, P., & Paquet, T. (2012). Logical segmentation for article extraction in digitized old newspapers. *Proceedings of the 2012 ACM Symposium on Document Engineering - DocEng '12*, 129–132. Doi: 10.1145/2361354.2361383.

Pan, S. J., & Yang, Q. (2010). A Survey on transfer learning. *IEEE Transactions on Knowledge and Data Engineering*, *22*(10), 1345–1359. Doi: 10.1109/TKDE.2009.191.

Peleato, R. A., Chappelier, J.-C., & Rajman, M. (2000). Automated information extraction out of classified advertisements. *International Conference on Application of Natural Language to Information Systems*, 203–214. Doi: 10.1007%2F3-540-45399-7_17.

Razavian, A. S., Azizpour, H., Sullivan, J., & Carlsson, S. (2014). CNN features off-the-shelf: an astounding baseline for recognition. *2014 IEEE Conference on Computer Vision and Pattern Recognition Workshops*, 512–519. Doi: 10.1109/CVPRW.2014.131.

Roy, P., Dutta, S., Dey, N., Dey, G., Chakraborty, S., & Ray, R. (2014). Adaptive thresholding: a comparative study. (2014). *International Conference on Control, Instrumentation, Communication and Computational Technologies (ICCICCT)*, 1182–1186. Doi: 10.1109/ICCICCT.2014.6993140.

Shaheen, F., Verma, B., & Asafuddoula, Md. (2016). Impact of automatic feature extraction in deep learning architecture. *2016 International Conference on Digital Image Computing: Techniques and Applications (DICTA)*, 1–8, IEEE. Doi: 10.1109/dicta.2016.7797053.

Sharma, S., Sharma, S., & Athaiya, A. (2020). Activation functions in neural networks. *International Journal of Engineering Applied Sciences and Technology*, *4*(12), 310–316. Doi: 10.33564/ijeast.2020.v04i12.054.

Torrey, L., & Shavlik, J. (2010). Transfer learning. In *Handbook of Research on Machine Learning Applications and Trends: Algorithms, Methods, and Techniques*, 242–264. IGI global. Doi: 10.4018/978-1-60566-766-9.ch011.

Vo, A. T., Tran, H. S., & Le, T. H. (2017). Advertisement image classification using convolutional neural network. *2017 9th International Conference on Knowledge and Systems Engineering (KSE)*, 197–202. Doi: 10.1109/KSE.2017.8119458.

Yosinski, J., Clune, J., Bengio, Y., & Lipson, H. (2014). How transferable are features in deep neural networks? *ArXiv:1411.1792 [Cs]*. http://arxiv.org/abs/1411.1792.

Zhang, L., Zhu, Z., & Zhao, Y. (2007). Robust commercial detection system. *2007 IEEE International Conference on Multimedia and Expo*, 587–590. Doi: 10.1109/ICME.2007.4284718

4 Evolutionary Computation Framework for Handling Resource and Optimization of Solar Energy Harvesting System for WSN

Anju Rani and Amit Kumar Bindal

CONTENTS

4.1 INTRODUCTION

A basic portion of the Internet of Things (IoTs) is the remote sensor organization (wireless sensor network (WSN)). These organizations have been generally utilized in different applications, for example, in the military, crisis recuperation, tolerant

DOI: 10.1201/9781003158165-4

well-being checking, and air quality observing (Taneja et al., 2012). Headway in late remote innovation has set off the requirement for gadgets to run on free force sources. This is clear particularly with the remote sensor organization (WSN). This can be refined by reaping energy from natural resources such as sunlight and wind. These energy-reaping gadgets can control the remote sensor hubs either straightforwardly or through a battery (Ibrahim et al., 2017). The WSN hubs experience the ill-effect of a significant plan imperative that their battery energy is restricted and hence they can work just for a couple of days relying on obligation pattern of activity (Sharma et al., 2018a). The sunlight-based energy is generally put away in sun-powered cells, so the energy putting away proficiency must be expanded. Sunlight-based energy framework can be changed over quickly into power-utilizing photovoltaic (PV) boards through the photovoltaic effect (Mohamed & Abd El Sattar, 2019). Be that as it may, the transformation productivity is low and the expense of intensity created is similarly high. PV age has numerous favorable circumstances: for example, it has low fuel costs, doesn't create contamination, requires little upkeep, and the PV framework has more different highlights (Ahmad et al., 2016). Sunlight-based PV energy collecting alludes to changing over sun-based light energy into electrical energy to work an electrical or electronic gadget. As applied to WSNs, sun-based light energy is changed over into electrical energy and is used to revive the battery of a WSN hub at the activity site itself (Sharma et al., 2018). In this manner, battery substitution is required over and over once the battery energy has been released. The electrical energy gathered from sun-based energy can likewise be utilized legitimately to control a WSN hub. Then again, the gathered energy might be warehoused in a battery-powered battery for future purposes. The SEH-WSNs are comprised of little self-sufficient WSN hubs connected to little measure sunlight-based boards for their energy-collecting needs. It is seen that the greatest conceivable gathered force from sun-powered energy at the outside is $15.0\,mW/cm^2$ with a productivity of up to 30% (Rasheduzzaman et al., 2016). Accordingly, we have picked sun-based energy gathering for providing substitute capacity to the WSNs as it has the most powerful thickness and great productivity. Various techniques to follow the most extreme force purpose of a PV module have been proposed (Choudhary & Saxena, 2014) to beat the restriction of effectiveness. Maximal power point tracking (MPPT) is utilized to remove the most extreme force from the sun-oriented PV module and move that capacity to the heap. The DC-DC converter goes about as an interface between the heap and the PV module as it effectively transfers the greatest force from the sun-powered PV module to the heap. By changing the obligation cycle, the heap impedance is coordinated with the source impedance to accomplish the most extreme force from the PV board (Veerachary & Saxena, 2011). Here, we compare the evolutionary computing techniques like INC-MPPT and P&O-MPPT with respect to efficiency and get the best result.

4.2 LITERATURE SURVEY

Power system networks take the PV-created energy by methods of matrix-associated inverters. Sometimes there is no coordination in the working particular of the heap and PV clusters which is a well-known concern in PV power frameworks. In particular, with various ecological states, the PV module exhibit shows a non-direct style for the V-I bend and the greatest force point on the V-P bend. The PV module productivity is in the scope of 10%–25%. This means that the greatest force point following (MPPT) calculations are gotten together with the whole framework to augment their capacity and lessen modules cost.

Different key research findings available in the literature are summarized in Table 4.1.

Table 4.1 has three columns: the first one shows the reference numbers, the second one shows the objective of the related work, and the third column shows the strategy of the work and how it performs.

4.3 PV CELL PRINCIPAL, CHARACTERISTIC, AND MODULE

The transformation of light energy into electrical energy depends on a marvel called PV impact (Mathew & Selvakumar, 2006). At the point when semiconductor materials are presented to light, a portion of photons of the light beam is consumed by the semiconductor gem, which causes a critical number of free electrons in the gem. This is the essential purpose behind delivering power because of PV impact. A photovoltaic cell is the fundamental unit of framework where PV impact is used to create power from light energy. Silicon is the most generally utilized semiconductor material for developing the PV cell. The silicon molecule has four valence electrons. In a strong precious stone, every silicon iota shares every one of its four valence electrons with another closest silicon particle, thus making covalent connections between them. Along these lines, silicon precious stone gets a tetrahedral cross-section structure. While light beam strikes on any materials, some part of the light is mirrored, some bit is sent through materials, and the rest is consumed by materials. Something very similar happens when light falls on a silicon precious stone. On the off chance that the power of episode light is sufficiently high, adequate quantities of photons are consumed by the gem, and these photons, thus, energize a portion of the electrons of covalent bonds. These energized electrons at that point get adequate energy to move from the valence band to the conduction band. As the energy level of these electrons is in the conduction band, the electrons leave from the covalent bond, leaving a gap in the bond behind each eliminated electron. These are called free electrons that move arbitrarily inside the precious stone structure of silicon. These free electrons and openings have a fundamental part in making power in PV cells. These electrons and openings are henceforth called light-created electrons and gaps separately. These light-created electrons and openings can't deliver power in the silicon gem alone. There ought to be some extra system to do that.

The electron-hole pair (EHP) is produced incident of a photon of light energy ($hv>Eg$) over a solar cell. The newly created EHP relates to the electric current termed light-induced current denoted by IL. The ideal equation of a solar cell with current-voltage (I–V) is as follows:

$$\text{Solar cell current } (I) = I_{\text{ph}} - I_0\left[\exp\left(\frac{qV}{kT}\right) - 1\right] \tag{4.1}$$

TABLE 4.1
Summarization of Different Key Approaches Proposed in Literature

Paper Ref.	Objective	Strategy
Baci et al. (2020)	Here, the authors have depicted that the length of daylight outperforms 2000 hours every year and can arrive at 3900 hours in the sunnier countries and the Sahara. The significance of this work depends on using sun-oriented energy to create power.	The model of the new artificial sun-based tree is proposed tentatively by utilizing material accessible in the nearby market: 25 sun-based panels, metal uphold, electrical lines, controller, and battery.
Amagai et al. (2019)	Energy-reaping innovation is standing out as "empowering innovation" that extends the utilization and chances of IoT usage, enhances lives, and improves social flexibility. This innovation harvests energy that disseminates around us, like electromagnetic waves, heat, vibration, etc. by converting it into simpler form to be utilized as electric energy.	Here, the creator depicts the highlights of the energy-reaping advancements, late subjects, and significant difficulties, and strongly predicts the future possibilities of the turn of events.
Sharma et al. (2018)	The WSN hubs experience the ill-effects of a significant plan limitation that their battery energy is restricted and can turn out just for a couple of days relying on obligation pattern of activity.	We propose another answer for this planning issue by utilizing surrounding sun-based photovoltaic energy. Here, we propose a profoundly productive and special sunlight-based energy-reaping framework for battery-powered WSN hubs.
Akinaga (2020)	The principal commitment of this exploration article is to propose an effective sun-based energy-gathering answer for the restricted battery energy issue of WSN hubs by using surrounding sun-based photovoltaic energy. Preferably, the Optimized Solar Energy Harvesting Wireless Sensor Network (SEH-WSN) hubs ought to work for an endless organization lifetime.	We propose a novel and effective sun-based energy-collecting framework with heartbeat width adjustment (PWM) and the greatest force point following (MPPT) for WSN hubs. The exploration center is to expand general collecting framework proficiency, which further relies on sun-powered board effectiveness, PWM productivity, and MPPT effectiveness.
Koech et al. (2019)	With an end goal to make sun-oriented energy bridling more effective and moderate, different advances have been created. The sun-based warm innovations have accomplished amazing sun-oriented transformation efficiencies and are completely popularized. Notwithstanding, PV innovation is as yet going through quick development with an end goal to accomplish high efficiencies and to lessen the expense.	New materials, ideas, and approaches in sunlight-based cell advancement have become the focal point of exploration in this field. This article gives a survey on the advancement of PV innovation with a distinct fascination with arising PV materials that hold the possibilities for accomplishing high efficiencies at low expenses.

(*Continued*)

TABLE 4.1 (*Continued*)
Summarization of Different Key Approaches Proposed in Literature

Paper Ref.	Objective	Strategy
Eseosa and Kingsley (2020)	The paper is on a reenactment of MPPT utilizing P&O and INC techniques. A numerical model of 100 kW PV framework was created utilizing MATLAB® M-document. The two models were planned and reenacted utilizing MATLAB/Simulink® *t* is shown that the PV framework yield power increments with ascend in sun-powered illumination and lower cell temperature. Accordingly, sun-based cell performs better in warm climates over cold climates.	It is suggested that the MPPT framework should comprise partial, three-point, temperature-based MPPT for more successful and improved examination. More so, the annoy and noticed technique ought to be enhanced by fluctuating the irradiance to keep and increment consistent voltage.
Liu et al. (2015)	To create cause-explicit mortality parts, we included new crucial enlistment and verbal examination information.	We utilized fundamental enlistment information in nations with satisfactory enrollment frameworks. We applied indispensable enlistment-based multi-cause models for nations with low (under-5) mortality but updated verbal autopsy-based multi-cause models for high mortality nations.
Kumar et al. (2017)	This paper deals with simulation/modeling and controlling of MPPT used in PV framework to boost the output of photovoltaic framework, irrespective of the temperature of VI attributes of load.	In this exploration, a significant greatest force point following method has been created, comprising a lift converter, which is controlling heartbeat given by a microcontroller-based unit.
Kinjal et al. (2015)	Of late, sustainable power innovation has had a critical impact on energy application. One commendable sort of sustainable energy will be energy from the sun that creates electrical power straightforwardly by utilizing PV modules helped by MPPT calculations to create maximum possible harvested energy.	More or less, by changing the yield force of the inverter, the objective of accomplishing MPPT in PV frameworks is to change the conceivable working voltage of PV boards to the voltage at MPPT.
Kumar et al. (2015)	The solar force differs chiefly based on the climate conditions. Numerous new calculations have been projected to follow the most extreme force point (MPPT) of the close planetary system. This paper presents a similar investigation of two astute control techniques to enhance the proficiency of the sun-powered PV framework.	This paper presents in subtleties near investigation between incremental conductance calculation and fluffy logic regulator calculation applied to a DC-DC boost converter gadget. The boost converter expands yield voltage based on the obligation pattern of the switch gadget.

where I is the solar cell output current, I_{ph} is the light produced by solar cell, I_0 is the reverse current of saturation because of reconjunction, q is the electron charge (1.6×10^{-19}C), V is the open-circuit voltage of solar cell, k is Boltzmann's constant (1.38×10^{-23}J/K), and T is the solar cell temperature (300 K).

The circuit model in Figure 4.1 represents the equivalent of the solar cell. It comprises light-produced source current (I_{ph}), a Shockley equation-modeled diode (D), and two series and parallel resistances.

The maximum power point (MPP) is a point on the power-voltage (P–V) characteristic of the solar cell, where the maximum power can be extracted from the solar cell as shown in Figure 4.2. Ideally, the solar cell efficiency should be high. But practically, it is limited to 5%–15% only (Green et al., 2018).

In Figure 4.2, the current law of Kirchhoff (KCL) can provide a characteristic equation of current for that corresponding circuit:

$$\text{Equivalent cell output current}(I) = I_{ph} - I_D - I_P \tag{4.2}$$

where I_P is the parallel resistance current, I_{ph} is the light-produced current, and I_D is the diode current.

$$\text{Diode current }(I_D) = I_o\left[\exp\left(\frac{V + IR_s}{nV_T}\right) - 1\right] \tag{4.3}$$

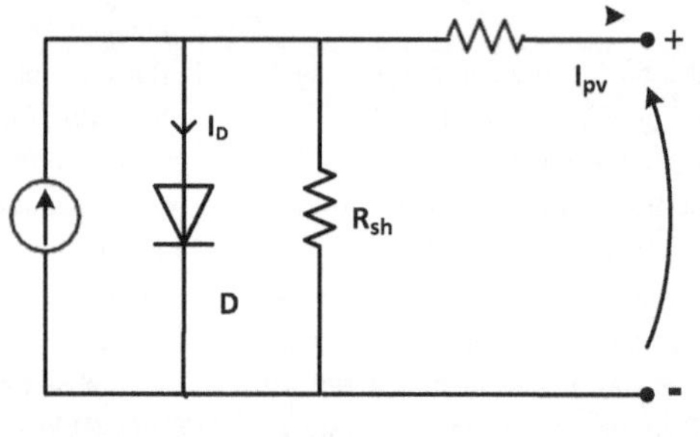

FIGURE 4.1 Single diode model of PV cell.

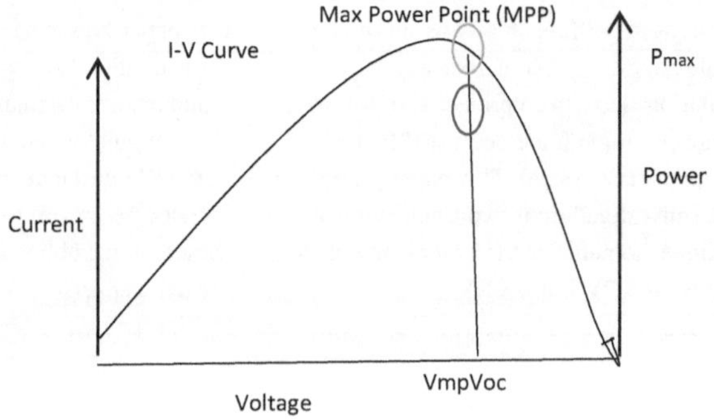

FIGURE 4.2 V-I and P-V characteristics.

where I_o is the reverse saturation current because of reconjunction, V is the solar cell open-circuit voltage, R_s is the series resistance, I_{pv} is the solar cell output current, n is the diode norm factor (1 termed as ideal, 2 termed as practical diode), k is Boltzmann's constant (1.38×10^{-23} J/K), V_T is the thermal voltage (kT/q), T is the solar cell temperature (300 K), and Q is the electron charge (1.6×10^{-19}C). The parallel resistance current is determined as follows:

$$\text{Current in parallel resistance}\left(I_p\right) = \frac{V + IpvR_s}{R_p} \tag{4.4}$$

Now, by placing the I_D and I_p values in Equation 4.2, we obtain complete equivalent circuit fourth equation of solar cell, under that all values are defined as connected with output current and voltage [9]:

$$\text{Solar cell current }(I) = I_L - I_0\left[\exp\left(\frac{q\left(V + IpvR_s\right)}{nkT}\right)\right] - \left(\frac{V + IR_s}{R_p}\right) \tag{4.5}$$

where R_p is the parallel resistance, and in Equation (4.3), the other parameters I_0, I_L, V, I, q, R_s, n, k, and T were already declared. The solar cell efficiency (η) is termed as follows:

$$\text{Solar cell efficiency }(\eta) = \frac{V_{OC} I_{SC} \text{FF}}{P_{in}} \tag{4.6}$$

where I_{SC} is the current short circuit, V_{OC} is the open-circuit voltage, FF is the fill factor, and Pin is the optical incident power. A solar cell's fill factor (FF) is given as follows:

$$\text{Fill factor }(FF) = \frac{P_{max}}{P_{dc}} = \frac{I_m V_m}{I_{SC} V_{OC}} \tag{4.7}$$

where V_m is the solar cell's maximum voltage and I_m is the maximum current. There are practically many kinds of solar cells, like amorphous silicon solar cells (a-Si), monocrystalline silicon solar cells (c-Si), thin-film solar cells (TFSCs), polycrystalline solar cells (multi-Si), etc. But the productivity of a-Si solar cells is greater than any other efficiency till 18%.

4.3.1 Solar Radiation Effect (G)

The efficiency of solar cells (η) is proportional to solar radiation's variations. The efficiency of solar cells (η) increases, if the solar radiation increases, and vice versa. Figure 4.3a displays the current-voltage (I–V) properties of a commercial solar panel of 10 W with varying values of irradiance. The solar panel of 10 W (Dow Chemical DPS 10-1000) is 232×546 mm in size and has a 0.13 m² module area. From Figure 4.3a, it is identified that the solar panel current increases with an increase in the degree of irradiance. Here, the solar cell current for solar irradiance of 1000 W/m² is optimum (6.2 A). Figure 4.3b shows the power-voltage properties of

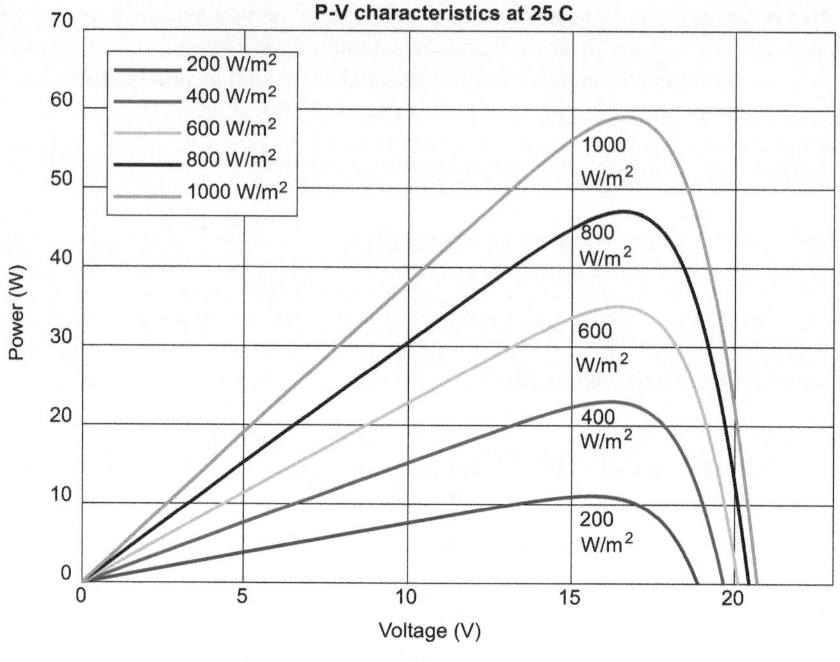

(a)

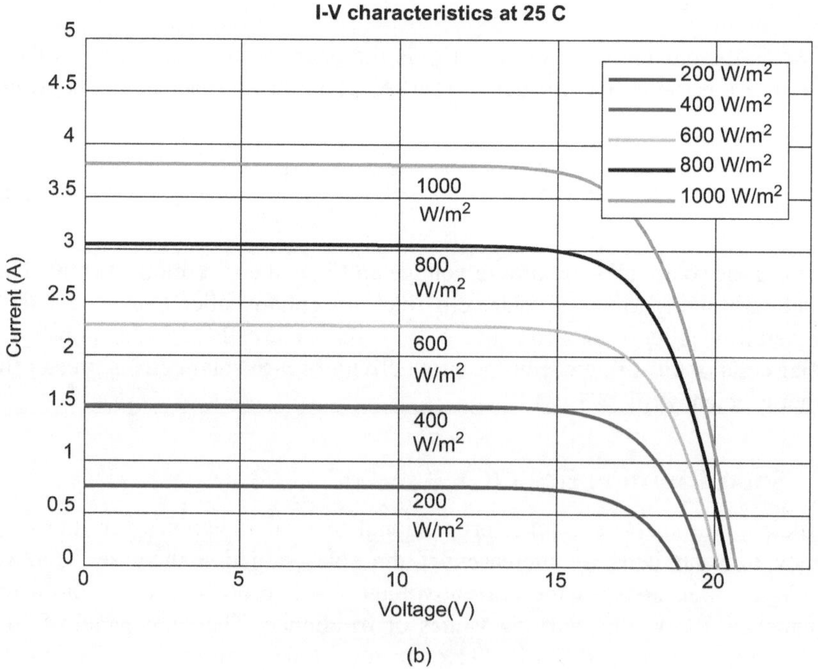

(b)

FIGURE 4.3 Solar panel characterization with irradiance-level variations (W/m²): (a) characteristics of (*I–V*) and (b) characteristics of (*P–V*).

solar panels in various radiation levels. For the highest solar irradiance like 1000, the extracted power is the optimum (9.8 W).

4.3.2 The Temperature Effect (T)

As shown in Figure 4.4a, if the temperature of the solar panel increases, then the production value decreases, and vice versa. The increase in output is in direct accordance with the fluctuations in temperature. Similarly, as the temperature in Figure 4.4b increases, output capacity decreases, and vice versa. Hence, the output power is inversely proportional to the variations of temperature.

4.4 SYSTEM FOR HARVESTING SOLAR ENERGY

A simple solar energy harvesting (SEH) system is a combination of rechargeable batteries, solar panels, DC-DC converter, battery management system (BMS) safety charging circuit, and DC-DC converter control unit. For DC-DC converters, control methods are generally MPPT control. The SEH unit in Figure 4.5 contains a rechargeable battery, DC-DC buck converter, MPPT solar panel and transmitter, and a WSN sensor node attached to the DC.

Solar energy from sunlight is collected in solar panels and transformed into electricity. The DC-DC buck converter is shut off, and this causes voltage magnitude to be controlled and transferred to the same rechargeable unit. An MPPT sensor controls the solar panel's current and voltage, changing the duty period as a Buck MOSFET DC-DC converter (Mathews et al., 2015).

Finally, the wireless sensor node is regulated by the voltage of the batteries. The WSN performs the role of detecting, analyzing, and interacting the same characteristics with other nodes. Thus, as with vibration, temperature, acceleration, and humidity, the SEH-WSN nodes can be used to track and control any physical phenomenon autonomously. In this scenario, the solar harvester circuit's efficiency exhibits a very significant function. If the solar power harvester's performance is low, the battery will not be recharged sufficiently, thereby reducing the lifespan of the WSN.

4.4.1 DC-DC Converter Modeling

In the configuration of a PV system, there are usually three kinds of DC-DC converters operationally, namely boost converter, buck converter, and buck-boost converter. Here, we used the DC-DC buck converter; their performance is very high as compared with those of two other converters. The DC-DC buck converter is an electric power generator with a lower output voltage than the source of electricity. The buck converter includes a DC voltage transmitter (V_{dc}), a MOSFET switch, an inductor (L), a condenser (C), and a diode (D), as shown in Figure 4.6. Upon closure of t_1 MOSFET (S), Vs input voltage is expressed via a load resistor. If for the time t_2, the MOSFET (S) is OFF, then the tension around the charging resistor is high. The output voltage amplitude (V_0) is lesser than the input voltage amplitude (V_o). The operating time (D) ranges marginally from 0 to 1. The running time of buck converter is $D = V_0/V_{in}$,

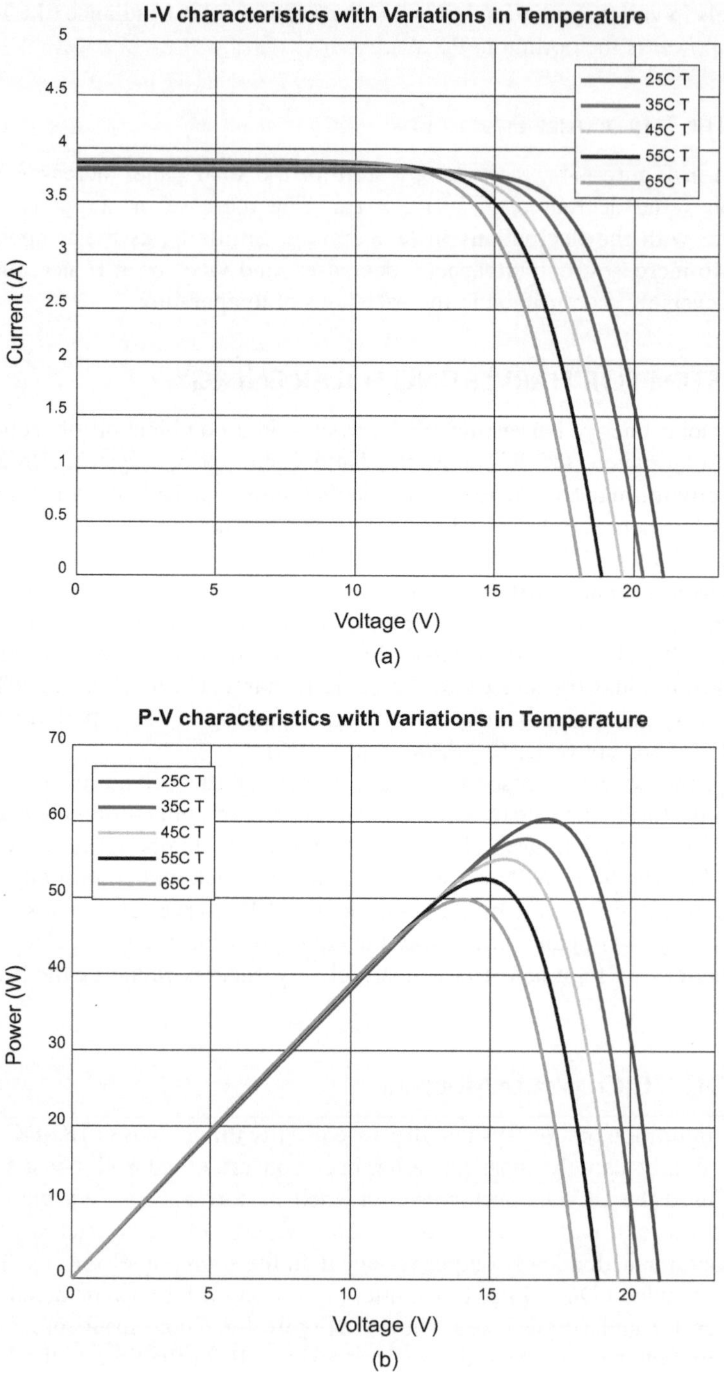

FIGURE 4.4 Characteristics of solar panels with temperature (°C) variations: (a) characteristics of (*I–V*) and (b) characteristics of (*P–V*).

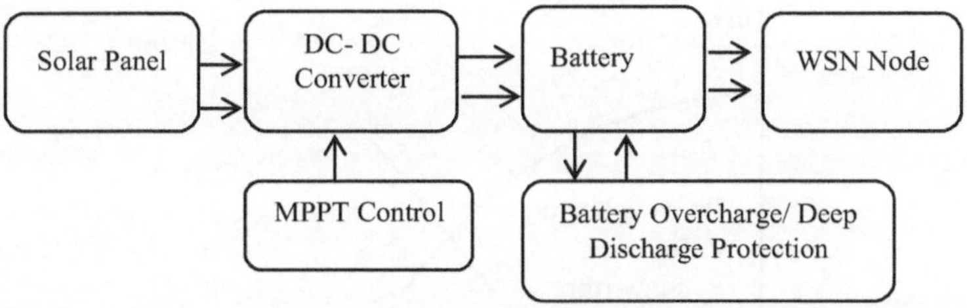

FIGURE 4.5 Solar energy recovery (harvesting) system block diagram, using input from MPPT capacity.

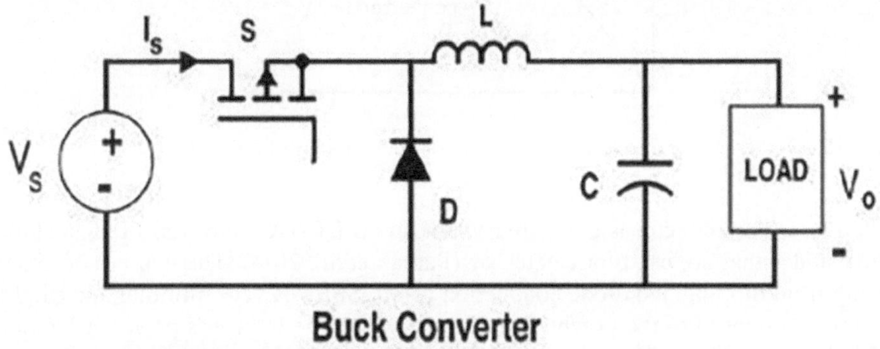

FIGURE 4.6 A DC-DC buck converter circuit.

depending on the time difference t_1. The average output voltage for buck converters is termed as follows:

$$V_0 = \frac{1}{T}\int_0^{t_1} v_0 dt = \frac{t_1}{T} V_{\text{in}} = f \cdot t_1 \cdot V_{\text{in}} = V_{\text{in}} \cdot D \tag{4.8}$$

where V_{in} is the input voltage, V_0 is the output voltage, t_1 is the MOSFET (S) ON time, T is the total time, f_s is the operating frequency, and D is the duty cycle.

The average load current is given at the output as follows:

$$I_0 = I_L = \frac{V_0}{R} = D \cdot V_{\text{in}} / R \tag{4.9}$$

where T is the period of chopping, f is the frequency of chopping, and duty cycle $D = t_1/T$.

4.4.2 DC-DC Buck Converter Power Losses

The DC-DC buck converter (Sanchez et al., 2013) has three major power dissipation components: inductor conduction loss, MOSFET conductivity loss, and MOSFET

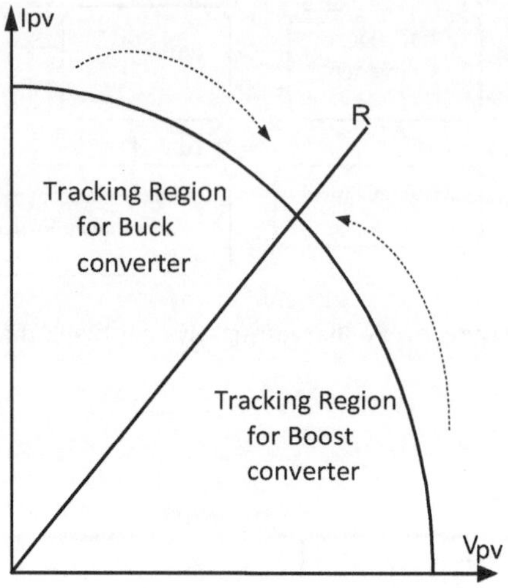

FIGURE 4.7 Tracking of optimal resistance for buck and boost converter.

switch loss. The inductor is the primary sink in all DC-DC converter (Taneja et al., 2018) modes that are used for electricity (Pathak et al., 2014). The number of losses arising from flipping and diode conduction of MOSFET is very minimal and can be ignored as opposed to the potential inductor losses. The inducer's power utilization deficit is indicated as follows:

$$P_L = I_{L(\mathrm{rms})}^2 \times R_{L(\mathrm{dc})} \tag{4.10}$$

where P_L is the inductor power loss (mW), $I_{L(\mathrm{rms})}$ is the RMS current of inductor, and $R_{L(\mathrm{dc})}$ is the inductor's DC resistance.

Figure 4.7 shows the tracking range for the buck and boost converter for solar energy to the electrical energy (Al-Bahadili et al., 2013). This is the graph that shows the value of resistance between the PV voltage and the current.

4.5 MAXIMUM POWER POINT TRACKING (MPPT) MODELING TECHNIQUE

MPPT techniques (Hauke, 2009) are mostly used in the formation of solar PV systems to get maximum extraction of power from the sun under various solar irradiance values and has a kind of algorithm that continually tests current (I_{PV}) and voltage (V_{PV}) from the solar panel and computes the quantity of the duty cycle (D) for supplying to DC-DC buck converter MOSFET switch. In PV applications, typically the following algorithms are used (Texas, 2018):

- Incremental conductance (INC)
- Fraction open-circuit voltage (OCV)
- Perturb and observe (P&O).

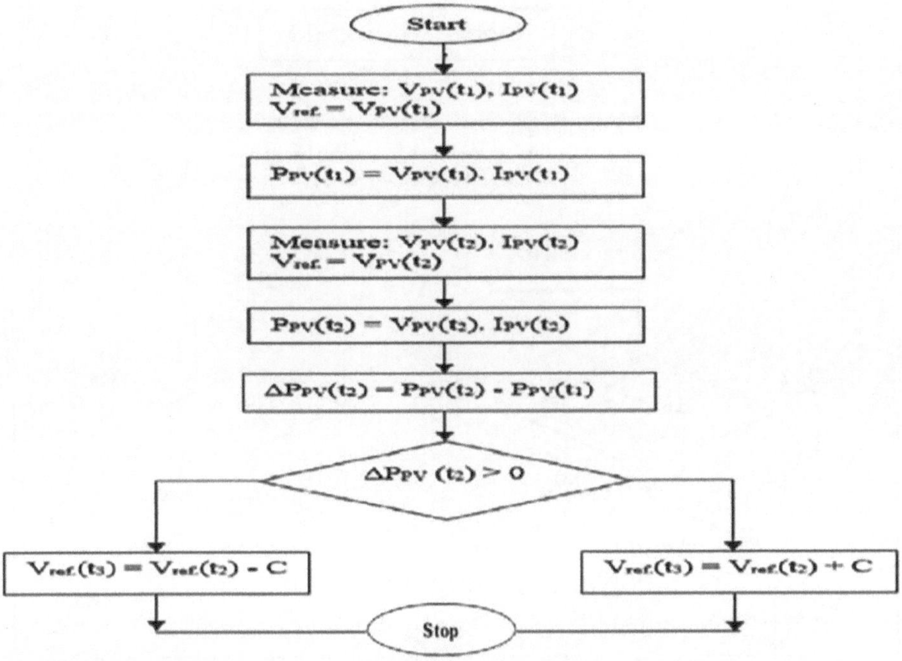

FIGURE 4.8 Flowchart for the P&O method.

The P&O method is mainly utilized in all forms of harvester systems with solar energy. Figure 4.8 is a flow diagram for the P&O algorithm. This algorithm's output is having a duty cycle (DD) variation that depends on the irradiance (W/m^2) input. If irradiance decreases, then the service cycle shifts and the voltage and current of the solar panel change (Haque, 2014). MPPT technique senses those alterations and adjusts solar panel impedance to the maximum point of power. So even if the irradiance changes, maximum power (P) may continue to extract by the solar panel. It produces a PWM (pulse width modulation) waveform of 0.7 initial duty cycle (ranging from 0 to 1) given as seed value for simulation.

The P&O method operates on the theory of balancing impedance between charge and solar panel. The impedance matching is important for optimum power transfer. Utilizing a DC-DC converter, this impedance matching is accomplished. Through utilizing a DC-DC converter, impedance is changed by adjusting the MOSFET switch's service cycle (DD). The relationship between input voltage, output voltage, and the duty cycle is given as follows:

$$V_o = V_{\text{in}} \cdot D \tag{4.11}$$

$$R_{\text{in}} = R_L / D^2 \tag{4.12}$$

But if the duty cycle varies (DD), then the output voltage of the solar energy harvester (V_o) changes. The output voltage (V_o) is therefore increased when the duty cycle (D) is

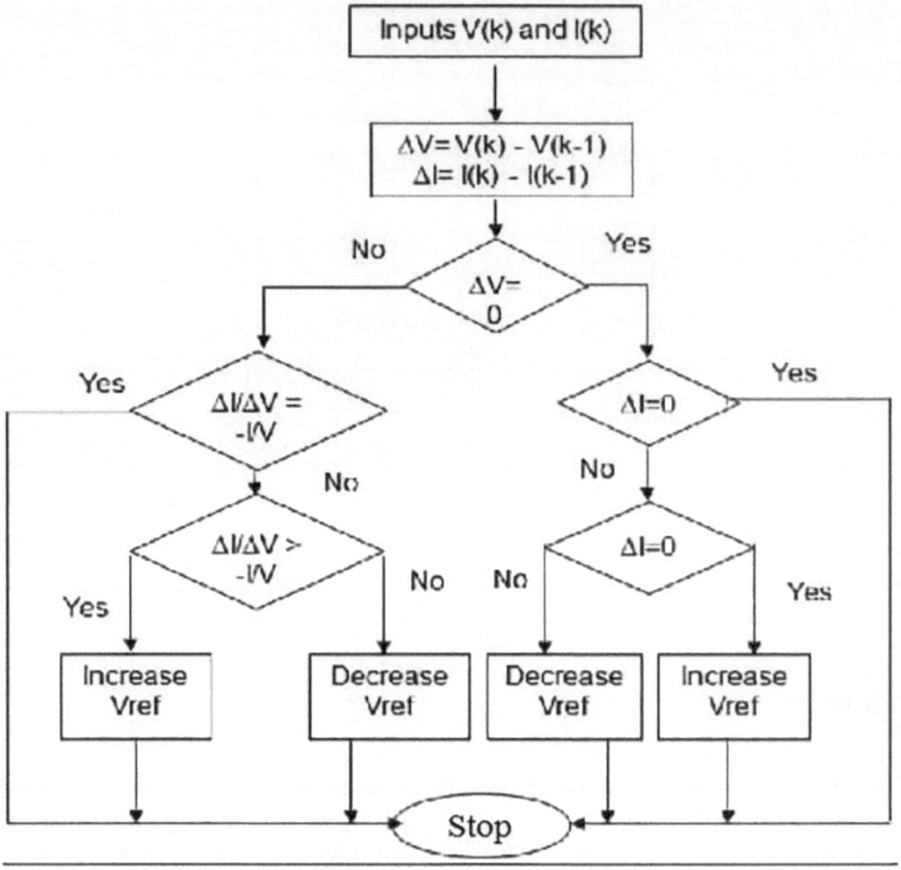

FIGURE 4.9 Flowchart for the incremental conductance method.

prolonged, and vice versa. The impedance of load resistance (R_L) may be compared to the impedance of the solar panel input by adjusting the duty cycle (D) for efficient transfer of power to the load for optimal output.

A flowchart displays the steps in the P&O algorithm, and MATLAB codes are depicted in Algorithm 4.1.

Algorithm 4.1

P&O-MPPT ALGORITHM

```
function D= PandO(Vpv, Ipv)
persistent DprevPprevVprev
if (isempty (Dprev))
Dprev=0.7;
Vprev= 190;
```

```
Pprev= 2000;i
end
deltaD= 0.0025;i
Ppv= Vpv *Ipv;
if (Ppv-Pprev)~= 0
if (Ppv-Pprev)> 0
if (Vpv-Vprev)> 0
D=Dprev– deltaD;
else
D= Dprev+deltaD;
end
else
if (Vpv-Vprev)> 0
D=Dprev+deltaD;
else
D=Dprev-deltaD;
end
end
else
D = Dprev;
end
Dprev = D;
Vprev = Vpv;
Pprev = Ppv;
```

4.6 INCREMENTAL CONDUCTANCE (INC) ALGORITHM

The process of INC is also used for PV systems. It monitors MPP by comparing the PV array's instant and INC. The INC system problem is close to P&O's. Usually, the fixed-step size is used, which computes MPPT's speed and accuracy of response. Therefore, the tradeoff between tracking speed and steady-state efficiency has to be made. Such architecture problems can be stable with MPPT strategies of variable-step duration.

The power with respect to voltage derivative (dP/dV) is utilized to change the MPPT phase scale. The INC technique surmounts the perturbation disadvantage and observes the method in monitoring peak power in rapidly changing atmospheric situations. This method will decide if MPPT has passed the MPP and also stops the perturbing point of service. If the condition is not true, it is possible to determine direction in the MPPT operating point using the relation between dI/dV and $-I/V$. This relation is found from the fact that when MPPT is on the right side of MPP, dP/dV is negative and is positive when on the left side of MPP. The phase size increases when the operating point is far from the MPP and progressively decreases when the operating point comes close to the MPP.

The quick tracking velocity and steady performance can be achieved simultaneously by changing the phase size. However, the MPPT algorithm convergence involves a scaling factor, and the factor significantly decreases response speed under abrupt changes in atmospheric situations. An MPPT algorithm with incremental resistance (INR) is to be tested with adjusted step variable size (Ibrahim et al., 2017). For moving between the fixed-step and variable-step mode, a threshold function is applied, and the variable-step phase is felt by scaling factor variation.

This method obtains quick response and precisely steady-state output, but its implementation is limited by the high computing load and the best nonlinearity of scaling factor. There are really two phase size alteration coefficients in Li and Shi (2015) to minimize the perturbation effects (duty ratio) underneath the drastic shift in irradiation with less computing, while the influence of the basic step size upon method efficiency is not considered.

This algorithm decides when MPPT hits MPP, while P&O toggles around MPP itself. This reflects a benefit over P&O. The INC could also monitor rapidly increasing and decreasing conditions of irradiance with greater accuracy than disturbance, and observe method (Praveen et al., 2016). The downside of the algorithm, compared to P&O, is that it is more complicated. The flow map shown in Figure 4.9 makes the algorithm simple to understand.

4.7 SIMULATION EXPERIMENT SETUP

As shown in Figure 4.10, we used MATLAB/Simulink R2017b to model a solar power boost converter control to load a WSN node package. Figure 4.11 displays the

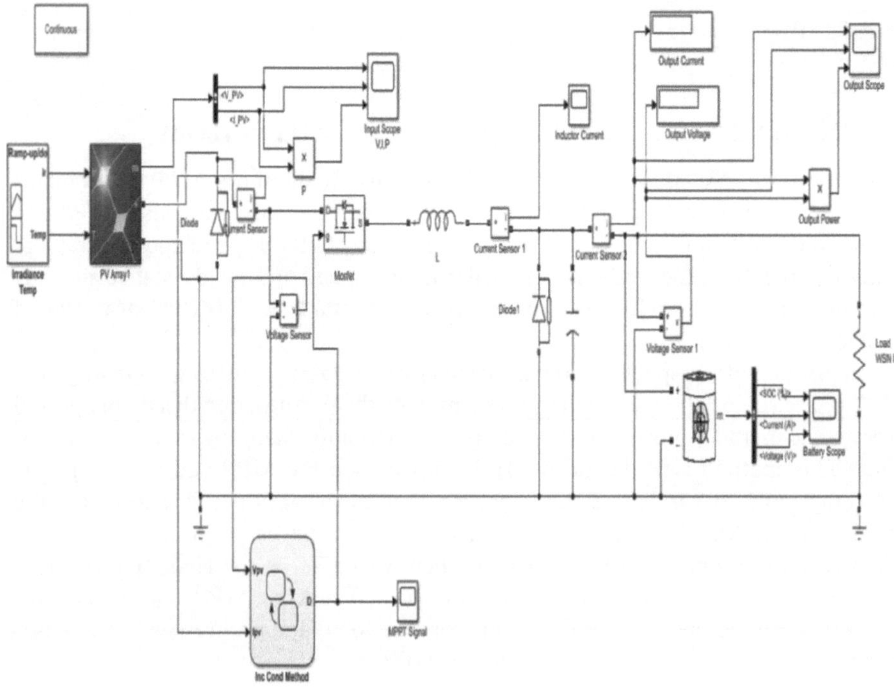

FIGURE 4.10 WSN node solar-powered buck converter with INC-MPPT battery charging control.

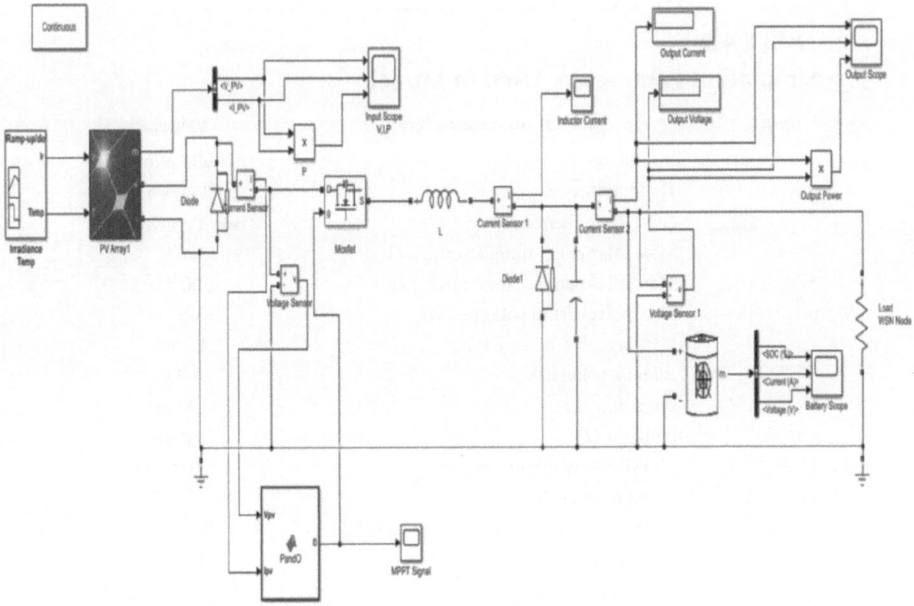

FIGURE 4.11 Solar-powered buck converter with P&O-MPPT control for battery charging of WSN node.

MATLAB/Simulink solar harvester model with MPPT functions. A solar irradiance of 1000 W/cm² occurs on the solar panel at a steady temperature of 25°C (Win, et al., 2010). Only this solar energy can be converted with an efficiency of 15% by the solar panel to 15 mW/cm² (LM2575, 1A, i3.3v-15v, 2018). For maximum irradiance, the performance of the simulation computer on the solar panel is 6.0 V, 500.0 mA, and 3.0 W (Castagnetti et al., 2012). That is why, the DC-DC boost converter provides this solar power, which increases the voltage output. The motor is operated by output voltage from a boost converter (Weddell et al., 2011). The WSN node operates on rechargeable batteries. Here, the WSN-load output along with a load resistance of 100.0 Ω is modeled against DC. Table 4.1 shows different simulation parameters such as irradiance, temperature, solar panel current (Moghadam et al., 2015), DC-DC configuration, voltage and capacity, duty cycle, battery type and battery size, WSN charging model, and loss of power.

Table 4.2 has three columns that are showing the simulation parameter for the used model in which the first one shows serial numbers, the second one shows simulation parameters' names, and the third one shows their values.

4.8 SIMULATION RESULTS

Simulation graphs of P&O-MPPT and INC-MPPT shown in Figures 4.12 and 4.13.

In Figures 4.12 and 4.13 above, for a simulation time of 10 seconds, three parameters of the regulated P&O-MPPT and INC-MPPT battery charger SEH mean that a state of charge (SOC), battery current, and voltage are obtained. The P&O-MPPT

TABLE 4.2
Simulation Parameters Used in Model

S. No.	Simulation Parameters	Values
1.	Irradiance	1000 (W/m^2)
2.	Temperature	25 (°C)
3.	DC-DC converter	Boost converter
4.	Max solar panel output voltage (V_m)	6.0V
5.	Max solar panel output current (I_m)	500 Ma
6.	Max power from solar cell (P_m)	3.0 W
7.	Rechargeable battery type	Li-ion
8.	Battery voltage	3.3V
9.	Capacitor (c)	100 uF
10.	Inductor (L)	200 uH
11.	MOSFET switching frequency (f)	5.0 KHz
12.	Initial duty cycle	0.5
13.	MOSFET switching power losses (P_{CW})	0.5 mW
14.	Switching voltage loss (V_{SW})	0.2V
15.	WSN-load model	10–Ω resistor
16.	Inductor conduction power loss (P_L)	50 mW

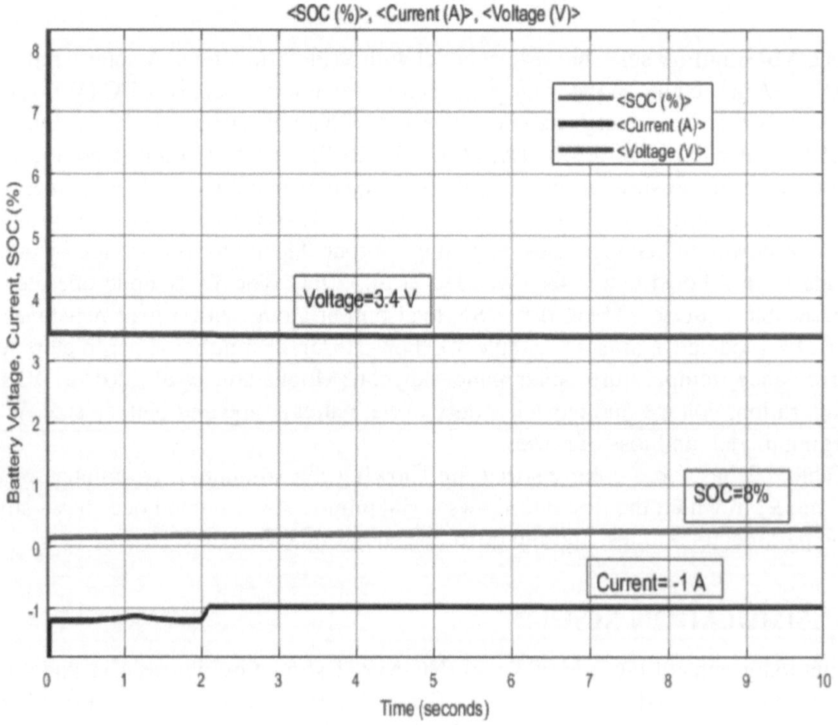

FIGURE 4.12 P&O-MPPT managed 10-second SEH system.

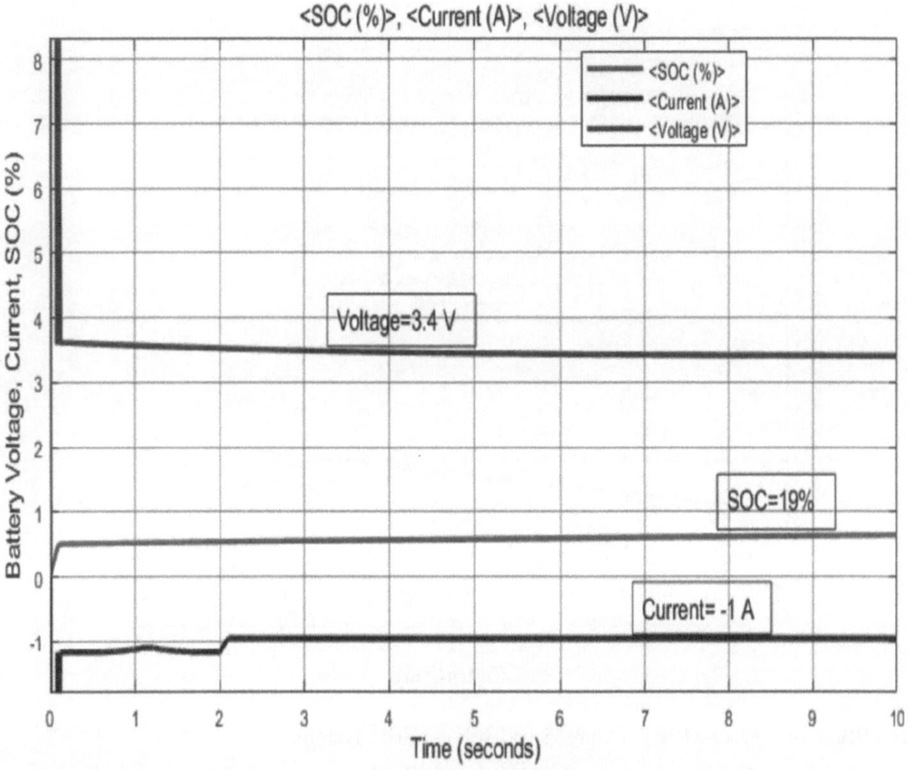

FIGURE 4.13 INC-MPPT managed 10-second SEH systems.

SOC dropped to 8%, while the INC-MPPT SOC dropped to 19%. SOCs view the INC-MPPT differently than P&O-MPPT.

Figure 4.14 shows the *x*-axis time of the P&O-MPPT managed 100-second SEH system, and the *y*-axis shows the battery voltage current. Figure 4.15 shows INC-MPPT managed 100-second SEH system, and the *y*-axis shows the battery voltage.

In Figures 4.14 and 4.15 above, again all the three parameters are obtained for a simulation time of 100 seconds for P&O-MPPT and INC-MPPT. Both show an increment in battery state of charge. P&O-MPPT SOC reaches 44%, while INC-MPPT SOC reaches 53%. Again INC-MPPT shows better increment than P&O-MPPT on different simulation times.

Figure 4.16 shows the *x*-axis time of the P&O-MPPT managed 200-second SEH system, and the *y*-axis shows the battery voltage current. Figure 4.17 shows INC-MPPT managed 200-second SEH system, and the *y*-axis shows the battery voltage.

In Figures 4.16 and 4.17, again all the three parameters are simulated for P&O-MPPT and INC-MPPT both for a simulation time of 200 seconds. Since the time increases, the SOC automatically increases. For this increased simulation time, the SOC of P&O-MPPT reaches 81%, and the SOC of INC-MPPT reaches 90%. The rate of increment of SOC is quite good in INC-MPPT in comparison with P&O-MPPT.

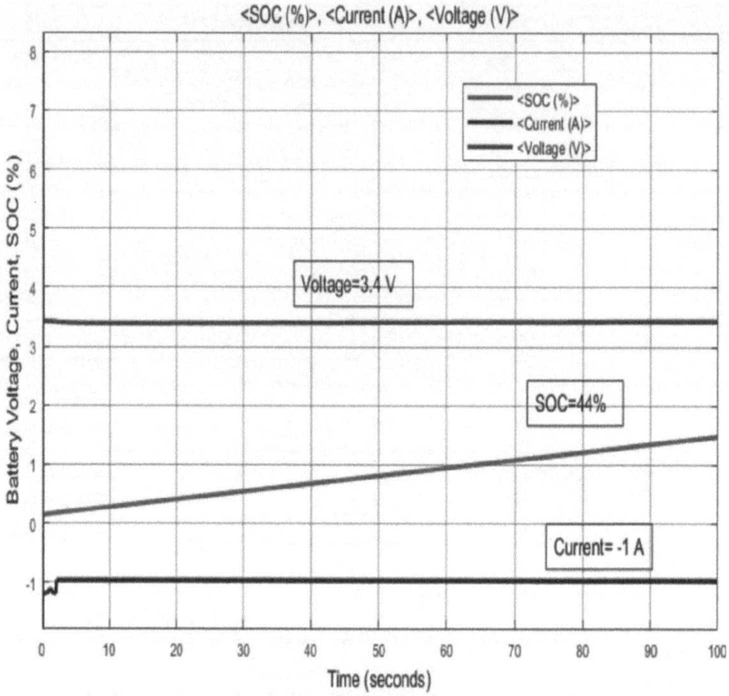

FIGURE 4.14 P&O-MPPT managed 100-second SEH system.

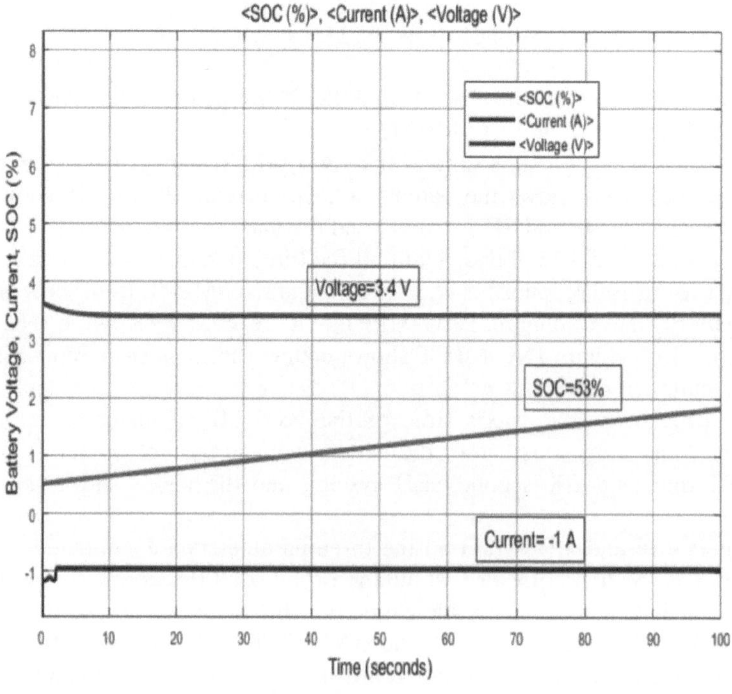

FIGURE 4.15 INC-MPPT managed 100-second SEH system.

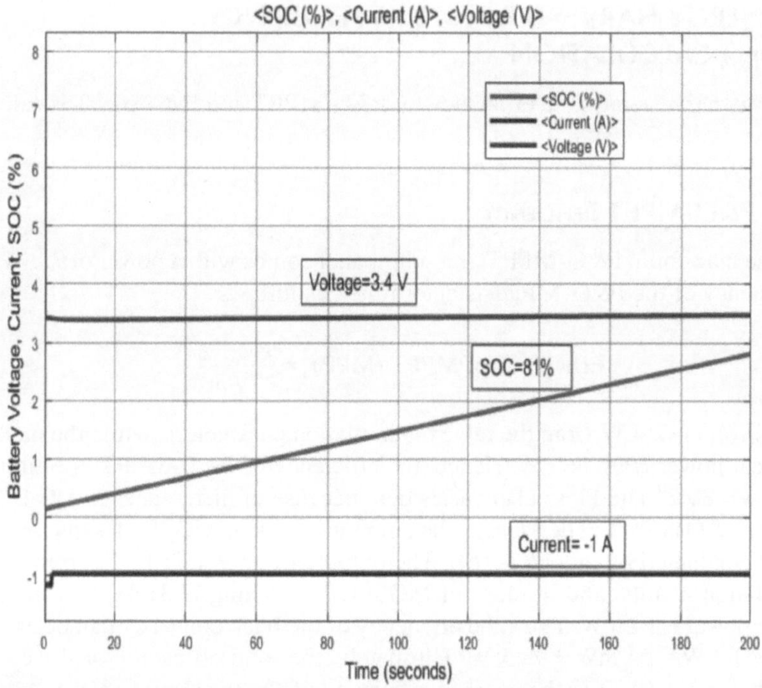

FIGURE 4.16 P&O-MPPT controlled 200-second SEH system.

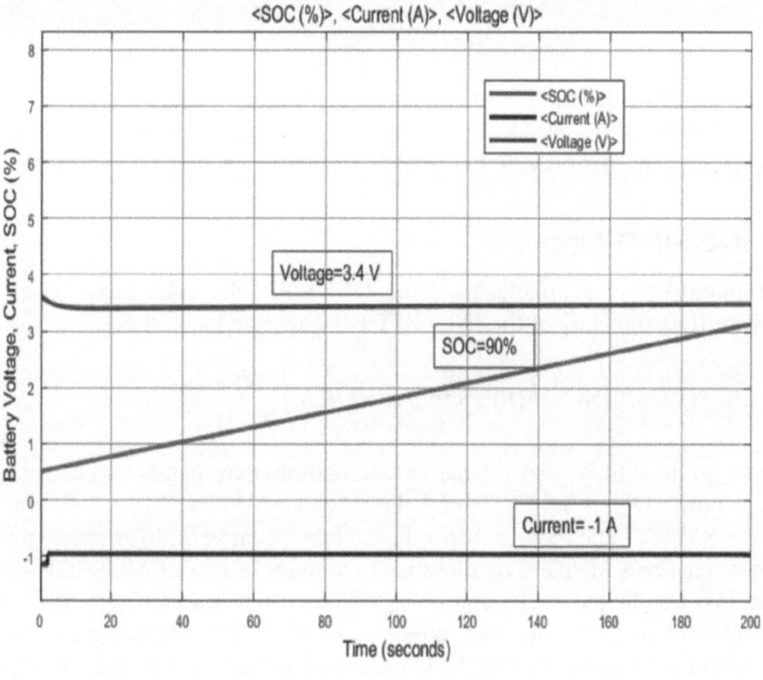

FIGURE 4.17 INC-MPPT controlled 200-second SEH system.

4.9 ENERGY HARVESTER SYSTEMS EFFICIENCY (η_{sys}) CALCULATION

The energy harvester system efficiency for P&O-MPPT and INC-MPPT is calculated below.

4.9.1 P&O-MPPT Efficiency

Using the maximum P&O-MPPT, the solar panel comes with a power of 2.8 W. Now, the efficiency of the P&O-MPPT is determined as follows:

$$\text{Efficiency of MPPT (MPP)} = \frac{P_{\text{P\&O MPP}}}{Pm} \tag{4.13}$$

The ($P_{\text{P\&O MPP}}$) is 2.4 W from the table of simulation parameters, while the maximum theoretical power (Pm) is 3 W. Hence, the efficiency of P&O-MPPT is estimated at 2.4 w/3 w = 80%. The Ploss also varies here because of differences in DC-DC buck converter P&O-MPPT. The Ploss is the amount of loss of MOSFET switching (P_{SW}) and loss conducted by inductors (P_L). The output power (P_O) is 1.7 W from the table of simulation results, and the loss of MOSFET switching is 3 mW, and the loss of inductor power is 22 mW. Thus, the efficiency of the buck converter can be estimated as 1.7 W/1.7 W+ 25 mW = 98.55%. Ultimately, the total efficiency of the circuit of energy harvester (η_{sys}) is the average efficiency of the buck converter and the efficiency of the P&O-MPPT.

Total energy harvester system efficiency utilizing MPPT is given by

$$\left(\eta_{sys}\right) = \frac{\eta_{\text{buck}} + \eta_{\text{MPP}}}{2} \tag{4.14}$$

By Equation (4.14), the total energy harvester efficiency (η_{sys}) estimated utilizing P&O-MPPT is (98.55% + 80%)/2 = 89.27%. The overall efficiency of the solar energy harvester $\left(\eta_{sys}\right)$ using P&O-MPPT is therefore 89.27%.

4.9.2 INC-MPPT Efficiency

The maximum power available, by using INC-MPPT by solar panel, comes with 2.6 W. Now, the efficiency of the INC-MPPT is computed as follows:

$$\text{INC} - \text{MPPT efficiency}\left(\eta_{\text{MPP}}\right) = \frac{P_{\text{INC MPP}}}{p_m} \tag{4.15}$$

The ($P_{\text{INC MPP}}$) is 2.6 W by a table of simulation parameters, while maximum theoretical power (P_m) is 3.0 W. Hereby, the efficiency of P&O-MPPT is estimated as 2.6 w/3 w = 86.66%. The Ploss is also adjusted here because of differences in DC-DC buck converter P&O-MPPT. The Ploss is the amount of loss of MOSFET switching (P_{sw}) and loss conducted by inductors (P_L). The output power (P_o) is 1.87 W from the table of simulation results, and the losses of MOSFET switching are 2 mW and the loss of inductor power is 18 mW. Thus, the efficiency of the buck converter can

TABLE 4.3
Simulation Results for the Controlled SEH System P&O-MPPT and INC-MPPT

S. No.	Energy Harvesting Parameters	P&O	INC
1.	Max solar panel output power (P_m)	2.4 W	2.6 W
2.	Average buck converter output voltage (V_m)	3.4V	3.4V
3.	Average buck converter output current (I_m)	500 mA	550 mA
4.	Buck converter output power	1.7 W	1.87 W
5.	Inductor loss	22.0 mW	18.0 mW
6.	MOSFET switching loss	3.0 mW	2.0 Mw
7.	Harvester system efficiency (%)	89.27	92.80

be estimated as 1.87 W/1.87 W + 20 mW = 98.94%. Ultimately, the energy harvester efficiency circuit (η_{sys}) on average is the efficiency of buck converter and the efficiency of P&O-MPPT.

$$\text{Total energy harvester systems efficiency utilizing } MPPT(\eta_{sys}) = \frac{\eta_{buck} + \eta_{MPP}}{2} \quad (4.16)$$

From Equation (4.16), the measured overall efficiency of the energy harvester device (η_{sys}) using INC-MPPT is (98.94% + 86.66%)/2 = 92.8%. The overall efficiency (η_{sys}) of the solar energy harvester using INC-MPPT is therefore 92.8%.

Table 4.2 shows the effects of the simulation on SEH systems operated by P&O-MPPT and INC-MPPT. Here, you will report the average buck converter voltage output (V_m), maximum solar output (P_m), buck converter performance, average buck converter output current (I_m), MOSFET switch breakdown, inducer failure, and harvester efficiency. No doubt, the INC-MPPT-regulated technique provides better results in view of output current, voltage, power, losses, and efficiency compared to P&O-MPPT control from Table 4.3.

Figure 4.18 shows the final result comparison in which the *x*-axis shows the P&O-MPPT and INC-MPPT values and the *y*-axis shows the efficiency.

4.10 CONCLUSION

In this work, modeling, simulation, and optimization are executed for SEH-WSN nodes. The two evolutionary computing control methods for harvester systems with solar energy, means, analysis, and comparison of P&O-MPPT and INC-MPPT have been performed by MATLAB simulation. On various simulation times, INC-MPPT and P&O-MPPT are simulated, and in every simulation time, INC-MPPT gives better results than P&O-MPPT. Also, the overall efficiency of P&O-MPPT is 89.27% and that of INC-MPPT is 92.8% which is better than the efficiency of the P&O-MPPT SEH-controlled technique. Thus, INC-MPPT is quite a promising technique. For future work, the other parameters will also be used to increase the efficiency. The other algorithm will also be used to increase the efficiency.

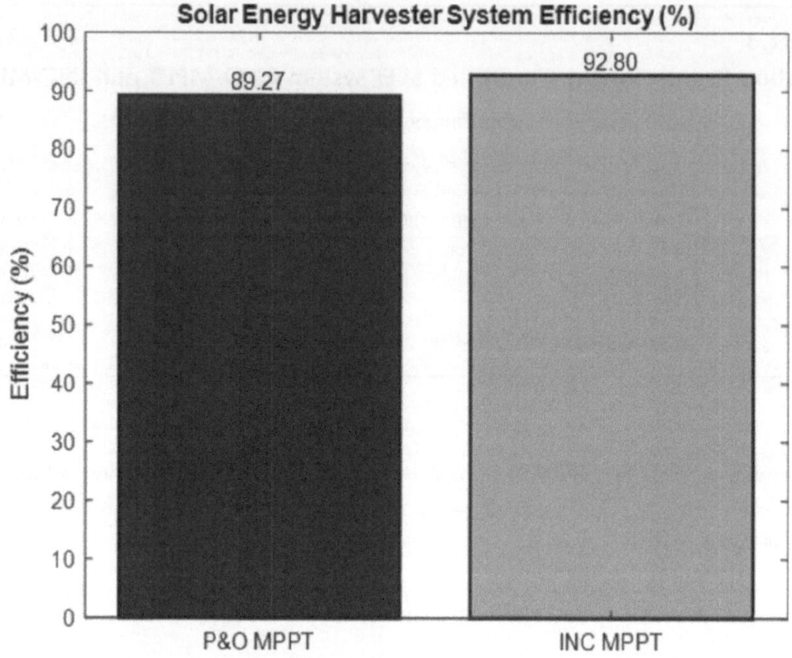

FIGURE 4.18 P&O-MPPT and INC-MPPT SEH-controlled techniques.

4.11 FUTURE SCOPE

All of the outcomes and comparisons for the proposed system are mentioned above, indicating that INC-MPPT is an efficient strategy. Other criteria can be used to maximize the productivity in future work as well. Some artificial intelligence-based algorithms can also be used to improve performance.

REFERENCES

Ahmad, T., Sobhan, S., & Nayan, M. F. (2016). Comparative analysis between single diode and double diode model of PV cell: concentrate different parameters effect on its efficiency. *Journal of Power and Energy Engineering*, *4*(3), 31–46.

Akinaga, H. (2020). Recent advances and future prospects in energy harvesting technologies. *Japanese Journal of Applied Physics*, *59*(11), 110201.

Al-Bahadili, H., Al-Saadi, H., Al-Sayed, R., & Hasan, M.A.-S. (2013), "Simulation of maximum power point tracking for photovoltaic systems", Applications of information technology to renewable energy processes and systems (IT-DREPS), *1st International Conference & Exhibition,* 2013, 79–84.

Amagai, Y., Shimazaki, T., Okawa, K., Fujiki, H., Kawae, T., & Kaneko, N. H. (2019). Precise measurement of absolute Seebeck coefficient from Thomson effect using AC-DC technique. *AIP Advances*, *9*(6), 065312.

Baci, A. B., Salmi, M., Menni, Y., Ghafourian, S., Sadeghzadeh, M., & Ghalandari, M. (2020). A new configuration of vertically connecting solar cells: solar tree. *International Journal of Photoenergy*, *2020*, 1–8. 8817440.

Castagnetti, A., Pegatoquet, A., Belleudy, C., & Auguin, M. (2012). A framework for modeling and simulating energy harvesting WSN nodes with efficient power management policies. *EURASIP Journal on Embedded Systems*, *2012*(1), 1–20.

Choudhary, D., & Saxena, A. R. (2014). Incremental conductance MPPT algorithm for PV system implemented using DC-DC buck and boost converter. *International Journal of Engineering Research and Applications*, *4*(8), 123–132.

Eseosa, O., & Kingsley, I. (2020). Comparative study of MPPT techniques for photovoltaic systems. *Saudi Journal of Engineering and Technology*, *5*, 12–14. Doi: 10.36348/sjet.2020.v05i02.002.

Green, M. A., Hishikawa, Y., Dunlop, E. D., Levi, D. H., Hohl-Ebinger, J., & Ho-Baillie, A. W. (2018). Solar cell efficiency tables (version 51). *Progress in Photovoltaics: Research and Applications*, *26*(1), 3–12.

Haque, A. (2014). Maximum power point tracking (MPPT) scheme for solar photovoltaic system. *Energy Technology & Policy*, *1*(1), 115–122.

Hauke, B. (2009). Basic calculation of a boost converter's power stage. *Texas Instruments, Application Report November, SLVA372C*, 1–9.

Ibrahim, R., & Chung, T. D. (2018). Solar energy harvester for industrial wireless sensor nodes. *Procedia Computer Science*, *105*, 112–119.

Ibrahim, R., Chung, T. D., Hassan, S. M., Bingi, K., & binti Salahuddin, S. K. (2017). Solar energy harvester for industrial wireless sensor nodes. *Procedia Computer Science*, *105*, 111–118.

Kinjal, P., Shah, K. B., & Patel, G. R. (2015, January). Notice of removal: comparative analysis of P&O and INC MPPT algorithm for PV system. In *2015 International Conference on Electrical, Electronics, Signals, Communication and Optimization (EESCO)* (pp. 1–6). IEEE.

Koech, R. K., Kigozi, M., Bello, A., Onwualu, P. A., & Soboyejo, W. O. (2019, August). Recent advances in solar energy harvesting materials with particular emphasis on photovoltaic materials. In *2019 IEEE PES/IAS Power Africa* (pp. 627–632). IEEE.

Kumar, M. Kapoor, S. R. Nagar, R., & Verma, A. (2015). Comparison between IC and fuzzy logic MPPT algorithm based solar PV system using boost converter. *International Journal of Advanced Research in Electrical, Electronics and Instrumentation Engineering*, *4*(6), 4927–4939.

Kumar, R., Choudhary, A., Koundal, G., & Yadav, A. S. A. (2017). Modelling/simulation of MPPT techniques for photovoltaic systems using MATLAB. *International Journal*, *7*(4), 178–187.

Li, Y., & Shi, R. (2015). An intelligent solar energy-harvesting system for wireless sensor networks. *EURASIP Journal on Wireless Communications and Networking*, *2015*(1), 1–12.

Liu, L., Oza, S., Hogan, D., Perin, J., Rudan, I., Lawn, J. E., & Black, R. E. (2015). Global, regional, and national causes of child mortality in 2000–13, with projections to inform post-2015 priorities: an updated systematic analysis. *The Lancet*, *385*(9966), 430–440.

LM2575, 1A, i3.3v-15v. (2018). Adjustable output voltage, iStep-down switching regulator. on semiconductor company datasheets. 2009. Available online: http://onsemi.com (accessed on 28 June2018).

Mathew, A., & Selvakumar, A. I. (2006). New MPPT for PV arrays using fuzzy controller in close cooperation with fuzzy cognitive network. *IEEE Transactions on Energy Conversion*, *21*(3), 793–803.

Mathews, I., King, P. J., Stafford, F., & Frizzell, R. (2015). Performance of III–V solar cells as indoor light energy harvesters. *IEEE Journal of Photovoltaics*, *6*(1), 230–235.

Moghadam, H. M., Khalili, A., & Mohammadinodoushan, M. (2015). Comparison study of maximum power point tracker techniques for PV systems in the grid connected mode. *International Journal of Review in Life Sciences*, *5*(10), 1175–1184.

Mohamed, S. A., & Abd El Sattar, M. (2019). A comparative study of P&O and INC maximum power point tracking techniques for grid-connected PV systems. *SN Applied Sciences, 1*(2), 174.

Pathak, G., Saxena, A. R., & Bansal, P. (2014). Review of dimming techniques for solid-state LED lights. *International Journal of Advanced Engineering Research and Technology (IJAERT), 2*(4), 108–114.

Praveen, K., Pudipeddi, M., & Sivaramakrishna, M. (2016, December). Design, development and analysis of energy harvesting system for wireless pulsating sensors. In *2016 IEEE Annual India Conference (INDICON)* (pp. 1–5). IEEE.

Rasheduzzaman, M., Pillai, P. B., Mendoza, A. N. C., & De Souza, M. M. (2016, July). A study of the performance of solar cells for indoor autonomous wireless sensors. In *2016 10th International Symposium on Communication Systems, Networks and Digital Signal Processing (CSNDSP)* (pp. 1–6). IEEE.

Sanchez, A., Blanc, S., Climent, S., Yuste, P., & Ors, R. (2013). SIVEH: numerical computing simulation of wireless energy-harvesting sensor nodes. *Sensors, 13*(9), 11750–11771.

Sharma, H., Haque, A., & Jaffery, Z. A. (2018a). An efficient solar energy harvesting system for wireless sensor nodes. In *2018 2nd IEEE International Conference on Power Electronics, Intelligent Control and Energy Systems (ICPEICES)* 2018 Oct 22 (pp. 461–464). IEEE.

Sharma, H., Haque, A., & Jaffery, Z. A. (2018b). Solar energy harvesting wireless sensor network nodes: a survey. *Journal of Renewable and Sustainable Energy, 10*(2), 023704.

Taneja, K., Taneja, H., & Ganesh, S. (2012, November). Computing issues in hybrid wireless networks. *International Journal of Applied Engineering Research, 7*(11), 2113–2115.

Taneja, K., Taneja, H., & Kumar, R. (2018, January). Multi-channel medium access control protocols: review and comparison. *Journal of Information and Optimization Sciences, 39*(1), 239–247.

Texas instruments application report on "calculating efficiency of PMP-DC-DC controllers". Available online: www.ti.com (accessed on 28June 2018).

Veerachary, M., & Saxena, A. R. (2011). Design of robust digital stabilizing controller for fourth-order boost DC–DC converter: a quantitative feedback theory approach. *IEEE Transactions on Industrial Electronics, 59*(2), 952–963.

Weddell, A. S., Merrett, G. V., & Al-Hashimi, B. M. (2011, March). Ultra low-power photovoltaic MPPT technique for indoor and outdoor wireless sensor nodes. In *2011 Design, Automation & Test in Europe* (pp. 1–4). IEEE.

Win, K. K., Wu, X., Dasgupta, S., Wen, W. J., Kumar, R., & Panda, S. K. (2010, November). Efficient solar energy harvester for wireless sensor nodes. In *2010 IEEE International Conference on Communication Systems* (pp. 289–294). IEEE.

5 Smart Systems for Global Sustainability with Enhanced Computing

Kritika Raj Sharma and Nitin Mittal

CONTENTS

5.1 INTRODUCTION

With the advancement in technology, human life is weaved into an intricate web of schedules that need his immediate attention and presence in many simultaneous situations at one time. Though distance in kilometers is still the same, it has significantly reduced in hours with new advancements in transport facilities and vehicles therein. Every man in his routine – whether for personal or professional purposes – is getting more and more dependent on vehicles. It initially started with a smaller ratio, but now the number of vehicles owned by one family is almost approaching to the same as the number of members therein. Usually, the size and dimension of a personal vehicle is

DOI: 10.1201/9781003158165-5

not merely selected by virtue of need and capacity, but it is also a showcase of status of the owner, hence making it bigger than what is actually required. Driving in modern times is one of the major routine activities of the present-day man. Traveling to near or far destinations is associated with botheration in driver's mind about the availability of suitable parking space, and thus, parking has emerged as one of the basic attentions seeking domain of researchers that is in need of effective management system and its implementation. Technological advancements in recent years have grown so fast with new emerging solutions presented by the fields of wireless sensor networks (WSNs), Internet of Things (IoT), Artificial Intelligence (AI), and others and thus bring the hope of better tomorrow with proper implementation of the proposed solutions in this regard. One such solution is proposed in this chapter, which takes into consideration the associated factors of safety, security, and fuel- and time-saving features in order to search for specific place to park a vehicle with prior information to the driver related to the same.

5.2 HISTORY AND BACKGROUND

In the early 20th century, the need and requirement of parking systems became more prominent with the increasing number of vehicles (Uddin, 2009) and hence their storage space. Major urban cities began to witness their drivers feeling troubled and worried because of inappropriate parking arrangements for their respective vehicles, and hence, many urban planners came up with a variety of solutions with varying technologies behind them. It also came in close look of businessmen to expand their business in providing this facility to the visitors ensuring safety and security and thus making them relaxed for the time they are away from their vehicles. In this way, the idea of ideal parking areas and spaces spread its branches to almost all public places such as shopping complexes, hospitals, and universities. It attracted many of the researchers' attention to design and propose more efficient parking system solutions based on the available space and affordable technology for its users (Atif et al., 2016; Chauhan et al., 2020; Funck et al., 2004; Takizawa et al., 2004; Teodorović & Lučić, 2006). Some of them are discussed in the following.

5.2.1 Parking Management System Based on WSN

WSNs-based parking management system is (Yang, Portilla & Riesgo, 2012) made up of four basic modules such as *wireless sensor module* for collection of data regarding the real-time slot status – whether occupied or empty denoted by big and small black dots, respectively – from the sensors placed in the parking regions and then sending the same information to the embedded web server for further work. The *web server* is embedded for transmitting and receiving the information back and forth from central web server and WSN. The *central web server* is for displaying the real-time information of available parking slots and sending it to the mobile application. The *mobile device module* displays the information fetched by the real-time monitoring of the parking zone with occupied or not occupied slots by the sensors deployed in those regions as shown in Figure 5.1.

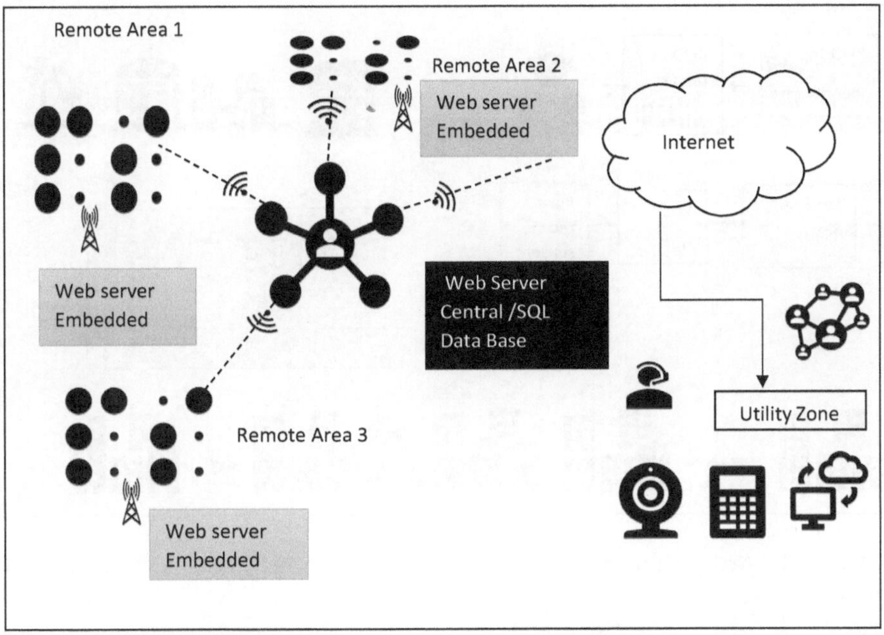

FIGURE 5.1 Architecture of Smart Parking Management System based on WSN.

5.2.2 Parking Management System Based on RFID

The parking system can be managed using radio-frequency identification (RFID) technology for ensuring the security of the parked vehicles such that the owners can do their work hassle-free without any segment of their mind worried about safety and security of their vehicles (Patil & Bhonge, 2013). The major components of the RFID network are as follows: *RFID Reader* from which emission of electromagnetic waves takes place to activate the corresponding *RFID tags*, which backscatter the received signals, and these are comprised of *antenna coil* and *transceivers* with encoder and decoder *transponder* in either active and passive states, keeping in view that tags in passive state are less costly, more environmentally friendly, and most energy-efficient due to low backscatter power, small and flat dimensions, and battery-related exemptions as compared to active tags (Decarli, Guidi & Dardari, 2018). The users of this system can be classified into two broad categories: the permanent users of that parking area can be given the permanent parking licenses, while the others who are the temporary users of the parking space – especially the visitors – can be issued the temporary recyclable passes which can be taken while entering the parking area and shall be deposited back while leaving that particular area of parking. In both of the above cases, the cards used can be the passive RFID cards and these cards contain the basic information regarding the vehicle and its driver which is entered into the system, while the entry by the vehicle is made using the card readers and the information related to the unoccupied/vacant parking slot is thereby passed to the driver, to help him to make his way to the parking space. In each array of the parking

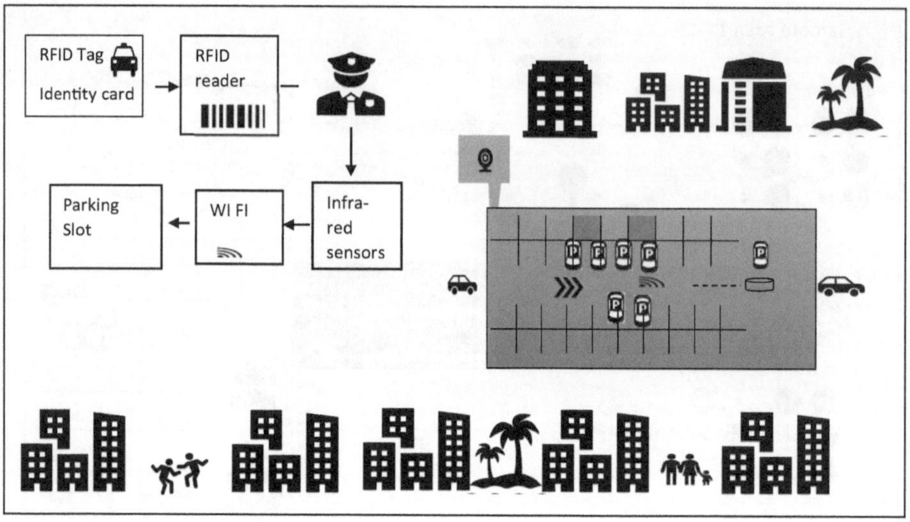

FIGURE 5.2 Architecture of the Smart Parking Management System based on the RFID system.

space, the RFID readers are placed to record and report the information related to the vacancy and the occupancy of the parking slots therein, and this continuous monitoring is the major feature in the said parking system that ensures the safety and security of the parked vehicle. The OCR (optical character recognition) cameras are basically used by the parking patrol for the purpose of continuous monitoring of the parked vehicles; however, this system involves higher expenditure on the equipment. Thus, the alternative of reader to tag communication is often substituted by the tag-to-tag communication in which this continuous monitoring is achieved by the RFID tags for the reduction in the operating cost of the security system (Decarli, Guidi, & Dardari, 2016). Figure 5.2 represents the architecture of the system mentioned in the paragraph.

5.2.3 Cloud-Based Parking Management System using VANETs

The inappropriate management in parking systems usually results in traffic congestion, which further leads to additional fuel consumption by vehicle, extra time involvement in searching the parking space, and increase in noise and air pollution levels of the area. One another way to find the solution in this regard is the use of IoT technology to manage the parking system of a place where the density of vehicles is on higher note. In 2018, Safi et al. proposed SVPS (Smart Vehicle Parking System), a cloud-based solution that provides the related information about the unoccupied parking space and other real-time recommendations with precision. The major components of this system are *centralized parking server* (*CPS*), i.e., cloud-based, which is responsible for running the central parking management system *Traffic Management Bureau* (*TMB*) to establish security and authenticity of the vehicle in association with Cloud Authentication Unit (CAU). The vehicles have their own on-board units (OBUs) for vehicle-to-vehicle or vehicle-to-infrastructure communication. *Parking side units* (*PSUs*) are there on the parking spaces, and with the help of some very intelligent and vigilant devices such as sensors, they regularly keep an eye on the parking slots and the information thus collected is passed to the central server for its update. *Road side units* (*RSUs*) take the information related to the available slots/spaces for parking from the central server and then passes this to the vehicles on road. It also assists by sharing the traffic congestion report on the road to the Smart Parking Management System (SPMS). The above scenario is diagrammatically represented in Figure 5.3. This whole system enhances the parking management system and is an appropriate stress reliever for the drivers of the urban cities where the parking problem has emerged as the major solution-seeking domain for the researchers, administrators, and the public. This system can be further attached with other technologies and features to make it more secure and safe. The vehicular communication can also be safeguarded from the various threats and attacks that can nullify the purpose of the said system. This gap needs to be bridged up with several encryption and network security methods. This system is helpful in preventing the environment from the poisonous emissions from the wandering vehicles in search of the parking area suitable for them. It also reduces the searching time and money of the end user.

5.2.4 Cloud-Based Parking Management System Based on IoT

The IoT is spreading its roots in each of the spheres and domain of the present-day world. The main aim of this technology is to let human be free from the unnecessary intervention in between the machines for smooth conduction of the work required from those machines in an effective, efficient, and seamless way. In recent times, we

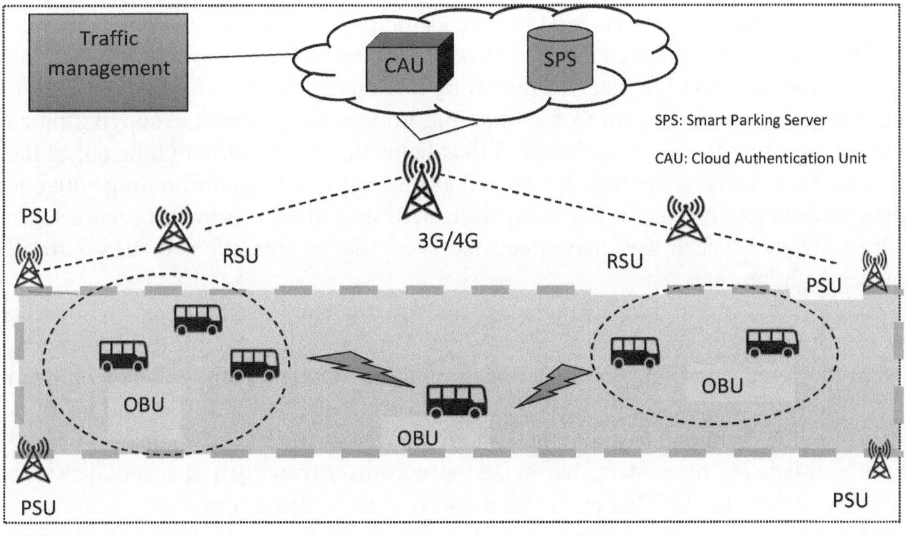

FIGURE 5.3 Architecture of the cloud-based smart parking system using VANETs.

are able to distinguish between the automatic and smart machines on work. Now, it's time to make the smart systems more intelligent than their predecessors with an inherent capabilities of responding faster and more decisive with greater parametric considerations behind these decisions. This technology weaves the machines in service of human being into a network that works for humans by not giving them the chance to bother for the operation and communication in between the network devices. The IoT has marked the new revolution in the industry standards and has stepped into Industry 4.0. The basic components of this field are the *sensing elements* that are used to sense the changes in the physical world that need immediate attention and response. These changes are converted into electrical signals as we have all the necessary circuitry to condition the electrical signals as per the need and requirement of the further stages. *IoT devices* are those physical everyday objects that can easily communicate and interact within the network when embedded with the compatible technology. These objects need to be remotely monitored and controlled within the network in order to achieve maximum derivable from these objects. These objects collect the data from the sensing elements; this data is in turn converted to the information, and through this information, the necessary response is generated within the network. *Network* is the interconnection between various devices that is responsible for back and forth of the information between the IoT devices which have their own gateways and service interfaces. In *cloud-based server*, all of the information can be stored in the cloud in order to save memory of local devices in the network, and this feature also enhances the utility by making the system more adaptable for remote operations, status check, and control. The system comprising the above units can resolve the issue of parking and its associated issues to very larger extent. The cloud-based SPMS using the IoT can be stated as follows.

In Figure 5.4, the architectural view is diagrammatically shown; the major units of this system are as follows. In 2015, Pham et al. (2015) have given the proposal of one such system and implemented it on a real-time scenario with some basic units of the system such as *Server* on cloud that takes the inputs from the local units at each parking space and stores it in the cloud that in turn helps the retrieval of information directly from the cloud in place of searching it from the local devices one by one. It becomes easy for the system to facilitate the information without occupying much of the memory in the local devices and thus helps the information to flow out of the boundaries of the local devices. It also makes the user free from the binding with one particular device to retrieve particular information; rather, it provides secure login to the network to each user and client that is not device specific and runs with all devices with internet.

Local unit is inside the parking space and is connected to the server for sending the information to it for retrieval and storage. This is responsible for the authentication of the vehicle entering and then assigning it the unique identity to be recognized on the network and also the display screen to show the status for its end users in order to guide them and to make the system more transparent. *Application software* is the accolade of this system that makes it more user-friendly and acceptable with the present-day clients. This application can run on Android operating system with its other variants on other commonly used operating systems with the 3G/4G internet connections. The overall system has significantly reduced the searching time of the

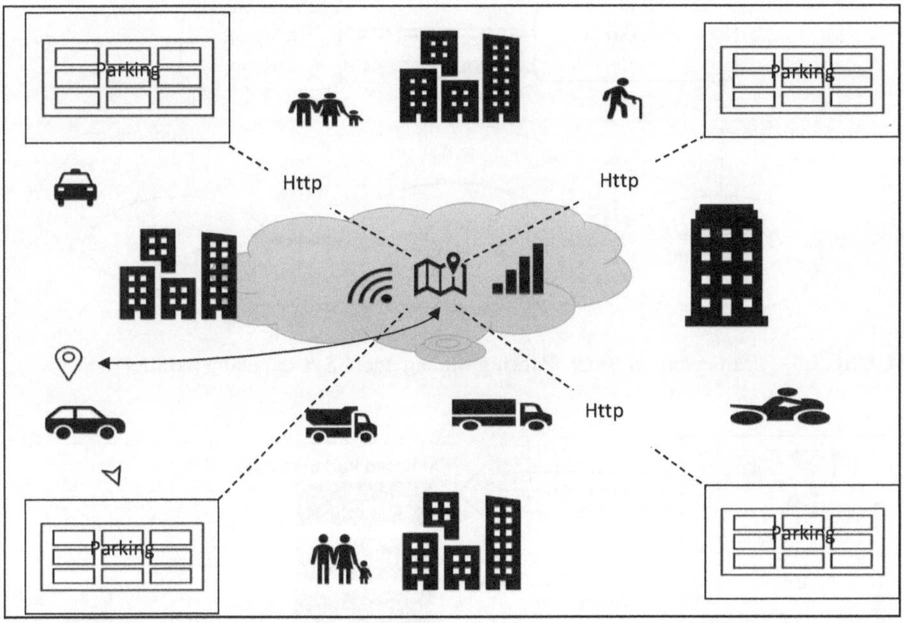

FIGURE 5.4 Architecture of the cloud-based smart parking system using IoT.

driver for the appropriate parking space but still needs to be improved to get shielded from the security threats.

5.2.5 Smart Parking Management System Based on Parallel Theory

In 2016, Wang et al. presented an idea of making the parking systems more intelligent for the prevailing circumstances caused by increasing density of vehicles in one particular area at one time, which has raised parking demand in significant measures. It is proposed that the system of parking can be managed using parallel parking systems based on ACP theory that is majorly used to solve complex system problems. The two main components of this system are *real systems* and *AI-based systems*. The coordination and communication in between the two above-mentioned entities helps to establish the management and control in the whole system. Figure 5.5 very well explains the framework of the system mentioned in the above paragraph.

In this system, the real system is connected with the AI-based system using the variety of the computational methods, and thus, the complex situations are tried to be resolved in the artificial scenario to pre-estimate about the reactions. Then, the appropriate solutions sorted thereby are introduced in the actual systems for better results. In this way, the planning about the parking under some complex situations can be done more effectively. This method also increases the efficiency of the parking management system. The architecture of the above system is shown in Figure 5.6.

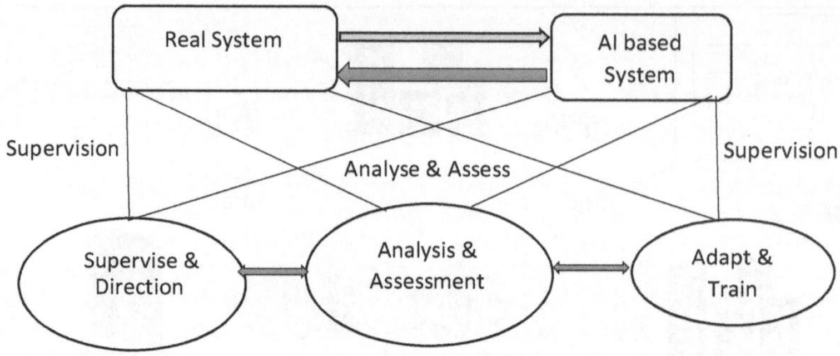

FIGURE 5.5 Framework of Smart Parking Management System using parallel theory.

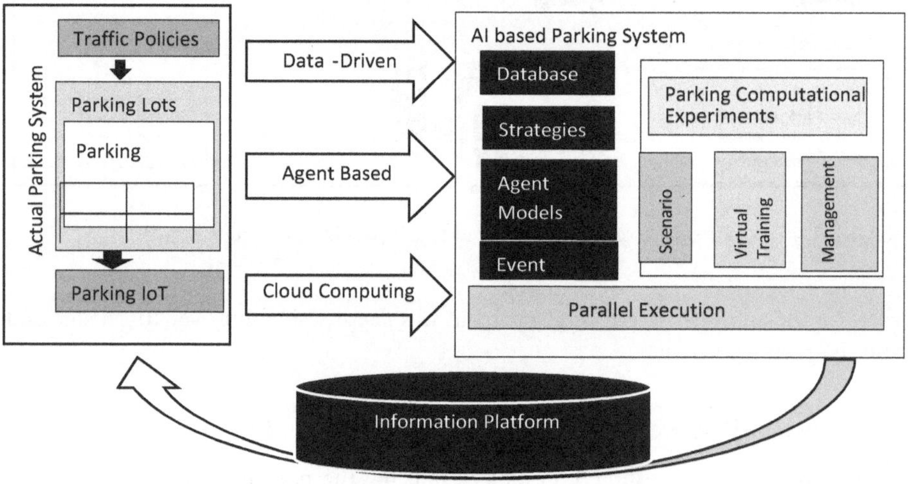

FIGURE 5.6 Architecture of Smart Parking Management System using parallel theory.

This method utilizes various input data-driven techniques, cloud computing, and parallel computing to make an artificial model for an actual model for the purpose of control and resource management: if, for example, the parking areas allotted for the working professionals are more occupied on the weekdays and the adjoining parking lot of the shopping complex is more occupied on weekends, then this trend can be observed with other different social complexities taken into consideration and both the resources can assist the need of each other during some complex situations where the demand increases in leaps and bounds to the actual available place. This system can also help to guide the driver about the most suitable parking space by observing their behavior and driving skills that can act as another facility entirely different from the existing technologies where the driver is given the option of a variety of parking slots, but as it is random, human behavior and the parking slots can also vary very frequently, so it is a kind of dynamic complex systems that can be resolved by the above parallel parking system. The information to and froth from the cloud and the system needs the security from the unnecessary threats and attacks; thus, various security techniques can be applied on the system to make it more secure from various threats and attacks.

5.2.6 Smart Parking Management System Using Image Processing and Artificial Intelligence

The parking system can also be managed by introducing the AI inside the system of parking utilizing other existing technologies in coordination to it. The concept of AI can go hand in hand with the image processing in making the existing parking system smarter and well managed. As presented by Ruili et al. in 2018, the parking slots in this system are equipped with the digital camera adjusted to capture the image of licensed plate number of the vehicle. Using various image processing techniques such as finding the RGB value and for unoccupied slots, the HSV (hue, saturation, value) image is obtained by the MATLAB® command rgb2hsv; it can be further extended by converting this image to grayscale image using command rgb2gray so that each pixel shall be compared to the threshold value using Equation 5.1 as follows:

$$\text{Gray} = (0.299 * r + 0.587 * g + 0.114 * b) \tag{5.1}$$

The vehicles while entering the parking lots are detected using the ultrasonic wave detection method. Cameras installed are used to read the number plates and fetch the required information from these plates and pass this information to the central processing system. The programming platform and the recognizing tool used over here are Node-RED and OpenALPR. To train the cameras, the Depth Recognizing Algorithm is used. The ALPR camera is used as the central device in this system. The driver is guided through the path using both the images and the voice. One of the good parts of this system is that it calculates the fee from the time driver has parked his vehicle in the parking lot to the time he makes his exit out of that parking lot, which means the time spent for searching the appropriate parking area is not included in the fee paid. That is why, it becomes the foremost duty of the managerial party to manage the system of parking in such a way that very less amount of the drivers' time gets wasted. The provision of penalty for the offenders of the management is also present in the system by first detecting the wrong parking or any other kind of the misconduct in the parking area by first warning the driver and then adding the fixed fine amount to the parking fees (Ruili, Haocong, Han, Connell & Sean, 2018).

5.3 STRENGTHS OF THE EXISTING TECHNOLOGIES FOR SMART PARKING MANAGEMENT SYSTEM

It has been observed from the above-discussed smart systems for parking management system that the various technologies can be integrated to manage the parking system: using WSN, the system can be arranged into the network transferring the information from one node to other utilizing the concepts of static and dynamic clustering, and the appropriate routing algorithms can be employed to let information reach to the destination in the minimum possible time; for this purpose, the routing optimization is required to be done. RFID cards and the associated system can be used to give the unique recognition to each vehicle entering the parking slot, which will help to recognize it on the system anywhere (Abdulkader et al., 2018; Abidin & Pulungan, 2020; Payal et al., 2014). The concept of VANETs helps to seek the reliable information from the road side units also that can give the information from the road to the parking unit as well as from the parking unit to the road, thus giving the real-time information to the driver related to the path guidance to the designated parking lot. The IoT utilizes the smart sensors to retrieve the information from the parking areas and to pass that information to the needy drivers on cloud (Fan et al., 2018). The AI system has introduced the AI inside the devices by helping the system to learn from its experiences, and the parallel parking system is a good framework for researchers to plan the actual system to resolve more and more complex situations (Perković et al., 2020; Schmidhuber, 2015; Singh et al., 2020).

5.4 SCOPES OF IMPROVEMENT IN THE EXISTING TECHNOLOGIES FOR SMART PARKING MANAGEMENT SYSTEM

The existing systems are very much dependent on the passage of information whether from road to the parking spaces or from the driving zones to the parking area. If this information is manipulated or played or mishandled, then it can raise some very serious issues; hence, there is the need to safeguard this information from various crypt-analysis attacks. It is also needed to identify the malicious nodes to find the exact source and destination. Thus, it is required to introduce security concepts hereby (Table 5.1).

5.5 PROPOSED SUSTAINABLE SMART PARKING SYSTEM WITH SECURITY

After studying the existing research work, it is observed that various technologies have proved themselves to be better in some or the other way, and the common point in which all of these still need to be improved for more reliable results is the identification of non-trustable nodes so that these nodes can be eliminated from the routes, in order to avoid the risk of message drops and also to enable the system to identify and detect the intruders and attackers. The identification of the malicious nodes will make the system more efficient and secure. It is also observed that much of the research work is being carried out to make system energy-efficient, but to make the system secure needs attention of the researchers as the work carried out under this aspect is still very limited. Hence, the proposed sustainable system of parking is using the Energy-Aware Trust-Based Secure Routing Algorithm (EASTRA) for route discovery and Firefly Algorithm (FA) with Artificial Neural Network (ANN)-based algorithms for route optimization. In addition to this, the

TABLE 5.1
Scope of Improvements in the Different Technologies for the Smart Parking Management Systems

S. No.	Technologies	Key Achievements	Improvement
1	Wireless sensor networks	Low-cost solution is proposed with wireless sensor networks along with the mobile communication.	Security algorithms can be used to protect the proposed work from the malicious nodes.
2	RFID	The proposed system is more energy-efficient method for information collection, exploiting tag-to-tag communication.	Security needs to be confronted against the strategically or non-strategically placed intruders in the proposed system.
3	Cloud-based VANETs	The cloud-based parking system processes big parking data with processing delegation in cloud, theft control, dynamic ITS management. Lessen the parking search time and fuel consumption.	Reliability of the proposed work can be enhanced by introducing the security features to protect it from attacks and intruders.
4	Internet of Things	The system adheres to Industry Standard 4.0. The status check and control features make it more convenient and acceptable solution.	The network needs to support high data rates, lower latency and enhanced mobility, and shall be enriched with the security features.
5	Parallel theory	The parallel system helps to optimize the actual parking management specially dealing in worst-case scenarios.	The parallel system needs to be secured by protecting the database from attackers and non-authorized users.
6	Image processing and Artificial Intelligence	The proposed system uses Raspberry pi as it is the best alternative because of its low cost and small form factor. Big Data analytics and neural networks form the basis.	Flexibility of the proposed system is compromised with increased hardware component demand and possession of the object. It is basic merger of multiple technologies.

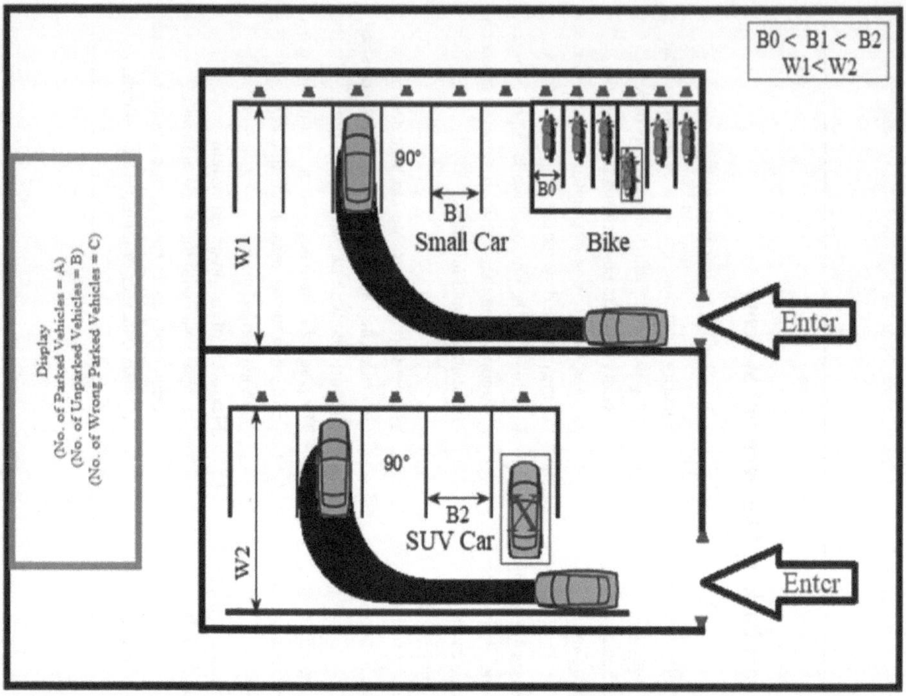

FIGURE 5.7 Layout of the proposed sustainable secure system of parking based on wireless sensor networks with customized features from the fields of IoT, ANN, and device-to-device communication.

device-to-device communication technology is also exploited to avoid communication gaps (if any) due to unavailability of traditional networks. The proposed system can be well understood from Figure 5.7, showing the detailed layout of the parking space further divided into various parking slots, each having a sensor deployed in it to pass the information to the subsequent node in the network and to transmit message further through the base station to the destination node. In this system, the parking slots will vary in size depending on the size of the vehicle to be parked in that; moreover, the sensors deployed in each of the parking slots will calculate the distance of the parked vehicle in relation to their own position of deployment, and if the vehicle is not in the range, then it will be considered as wrongly parked vehicle with an initiation of buzzer or alarm in response to this inappropriate action of the driver.

The flow of the proposed work can be understood using the flowchart in Figure 5.8.

Further implementation of the proposed work can be made clearer by distributing the process into major steps as follows, with the supporting diagrams taken from the simulator screen for better understanding of the process:

Step 1: The parking area is selected on the simulator and divided into the various parking slots; here, we have divided it into 15 parking slots of variable

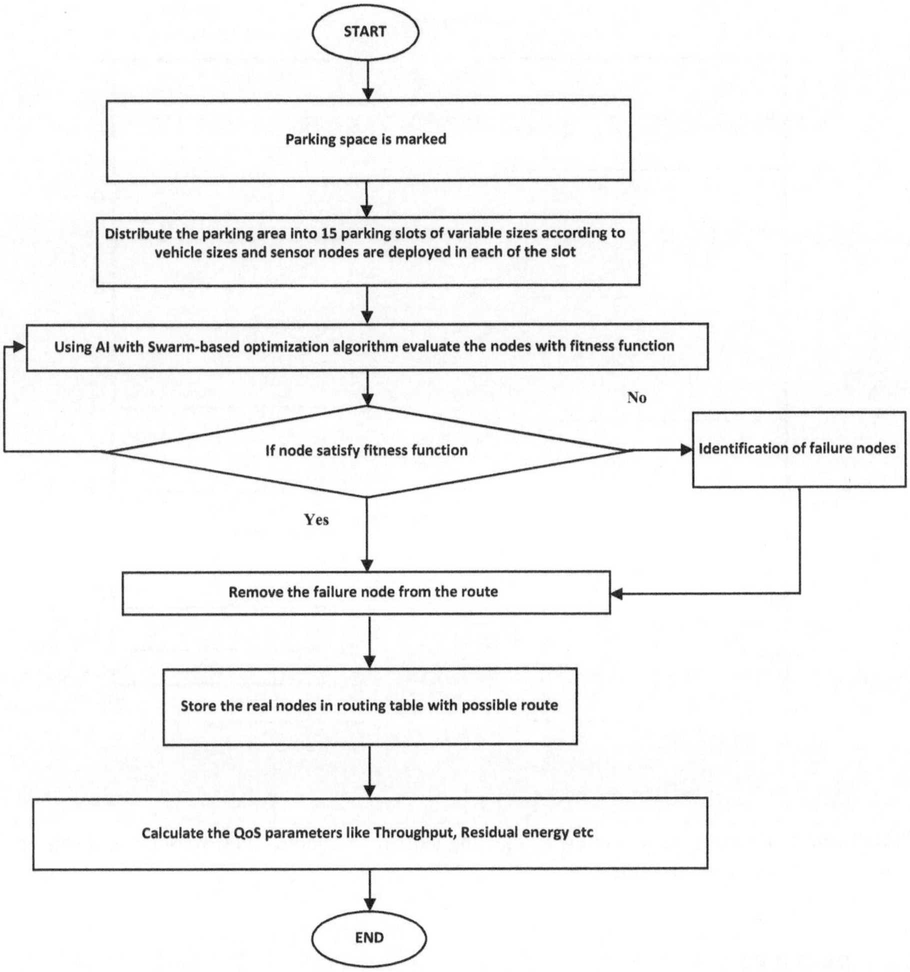

FIGURE 5.8 Flowchart of the proposed work.

sizes covering all three sizes with different widths, such as B0 < B1 < B2. In each of the slot, a sensor is deployed (Figure 5.9).

Step 2: During the entry of the vehicle, the identification of wrongly parked vehicle is taken care of and is prompted through message and alarm (Figure 5.10).

Step 3: The route is discovered as there are no suspicious/malicious nodes in between the route using the EATSRA algorithm to identify such nodes. The result after the first iteration from the available slot to the base station is made prominent in Figure 5.11.

Step 4: The nodes with low trust score are identified as non-trustable nodes and eliminated from the route. The alternative route is established to transmit the information comprised of the trustable nodes only (Figure 5.12).

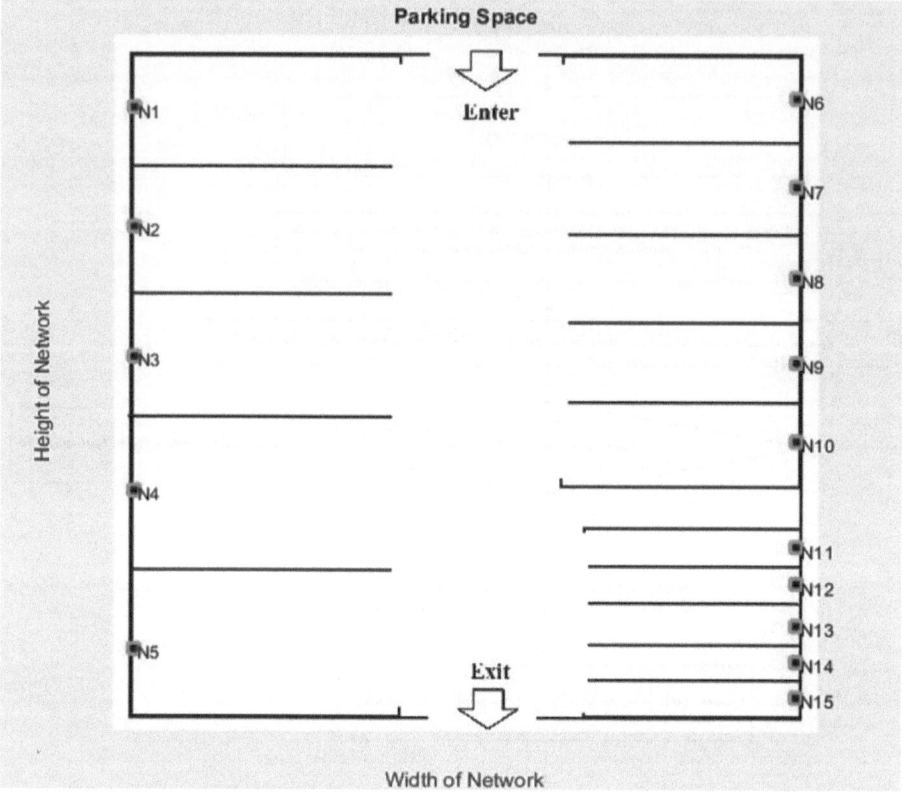

FIGURE 5.9 Parking space having 15 parking slots for variable-sized vehicles and sensors deployed on each of the parking slot.

5.6 RESULTS

From the implementation of the proposed work on the simulator, the multiple iterations have resulted in multiple routes and have also identified the malicious nodes, which are highlighted as attackers; these are the nodes for which the trust score has not passed the desired value and are thus not included in the routes to transmit data; information about the screenshot of the same is shown in Figure 5.13.

The values already set for the above simulation and the values obtained from the simulation are mentioned in Table 5.2 for the reference of the readers.

To estimate the quality of service of the proposed work, various parameters are plotted to ensure that the proposed work is better than the existing systems.

5.6.1 Throughput vs Number of Rounds

For any system to function well, the measure of the throughput should be high. It is a positive parameter for the analysis of any of the communication network. It can be

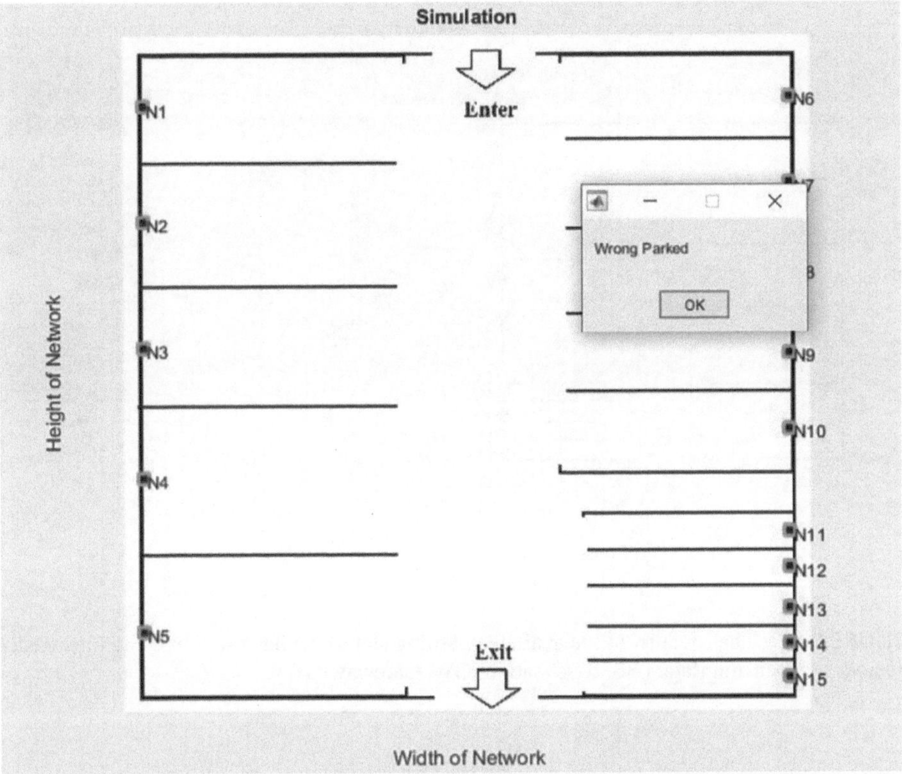

FIGURE 5.10 Identification of the wrongly parked vehicle by the simulator.

defined as the rate of the information processed in given time. In the plot shown in Figure 5.14, the throughput vs. number of rounds is measured for the two systems: one without prevention and the other with prevention, that is, our proposed system. It is observed that the throughput is decreasing with the increasing number of rounds, but there exists significant difference in the value of throughput for both the systems. The proposed system is exhibiting far more better results in comparison with the previously existing system, that is, the system without prevention.

5.6.2 Residual Energy

At the end of each iteration, the amount of energy left with each of the node is a prime concern for any of the communication network; the average of this energy is termed as the average residual energy. The greater the average residual energy, the better the system. We can see in Figure 5.15 that the proposed system is left with more residual energy in comparison with the system without the proposed prevention scheme; hence, we can consider our proposed system in this regard to be better in terms of the residual energy.

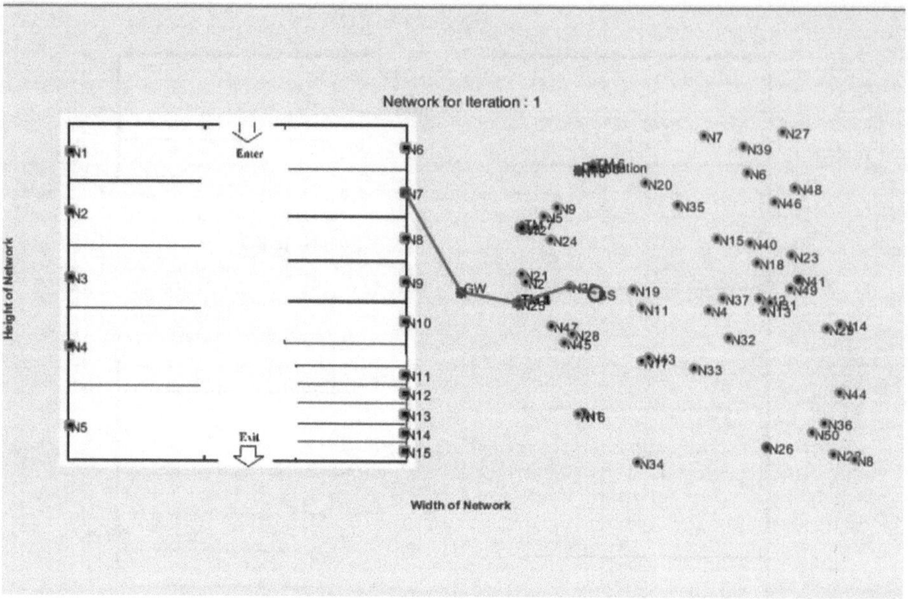

FIGURE 5.11 The route from the available parking slot to the base station using intermediate nodes to transmit data. (BS, base station; GW, Gateway.)

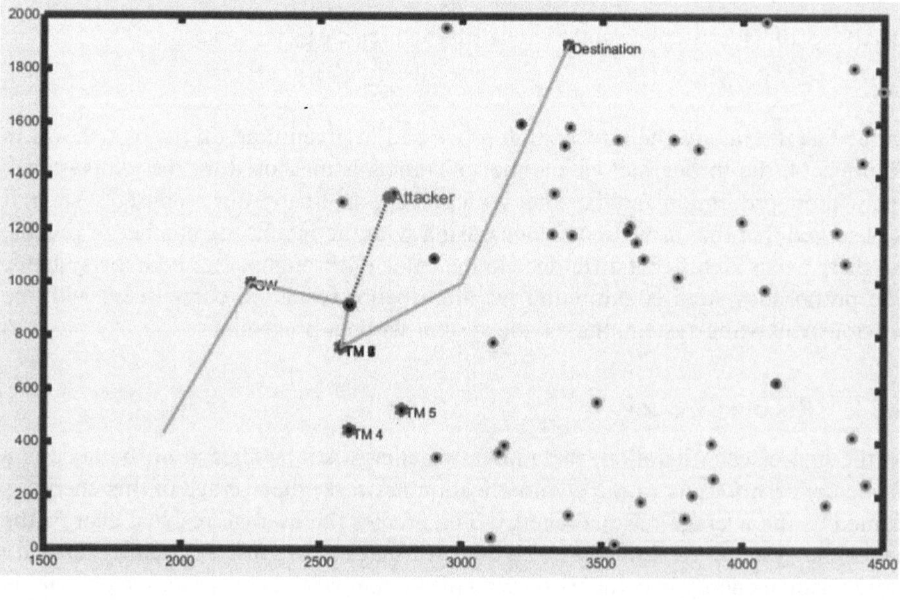

FIGURE 5.12 Identification of nodes with low trust value and thus not included in the route.

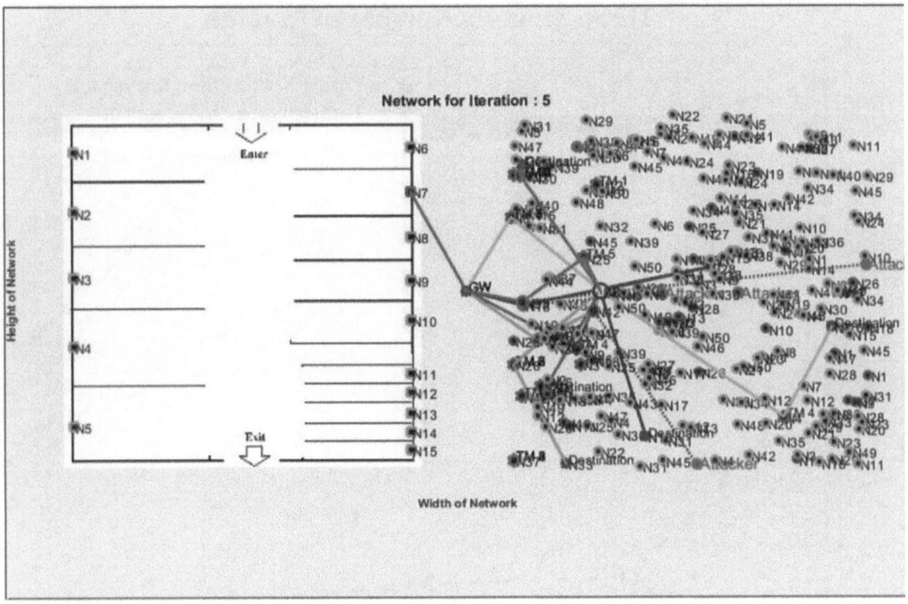

FIGURE 5.13 Routes discovered by the multiple iterations and detection of the attacker nodes.

TABLE 5.2
Specifications of the Parking Slot

Total number of parking slots	15
Total number of the parked vehicles	1
Total number of the unparked vehicles	14
Total wrongly parked vehicles	0
Routing Mechanism	
Destination node	N17
Route nodes	52 33 25 51 36 17
Optimized route nodes	52 18 25 51 36 17
Route is affected by attackers!!!	

5.7 CONCLUSION

The parking management system is the present-day need of the drivers, management, and the other entities of the society. Thus, it is required to build up the smart and more efficient parking management system, which can also be a merger of all

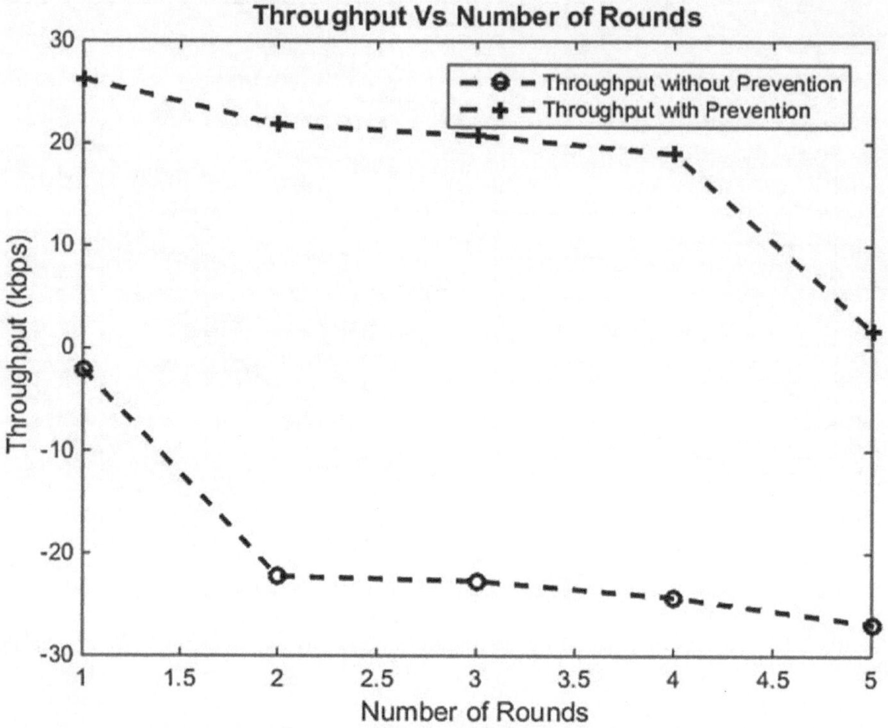

FIGURE 5.14 Throughput of the system without prevention vs. throughput of the system with prevention.

the above-discussed existing solutions to plan to make as one system. This will form the support of ITS, that is, intelligent transport system, and will help to reduce the driver's stress to a greater extent. Further, the study can be extended by adding the security algorithms to the system to make the shared information flow fearlessly within the network. In this chapter, one such system for efficient parking management is proposed, which is based on WSN and is enriched with add-on technologies such as AI and device-to-device communication with the traditional communication networks. The participating nodes in the route identified for the exchange of information will be selected if and only if they qualify valid trust score or they will not participate in the discovered route, and this is ensured using EATSRA algorithm for efficient routing mechanism. The energy consumption and delay are also minimized using the swarm-based algorithm for the proposed system. The system also has the capability to detect wrongly parked vehicles, which is again a preventive step from fatal accidents and unusual chaos among drivers. The proposed research work can undoubtedly prove to be better in terms of effective traffic management and minimizing the accidents on roads.

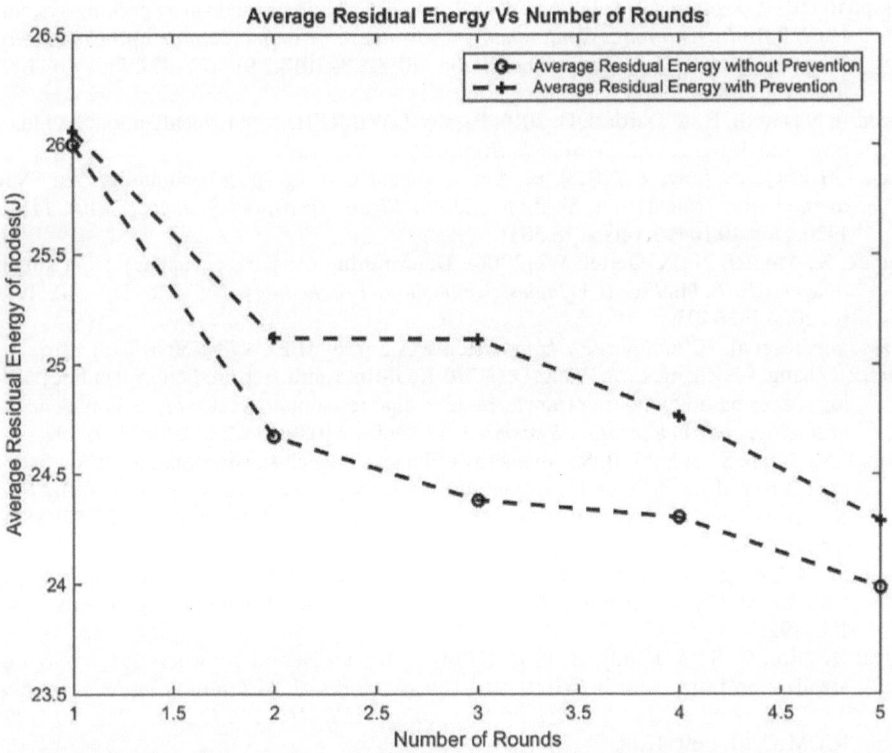

FIGURE 5.15 Average residual energy of the system without prevention vs. the system with prevention.

REFERENCES

Abdulkader, O., Bamhdi, A. M., Thayananthan, V., Jambi, K., & Alrasheedi, M. (2018). A novel and secure smart parking management system (SPMS) based on integration of WSN, RFID, and IoT. *2018 15th Learning and Technology Conference, L and T 2018*, 102–106. Doi: 10.1109/LT.2018.8368492.

Abidin, M. Z., & Pulungan, R. (2020). A systematic review of machine-vision-based smart parking systems. *Scientific Journal of Informatics*, *7*(2), 213–227. Doi: 10.15294/sji.v7i2.25654.

Atif, Y., Ding, J., & Jeusfeld, M. A. (2016). Internet of things approach to cloud-based smart car parking. *Procedia Computer Science*, *58*, 193–198. Doi: 10.1016/j.procs.2016.09.031.

Bahrami, S., & Roorda, M. J. (2020). Autonomous vehicle relocation problem in a parking facility. *Transportmetrica A: Transport Science*, 16(3), 1604–1627. Doi: 10.1080/23249935.2020.1769226.

Chauhan, V., Patel, M., Tanwar, S., Tyagi, S., & Kumar, N. (2020). IoT enabled real-time urban transport management system. *Computers and Electrical Engineering*, *86*. Doi: 10.1016/j.compeleceng.2020.106746.

Chippalkatti, P., Kadam, G., & Ichake, V. (2018). I-SPARK: IoT based smart parking system. *IEEE* 2018 *International Conference on Advances in Communication and Computing Technology (ICACCT)*, Sangamner, India (2018.2.8–2018.2.9), 473–477. Doi: 10.1109/ICACCT.2018.8529541.

Decarli, N., Guidi, F., & Dardari, D. 2016. Passive UWB RFID for tag localization: architectures and design. *IEEE Sensors Journal*, *16*(5), 1385–1397.

Fan, J., Hu, Q., & Tang, Z. (2018). Predicting vacant parking space availability: An SVR method with fruit fly optimisation. *IET Intelligent Transport Systems*, *12*(10), 1414–1420. Doi: 10.1049/iet-its.2018.5031.

Funck, S., Möhler, N., & Oertel, W. (2004). Determining car-park occupancy from single images. *IEEE Intelligent Vehicles Symposium, Proceedings*, 325–328. Doi: 10.1109/ivs.2004.1336403.

Hassoune, K. et al. 2016. Smart *Parking* Systems: A Survey IEEE 978-1-5090-5781-8/16.

Mei, Z., Zhang, W., Zhang, L., & Wang, D. (2019). Real-time multistep prediction of public parking spaces based on Fourier transform–least squares support vector regression. *Journal of Intelligent Transportation Systems*, 1–13. Doi: 10.1080/15472450.2019.1579092.

Mittal, N., Singh, S., & Sohi, B. S., An energy efficient stable clustering approach using fuzzy type-2 neural network optimization algorithm for wireless sensor networks. *Ad Hoc & Sensor Wireless Networks, June 2020.* SCI-Indexed IF: 0.948 ISSN: 1551-9899 (print), 1552–0633.

Patil, M., & Bhonge, V. N. 2013. Wireless sensor network and RFID for smart parking system. *International Journal of Emerging Technology and Advanced Engineering*, *3*(4), 188–192.

Payal, A., Rai, C. S., & Reddy, B. V. R. (2014). Artificial Neural Networks for developing localization framework in Wireless Sensor Networks. *2014 International Conference on Data Mining and Intelligent Computing, ICDMIC 2014*, 0–5. Doi: 10.1109/ICDMIC.2014.6954228.

Perković, T., Šolić, P., Zargariasl, H., Čoko, D., & Rodrigues, J. J. P. C. (2020). Smart parking sensors: state of the art and performance evaluation. *Journal of Cleaner Production*, *262*. Doi: 10.1016/j.jclepro.2020.121181.

Pham, T. N., Tsai, M. F., Nguyen, D. B., Dow, C. R., & Deng, D. J. (2015). A cloud-based smart-parking system based on internet-of-things technologies. *IEEE Access*, 3, 1581–1591. Doi: 10.1109/ACCESS.2015.2477299.

Ruili, J., Haocong, W., & Han, W., O'Connell, E., McGrath, S. (2018). Smart parking system using image processing and Artificial Intelligence. *Twelfth International Conference on Sensing Technology (ICST)* Doi: 10.1109/ICSensT.2018.8603590.

Safi, Q. G. K., Luo, S., Pan, L., Liu, W., Hussain, R., & Bouk, S. H. 2018. *SVPS:* cloud-based smart vehicle parking system over ubiquitous VANETs. *Computer Networks*, S1389128618301531. Doi: 10.1016/j.comnet.2018.03.034.

Schmidhuber, J. (2015). Deep learning in neural networks: An overview. In *Neural Networks* (Vol. 61, pp. 85–117). Elsevier Ltd. Doi: 10.1016/j.neunet.2014.09.003.

Singh, R., Dutta, C., Singhal, N., & Choudhury, T. (2020). An improved vehicle parking mechanism to reduce parking space searching time using firefly algorithm and feed forward back propagation method. *Procedia Computer Science*, *167*, 952–961. Doi: 10.1016/j.procs.2020.03.394.

Takizawa, H., Yamada, K., & Ito, T. (2004). Vehicles detection using sensor fusion. *IEEE Intelligent Vehicles Symposium, Proceedings*, 238–243. Doi: 10.1109/ivs.2004.1336388.

Teodorović, D., & Lučić, P. (2006). Intelligent parking systems. *European Journal of Operational Research*, *175*(3), 1666–1681. Doi: 10.1016/j.ejor.2005.02.033.

Tsiropoulou, E. E., Baras, J. S., Papavassiliou, S., & Sinha, S. 2017. RFID-based smart parking management system, *Cyber-Physical Systems*, Doi: 10.1080/23335777.2017.1358765.

Wang, F. Y., Yang, L. Q., Yang, J., Zhang, Y., Han, S., & Zhao, K. (2016). Urban intelligent parking system based on the parallel theory. *IEEE 2016 International Conference on Computing, Networking and Communications (ICNC),* Kauai, HI, USA (2016.2.15–2016.2.18)], 5. Doi: 10.1109/ICCNC.2016.7440708.

Uddin, A. 2009. Traffic Congestion in Indian Cities: Challenges of a Rising Power. Kyoto of the cities, Naples.

Yang, J., Portilla, J., & Riesgo, T. 2012, October. Smart parking service based on wireless sensor networks. In *IECON 2012–38th Annual Conference on IEEE Industrial Electronics Society*, 6029–6034, IEEE.

6 Intelligent Systems
Techniques for Optimized Decision Making

J. Senthil Kumar and G. Sivasankar

CONTENTS

6.1 INTRODUCTION

Robots are becoming part of our lives and are operating and assisting us in environments that are inherently social. Such robotic systems need to possess social intelligence for better navigation in an environment. People while moving around the roads amidst their busy schedule are supposed to have thousands of interactions happening at the same time, and they are able to cross the roads from one side to the other while not being collided with each other. Everyone does this naturally without thinking too much, while someone is talking with a mobile and doing other things while moving around. The primary motivation for us in this chapter is whether the same aspects can be performed with robots in a social environment and if it is feasible what challenges to be dealt with. The dynamic constraints are not only just a perception problem but also a planning problem where numerous interactions are taking place. When we use planner the dynamic window approach, it basically treats

DOI: 10.1201/9781003158165-6

obstacles just as static obstacles. Once you start adding dynamic constraints, it is not really able to deliver a good plan. All these issues need to include, and we need to think about these dynamic obstacles.

The intelligent navigation in the mobile robotics environment is gaining its importance because of its real need across all domains of robot usage. Navigation requires appropriate path planning that can be achieved by finding a feasible optimized path from the starting position to the desired goal position of the robot by avoiding obstacles. This in turn drastically reduces the chances of collision drastically. Emotion-aware navigation of a mobile robot is also possible among a group of people that can be carried out using the real-time algorithms (Bera et al., 2019). With common sense, metal models are characterized for exhibiting key decisions and strategies for cognitive robots to successfully share their space and tasks among the human in social environments (Lemaignan et al., 2017). Configuring social robots to play its part as caregivers are done with the user perception using computers as a social actors paradigm (Kim et al., 2013). Tool based on human neuroscience and neuroimaging has also started to play a crucial role for collaborating with social robots (Henschel et al., 2020).

Researches toward socially compatible robots are focused on social learning, learning by observation, people detection, and tracking. They are carried out by effective means of motion planning and navigation with better human–robot interaction. By modeling the human behavior in the social environment using spatiotemporal models, robots can be subjected to social isolation in the environment.

This chapter discusses the implementation of the path planning algorithm in TurtleBot3 in both real time and simulation. While considering path planning algorithms, many path planning algorithms already exist to solve the path planning problem. Initially, graph-based path planning algorithms such as Dijkstra's et al. (1959) and A* (Hart et al., 1968) were developed. These algorithms provide the shortest path from the starting position to the ending position and provide collision-free position, but the computational complexity increases with an increase in the dimension of the space. Therefore, the performance of these search algorithms degrades with an increased dimension of the workspace. This challenge is overcome by sampling-based path planning algorithms, one of the popular path planning methods. Sampling-based planners take random samples from the workspace instead of discretizing the workspace and connecting it to form a tree from the starting position to end.

The rest of this chapter will discuss the following issues related to robot navigation elaborately:

- A brief introduction to the motion prediction strategy among different environments.
- A brief discussion on Autonomous Mobile Robot Navigation Framework
- Navigation and control aspects of mobile robots among social environments.
- Usage of TurtleBot3 robot via Robot Operating System (ROS) for navigation tasks.
- Using Q-learning strategy for TurtleBot3 robot navigation with the map using Gazebo and RViz tools for robot navigation

6.2 MOTION PREDICTION

Motion prediction is one of the most primary challenges to be addressed in dynamic environments of social robots. Human–robot interaction requires the reaction of robot with respect to human movement (Lisetti et al., 2004). Coordinated motion planning models are highly important to assess the coordination of agents explicitly in prediction, planning, and social cooperation among human and robots. Trajectory optimization is one of the vital parameters to be assessed for imparting space exploration features in the robotic systems. It includes numerous submodules such as trajectory selection, optimization of trajectory, parallel as well as dynamic class optimizations, and coordinated motion planning. In addition, along with those parameters, environments with social encountering options need to be considered for simulating the developed planning approaches. Planning algorithms are helpful to explore topological-based alternate trajectories using time elastic band frameworks. With the inputs on goal pose, start pose, and set of obstacles in the environment, the trajectory optimization techniques are aimed to provide best topology-based alternate paths in the environment. Most of the exploration algorithms in motion prediction tasks are closed-loop kind with feedback. They are helpful to extend the initial trajectory from the class of alternate solutions captured through the feedback in the path. Learned trajectory tracking techniques in the motion prediction environments are helpful to predict the trajectory of human in social environment with the best set of trained features from the gathered data.

In few cases, the multiagent platform predicts the coordination of robotic agents with the environment and human being. Here, each agent predicts its own best trajectory, which in turn is helpful for estimating the motion of the robotic systems amidst the human being. Also, joint collision avoidance techniques are gaining popularity, which are used to predict the pairwise intersection points and rate them as obstacles based on the learned set of features they encounter. This feature enables to estimate the interaction points to find the closest spatial-temporal position on the neighbor's future trajectories in the environment.

Especially in the social environment with pedestrians encountered with robotic systems, the predicted trajectories are estimated along with the topological alternate paths in the environments. In such use cases, the start and goal position of human being navigating along with social robots needs to be updated for estimating the probability of both sides. The oscillations in the probabilistic condition indicate that both the predicted and topological paths are similar. In such cases, the current orientation of the robotic system in the environment significantly influences the decision-making chances of the motion prediction modules. In addition, alternative optimized paths could also be estimated by the motion prediction modules based on the quality of the data existing in the environment. Social experimentations under environmental conditions greatly affect the performance of the algorithm; they may have a large impact on the trajectory shift due to the effect of cooperation between human and robots in the chosen environment.

Researchers have put forward several such models for motion prediction. The Extended Social Force Model (ESFM) is one of the popular models used for motion prediction in robots subjected to navigation in social environments.

6.2.1 The Extended Social Force Model (ESFM)

Social Force Model (SFM) is one of the popular approaches to model human motion in the social environment. The ESFM model is the extended version of SFM, in which it treats the objects in the dynamic environment just as points with two-dimensional representation, and the trajectory is described by these points. The extended forces are the attraction forces that are enforced to reach the goal position of the robot. There are several goals needed to estimate because of the dynamic environment with fluctuating trajectories. Unlike the analysis on the forces of interaction between people moving in an environment, determining the appropriate trajectory for robots needs to be quantified for better interaction of the robots in dynamic environments. Robots cannot be treated with the same set of parameters used for human interactions with each other and within the environment. The fitting parameters to predict the difference between dynamic objects or people present in the environment with respect to the robot's position is the superposition and all the sum of all the forces acting on the environment as shown in equation 6.1.

$$f_n(t) = f_n^{\text{goal}}\left(t, D_n(t)\right) + \sum_{j \in P \setminus n} f_{n,j}^{\text{int}}(t) + \sum_{o \in O} f_{n,o}^{\text{int}}(t) + \sum_{r \in R} f_{n,r}^{\text{int}}(t) \tag{6.1}$$

Figure 6.1 shows the ESFM in which humans navigate with the corresponding forces applied over the dynamic environment. When we have a trajectory and while we are trying to estimate where we need to move toward next successive steps, for a

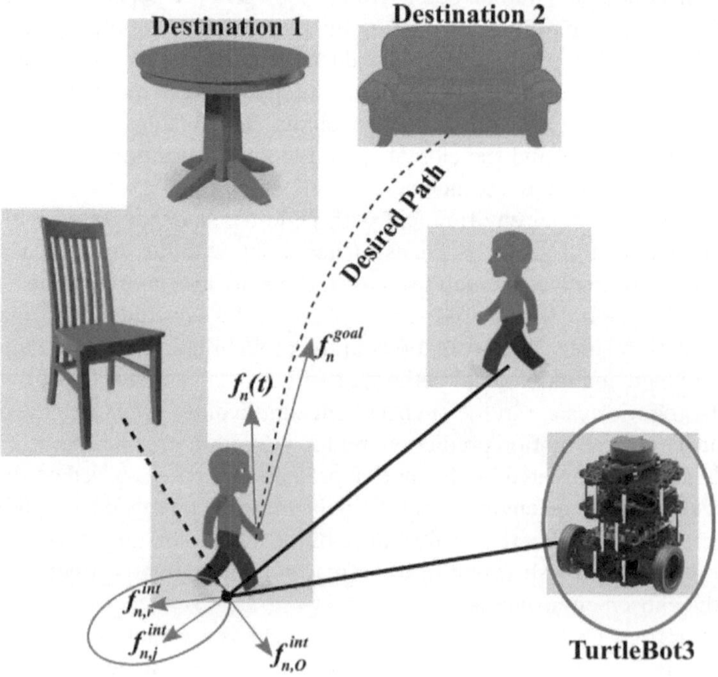

FIGURE 6.1 Scheme of forces centered around a person during social interaction with robots.

particular agent, it just tries to reach this destination. When a second agent is introduced in the scene, the trajectory changes and it is a vital parameter as this model is faster to calculate, and we need to capture those interactions. When these features are imparted in mobile robotic systems for social interactions, three types of behavior are observed in the human motion prediction such as aware behavior, balanced behavior, and unaware behavior. The social forces developed with the help of prescribed model with the prediction performance will lead to a better accuracy based on the environment and the time horizon of prediction.

6.2.2 Robot Navigation Using the Extended Social Force Model (ESFM)

Reactive navigation strategy of mobile robots based on ESFM can be obtained by a straightforward solution. In this scenario shown in Figure 6.2, the TurtleBot3 mobile robot navigates toward global goals by solving intermediate local goals avoiding all the potentially dynamic and static obstacles just using the interaction forces. The navigation of the robot is supported by steering and interaction forces as described in Equations 6.2 and 6.3. This, in turn, provides the resultant force for navigation of the TurtleBot3 robotic system.

$$f_n^{\text{goal}}\left(t,D_n(t)\right)=k_r\left(v_r^0,D_r(t)-v_r(t)\right) \tag{6.2}$$

$$f_r(t)=\propto f_r^{\text{goal}}\left(t,D_r(t)\right)+\gamma f_r^{\text{per}}(t)+\delta f_r^{\text{obs}}(t) \tag{6.3}$$

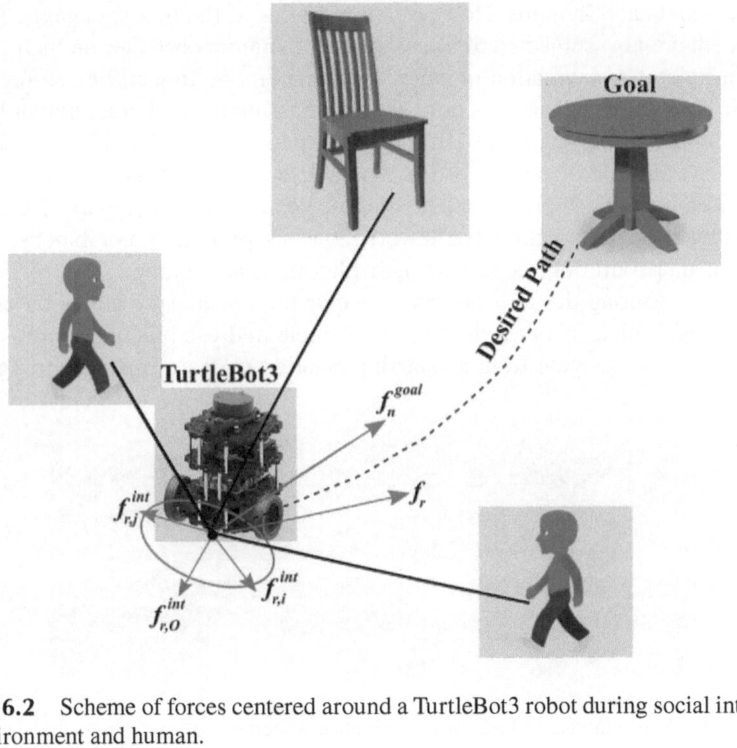

FIGURE 6.2 Scheme of forces centered around a TurtleBot3 robot during social interaction with environment and human.

6.3 AUTONOMOUS MOBILE ROBOT NAVIGATION AND CONTROL

In this section, we will discuss on the details of the autonomous navigation stack using TurtleBot3 robot and ROS. Human-aware navigation of robots in social environments opens up a new range of possible interactions of the robotic systems with the environment. Performance measurement of human-centered robots is different from those of the conventional industrial and mobile robotic systems. They need to possess a high level of adaptability, flexibility, and autonomous capability depending on the deployed environment. Moreover, compared with the traditional robots such as industrial robots, human-centered robots require completely different performance measurement requirements.

Navigation of social robots in constrained and dynamic environments needs to be modeled as cooperative activity. Multiple robots engaged in performing an activity need to form cooperative network. Such configurations of robotic systems are widely used in many use cases and end applications such as military, particularly for enemy detection. For carrying out such crucial tasks, connectivity among the robots plays an important role for seamless exchange of information. In addition, the navigation planning approaches in such dynamic and complex environments cannot be directly adopted while carrying out mission-critical tasks. Dynamic connectivity among social robots engaged in the mission needs perfect coordination with human for making dynamic decisions based on their own states. This could also be effectively configured and applied to preserve the issues in establishing connectivity among the social robots and humans. However, in most cases, the robots engaged in social environment are also subjected to issues such as dynamic obstacles on their path and human intruders. A navigation planner with cooperative trajectories enforced with kinematic constraints of robots is done for avoiding human and other dynamic obstacles (Khambhaita and Alami, 2020). From the perspective of mobile social robots, better relative motion of people in the dynamic environment is observed by using unsupervised learning approach on surveillance data sets (Luber et al., 2012).

An autonomous navigation framework consists of four major blocks, namely, perception, map building, cognitive path planning, and motion control as shown in Figure 6.3. Among the four blocks, this chapter's primary focus is on cognitive path planning, which involves finding an obstacle and establishing a collision-free path for a robot to navigate from its starting position to its destination through many

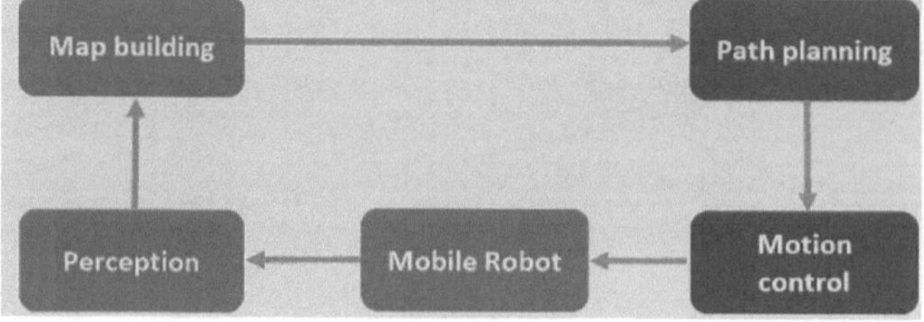

FIGURE 6.3 Autonomous mobile robot navigation scheme.

intermediate points. In general, path planning is performed after the perception and map building phases. Perception represents sensing of the environment, which is done by getting raw data from sensors like LiDAR. The collected raw data are further processed by certain transformations and visualized as a map using ROS topics. Localization of the robot can be done using various localization algorithms like Adaptive Monte Carlo Localization (AMCL). RViz tool is used for visualizing the map and the robot.

6.3.1 Perception

An Autonomous Mobile Robot Navigation scheme is used to move the robot from the starting position to the specified destination location on the map using the sensor data. Sensors mounted on autonomous mobile robots are helpful in capturing the status of their environments, and they generate numerous raw information. Certain sensor data could capture the internal status of robotic parameters as well. Based on the perception data for the exteroceptive and proprioceptive sensors mounted on robotic systems, accuracy in the perception modules could be ensured by the learning algorithms for assisting the robotic systems toward desired direction. For example, from the odometric sensors mounted on the TurtleBot3 robot, its pose and orientated direction of the robot could be estimated using the odometry data. Apart from the odometry data, other sensor data also could be fused to estimate the accurate position of the robot in the environment. The odometry information is updated in the map using wheel optical encoders, IMU sensor, and distance sensor in TurtleBot3 robot. The ROS topic meant for odometry is used to measure the estimated pose of TurtleBot3 robots. Figure 6.4 shows the ROS navigation architecture for the proposed social navigation application.

6.3.2 Localization/Pose Estimation

Pose of the robot sensor changes based on the robot hardware configuration, ROS with relative coordinate transformation (TF), most useful parameters to describe the robot parts as well as objects and obstacles. The pose of a robot can be described as a combination of positions and orientations. Robot position estimation is performed using the given map and based on the encoder, IMU sensor, and the distance sensor. The Monte Carlo localization (MCL)-based pose estimation algorithm method, namely, particle filter, is widely used in the field of robot pose estimation. AMCL, an improved version of Monte Carlo pose estimation method, is used in TurtleBot3 to estimate the robot's current pose. Various AMCL parameters are found in "amcl.launch.xml" file.

6.3.3 Map Building

Navigation of robotic systems demands appropriate map-building strategies that could enable exploration amidst strange paths. Particularly high-accuracy GPS systems could assist map building in robotic systems to a larger extent. It also helps to navigate the robots inside buildings, outdoors, and also users to build maps in real time as they are involved in exploring their environment during the navigation.

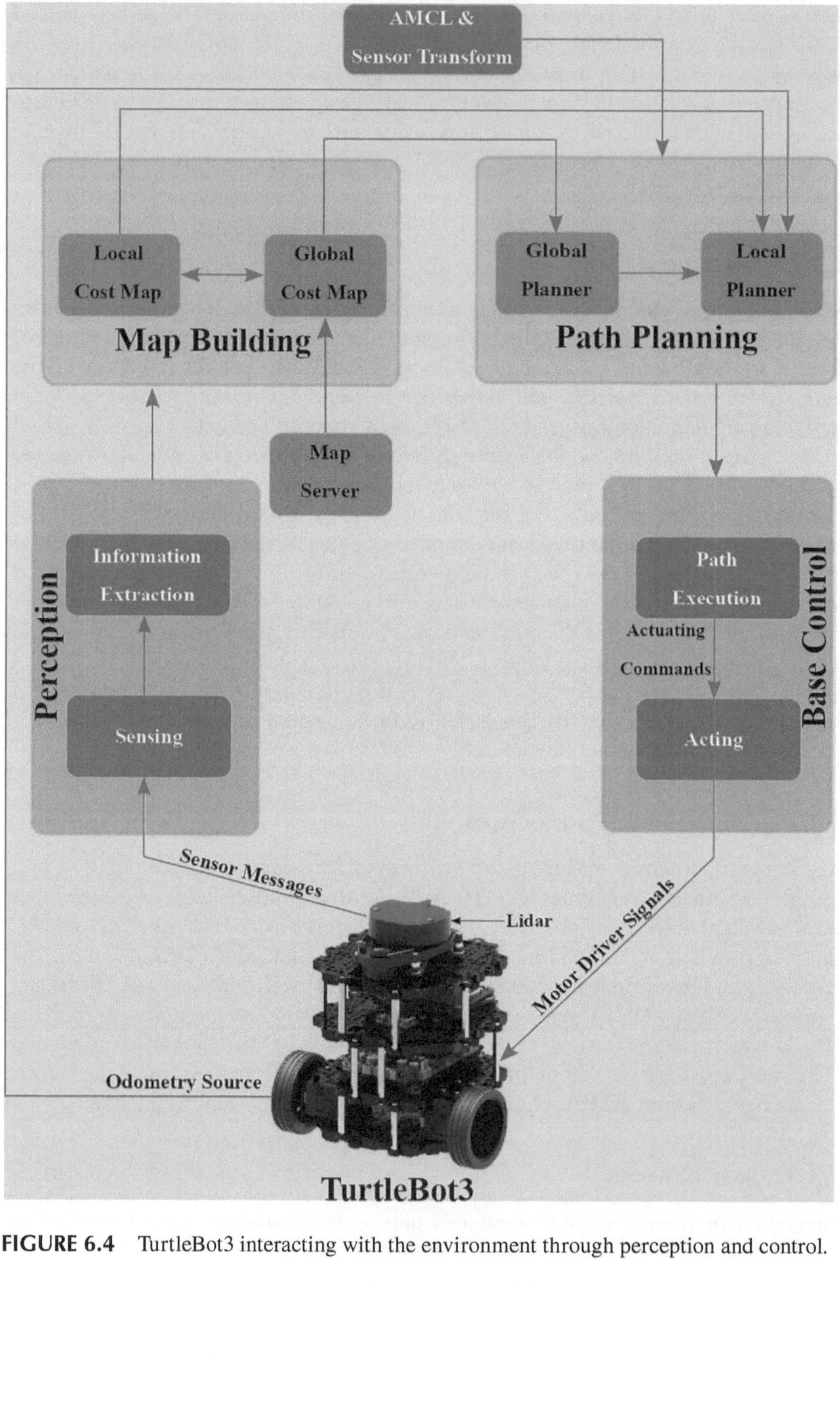

FIGURE 6.4 TurtleBot3 interacting with the environment through perception and control.

Sensors capable of performing on-board processing are in demands, since quick decision-making about the environment could result in a better strategy formation for the path planning approaches. However, usage of kinetic depth sensors, inertia sensors, ranging LiDAR, or laser rangefinder could require high computation resources for imparting quick and accurate map building. Cost map calculates the possible collision area, obstacle area, and movable area, which is used to set up a path plan for navigating in the fixed map. SLAM is used to build a map based on the robot pose using laser sensor (LiDAR) scan information and the position information. Incremental LiDAR scan matching principles are used by the robotic systems to navigate along the strange paths in the environment they are being deployed. The LiDAR system operates on the principle of measuring the time taken by the emitted light pulses to return from the obstacles for building a SLAM-based map.

Reproduction of LiDAR scans produces a continuously expanding map of the environment, in which they are deployed. However, motion drift could possibly cause some errors in the map building for the robot. Those errors can be rejected when the robotic system returns to a location that has been previously observed. In addition, it is important that these scans be corrected for the robot's path planning modules. Inertial sensor helps in this regard to carry out the entire process of map building, and the path planner provides the complete map of the environment in real time. All the computations are carried out in the on-board computer in the robot and a camera module could be integrated to observe the synchronous coordination between the map built using the SLAM through the LiDAR and camera feed. This also helps to give feedback to the robot to ensure that the locations are previously visited or new locations are on the way of the robot.

In a social environment during larger excursions, a significant drift could occur. These drift could potentially corrupt the visibility of the robot in the map. For instance, obscuring doorways or in exploration of new areas should be potentially cross-checked before the LiDAR feed is built into the map using the SLAM. When a previously visited location is determined, map smoothing can resolve this inconsistency. For easy navigation in the dynamic environments, the robot can also be configured to inject tags into the maps, by labeling interested and important locations on the way. The robot could also be annotated in the map with higher levels of information, such as directions or detected obstacles on its path. These maps could also be transmitted through wireless in real time back to remote stations. This mechanism enables perfect situational awareness of the robot in a social environment amidst humans. It could also be potentially used in search and rescue operations as well.

6.3.4 Path Planning

Autonomous mobile robot navigation systems are divided into three different subsystems: sensor data perception, decision making based on reasoning, and robot navigation and control. Basis of mobile robot navigation and control is path planning (Orozco-Rosas et al., 2019), which finds a path from the given current position to the goal position. Generated path should be collision-free, less processing time, minimum path length, less energy consumption, and meet the dynamics of the mobile robot (Han & Seo, 2017).

Human safety and comfort are also important measures in social robot navigation. Since the path quality affects the robotic application, an efficient path planning is needed to achieve effective and safety navigation. Based on the environment information, path planning is divided into local path planning and global path planning (Wang et al., 2017). In the local planner, the robot knows only the information it is nearby and whereas in the global planner, the robot knows the global environmental information. When the robot plans its path, it requires the map information, current position of the robot, and the goal position.

To identify where the mobile robot is moving around, it requires a map of its environment. It helps the mobile robot to know the directions and locations. The map can be generated either manually or gradually, while the robot discovers the new environment. Map can be generated using either SLAM algorithm or frontier exploration through mobile robot teleportation. Current position and orientation of the mobile robot is determined using numerous sensing methods such as laser rangefinder, ultrasound sensors, cameras, and GPS absolute coordinate (e.g., longitude, latitude, altitude) in outdoor environments (Koubaa et al. 2018). Current position and orientation of the mobile robot is computed using AMCL algorithm based on the sensors data received from the mobile robot using ROS message. Based on the information about mobile robot, path planning can be generated and motions can be visualized using Rviz tool.

Path planning algorithm creates a trajectory to move the robot from the starting position to the target position on the map. The generated path plan includes the global path planning and the local path planning. In this ROS package, sampling-based path planning algorithm (RRT with RL) is used as a global path planning. Figure 6.5 shows the exploration and search for a path from starting to the target point for the TurtleBot3 robot in a cluttered environment using a sampling-based path planning algorithm. Sudden moving objects or obstacles around the robot can be avoided by using Dynamic Window Approach (DWA), as an obstacle avoidance algorithm, and as a local path planning solution.

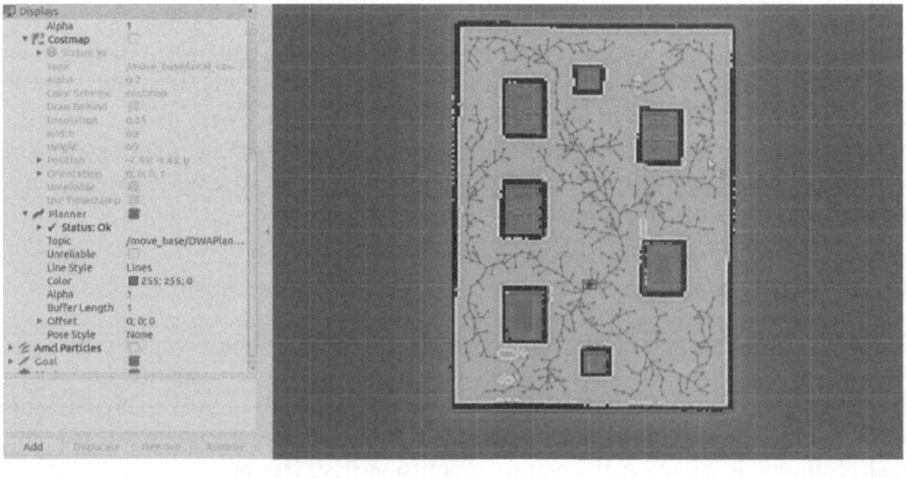

FIGURE 6.5 TurtleBot3 interacting with the environment in cluttered environment using sampling based path approach.

TurtleBot3 plans its path using map information, current, and goal position of the TurtleBot3. To identify where the TurtleBot3 is moving, it requires a map of its environment. It helps the TurtleBot3 to know the directions and locations. The map can be generated either manually or gradually while it discovers the new environment. Map is generated using SLAM algorithm through TurtleBot3 teleportation package. Current position and orientation of the robot is determined using sensor data. Current position and orientation of the TurtleBot3 is computed using AMCL algorithm based on the odometry, laser rangefinder, and IMU sensor data received from the TurtleBot3 using ROS topic /odom and /scan. Generated path in TurleBot3 is visualized using Rviz tool.

6.3.5 State-of-the Art Path Planning Approaches

Rapidly Exploring Random Search Tree (RRT) (LaValle, 1998) is one of the most popular among the sampling-based planning algorithms. Sampling-based path planning algorithms are probabilistic complete algorithms because they can provide a solution if one exists when the time tends to infinity, i.e., probability of finding the solution becomes one as the running time approaches infinity. RRT takes a random point in space and checks whether the point is present in the obstacle. Since it is a tree-based algorithm, it finds another point between the starting point and the random point with specific branch length. If the new point meets all the necessary constraints, it will be added to the tree as a node and the starting point becomes the root of the tree and also parent for the initially added node. This process continues until the search reaches the goal point. RRT provides the solution, which is not necessarily an optimal solution but a suboptimal solution.

There are variants of RRTs which claim to attain optimality reported in the literature (Elbanhawi & Simic, 2014). RRT* (Karaman & Frazzoli, 2011), a modified form of RRT, employs rewiring procedures to find the optimal solution. RRT* samples the space as that of RRT that performs, but it can modify the parent of the nodes by rewiring procedure and choose parent procedure. By calculating the distance using any of the distance functions available, the parent of the nodes can be changed so that the cost to traverse from one node to another node is reduced. Therefore, the solution automatically reaches optimality as the time increases. Even though a variant of RRTs is available in the literature, robot autonomous navigation using RRT is presented by Chang-an et al. (2008). Kuwata et al. (2009) proposed an RRT-based real-time motion planning algorithm for autonomous vehicles in urban environments. ROS-based autonomous navigation implementation issues in a Pioneer 3-DX robot are discussed in Zaman et al. (2011). In Crépon et al. (2018), the authors presented the implementation of the RRT navigation planner in a real mobile robot DynIBEX. ROS-based autonomous navigation scheme using the RRT algorithm in the TurtleBot robot is presented in Zhang et al. (2020).

6.4 *Q*-LEARNING STRATEGY

This section deals with the usage of *Q*-learning strategy for TurtleBot3 to effective navigation and path control in a social environment.

Robotic systems interacting with the environment are mostly based on knowledge-based methods, in which the robotic systems learn from the environment by learning about static and dynamic obstacles in their path. The major problems to be addressed in the dynamic environments are motion prediction, navigation, control by appropriate path planning, and considering the uncertainties during the real-time execution of the robotic system while navigating in social environments.

To handle human–robot interaction, RRT-based path planning alone is not enough to handle the situation, so we are proposing reinforcement learning (RL)-based RRT, where RL is used to learn the human–robot interaction through the trial-and-error method to achieve the desired performance. In RL, the agent earns positive and negative rewards for actions for each state. Markov decision process (MDP) is defined as a state, action, and the reward pair $(s,\ a,\ r)$ where the state space is $S \subseteq R^N$, the action space is $A \subseteq R^N$, and the reward function is : $S \times A \rightarrow$ R. If action a to the state s, then the probability of the next state s' and the next reward r is given in Equation 6.4.

$$p(s,\ a,\ s') = P_r(S_{t+1} = s' | \ S_t = s, A_t = a) \tag{6.4}$$

The exploration, delayed reward, partially observable state, and lifelong learning differ in the RL problem from other function approximations.

Q-learning is a basic off-policy RL method, where the agent learns to find the best action pair to the current state by taking the action randomly. The random action value is obtained from the epsilon greedy method, where epsilon (ε) balances the exploration and exploitation as given in Equation 6.5.

$$a = \begin{cases} \text{random } a \in A & \text{if } \xi < \varepsilon \\ \arg\max_{a \in A} Q(s,a) & \text{otherwise} \end{cases} \tag{6.5}$$

Q-learning is influenced by certain variables like learning rate α, discount factor γ, and reward r. The learning rate is referred to as alpha $\alpha \in [0,1]$, which defines the acceptance between the new values to the old value. It controls the model convergence. The learning rate is kept small that requires more training episodes, whereas the larger learning rate requires fewer training episodes and results quickly. However, if we choose too large a learning rate, it may converge the model to a suboptimal solution too quickly. Hence, the selection of is important in Q-learning. Gamma is a discount factor $\gamma \in [0,1]$, which is used to balance immediate and future rewards. The discount factor tells about how RL agents care about distant future rewards to the immediate future rewards. If $\gamma = 0$, then the RL agent evaluates its actions based on immediate reward. If , then the RL agent evaluates its actions based on the future rewards.

Policy, reward, and value function are the main elements of any RL system. A policy is a mapping of the perceived state to the actions taken on the state. An RL agent uses policy to select the actions. The main objective of the RL agent is to learn the stochastic optimal policy $\pi^* : S \times A \rightarrow \mathbb{R}$ that maximizes the value as given in Equation 6.6.

$$\pi^* = \arg\max_{\pi} \mathbb{E}\left[\sum_{k=0}^{T} \gamma^t r_{t+k+1}\right] \tag{6.6}$$

The reward is the value received after completing a certain action at a given state. The reward is defined as Equation 6.7.

$$r = \begin{cases} r_{\text{neg}}; & \text{if the robort collides with obstracles} \\ r_{\text{pos}}; & \text{if the robort reaches the goal} \\ r_{\text{default}}; & \text{otherwise} \end{cases} \tag{6.7}$$

Most of the RL algorithms are based on the value functions of the states. It tells about the estimate which is based on future expected rewards or expected return. The state-value function $V_\pi(s)$ for the policy π is given in Equation 6.8.

$$V_\pi(s) = \mathbb{E}_\pi\left[\sum_{k=0}^{T-t-1} \gamma^k r_{t+k+1 \mid S_t = s}\right] \tag{6.8}$$

The Q-agent of the proposed algorithm is interacting with the environment as shown in Figure 6.6. The agent decides to select a safe boundary using the learned policy which is observed from the environment. A reward is used to evaluate whether the obtained action is good or not. The state-action pairs Q-values for the agent are recorded. The Q-value update is done using the rewards received from the environment:

6.5 RESULTS AND DISCUSSION

The proposed algorithm selects the parameters for carefully avoiding the obstacles based on the given environment using the Q-learning training. The choice of parameters chosen in the proposed experimental setup for the Q-learning technique in the proposed system is as $\varepsilon = 0.9$, discount factor $\gamma = 0.8$, learning rate $\alpha = 0.8$; reward is chosen as $r_{\text{negative}} = -2.5$ and $r_{\text{positive}} = +0.5$.

The map information is loaded using the ROS nodes in the TurtleBot3. After map creation, the TurtleBot3 navigation stack is done using the launch file. On execution of the launch file, it runs the navigation stack and automatically opens RViz tool for visualization. We can see the robot pose with the cost map shown in blue color for the given cluttered map as shown in Figure 6.7.

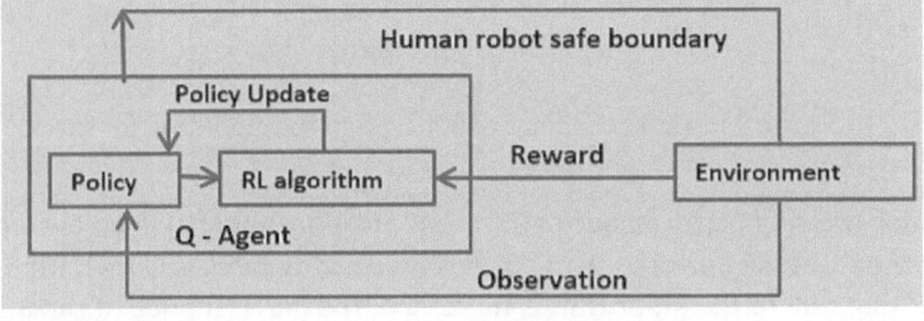

FIGURE 6.6 Block diagram of Q-learning process for adaptive to environment.

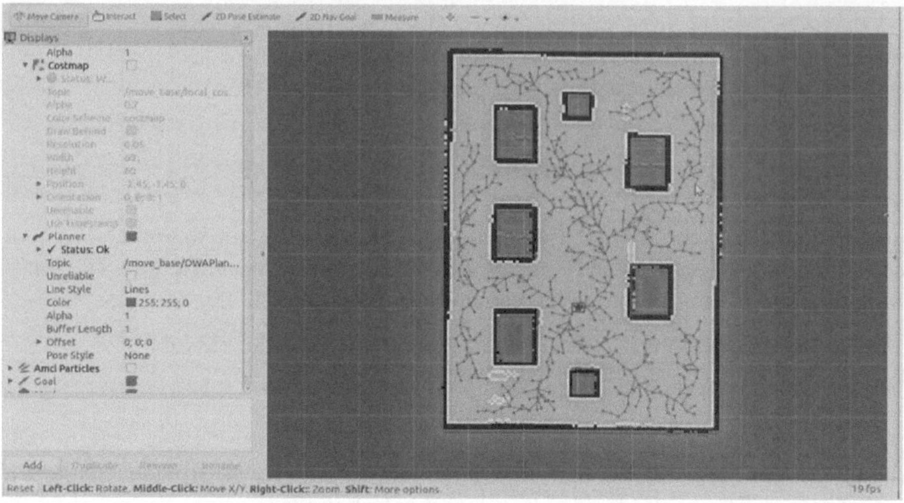

FIGURE 6.7 TurtleBot3 cost map in cluttered environment.

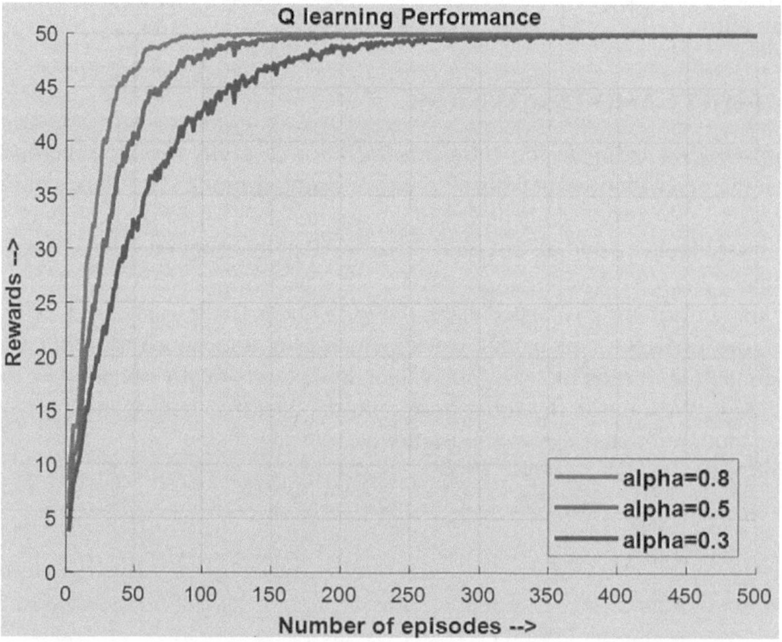

FIGURE 6.8 Q-learning reward performance of the proposed algorithm for various episodes.

Q-learning training performance of the proposed algorithm between reward per episode and the number of episodes using different learning rates is shown in Figure 6.8. As the learning rate of the algorithm is increased, the performance of the algorithms improves with increase in the rewards. As the number of episodes increases, the impact of the rewards converges and saturates at 50 in the proposed scheme.

6.6 CONCLUSION

To summarize, this chapter provides the reader with useful information on autonomous navigation schemes of robotic systems in social environments. The discussions are focused on the path planning algorithms implemented for the TurtleBot3 robot. Among the different navigation schemes, *Q*-learning algorithm is used for implementation and its performance is well suited for autonomous navigation and social human interaction tasks. The perception and control tasks are effectively managed by the developed algorithm, and it belongs to a better contestant among the family of existing navigation schemes. Finally, based on the learning parameters of the *Q*-learning strategy, the more time the algorithm spends on learning on the social environment of human-robot interaction, the better the reward from the environment and enhanced performance. This overcomes the challenges existing in the social robotic interactions with better navigation mechanisms.

ACKNOWLEDGEMENT

The authors gratefully acknowledge the Management and Faculty of Mepco Schlenk Engineering College, Sivakasi, India, for their support and extending necessary facilities to carry out this work.

DISCLOSURE STATEMENT

The authors of this manuscript are not involved and not availed any financial assistance from any funding agencies for carrying out the research work shared with this chapter.

REFERENCES

Bera, A., Randhavane, T., Prinja, R., Kapsaskis, K., Wang, A., Gray, K., & Manocha, D. (2019). The emotionally intelligent robot: Improving social navigation in crowded environments. *arXiv preprint arXiv:1903.03217.*

Chang-an, L., Jin-gang, C., Guo-dong, L., & Chun-yang, L. (2008). Mobile robot path planning based on an improved rapidly-exploring random tree in unknown environment. *2008 IEEE International Conference on Automation and Logistics*, IEEE, 2375–2379.

Crépon, P.-A., Panchea, A. M., & Chapoutot, A. (2018). Reliable navigation planning implementation on a two-wheeled mobile robot. *2018 Second IEEE International Conference on Robotic Computing (IRC)*, IEEE, 173–174.

Dijkstra, E. W. et al. (1959). A note on two problems in connexion with graphs. *Numerische Mathematik*, 1 (1), 269–271.

Elbanhawi, M., & Simic, M. (2014). Sampling-based robot motion planning: A review. *IEEE Access*, 2, 56–77.

Han, J., & Seo, Y. (2017). Mobile robot path planning with surrounding point set and path improvement. *Applied Soft Computing*, 57, 35–47.

Hart, P. E., Nilsson, N. J., & Raphael, B. (1968). A formal basis for the heuristic determination of minimum cost paths. *IEEE Transactions on Systems Science and Cybernetics*, 4 (2), 100–107.

Henschel, A., Hortensius, R., & Cross, E. S. (2020). Social cognition in the age of human–robot interaction. *Trends in Neurosciences.*

Karaman, S., & Frazzoli, E. (2011). Sampling-based algorithms for optimal motion planning. *The International Journal of Robotics Research*, 30 (7), 846–894.

Khambhaita, H., & Alami, R. (2020). Viewing robot navigation in human environment as a cooperative activity. *Robotics Research* (pp. 285–300). Springer, Cham.

Kim, K. J., Park, E., & Sundar, S. S. (2013). Caregiving role in human–robot interaction: A study of the mediating effects of perceived benefit and social presence. *Computers in Human Behavior*, 29 (4), 1799–1806.

Koubaa, A., Bennaceur, H., Chaari, I., Trigui, S., Ammar, A., Sriti, M.-F., Alajlan, M., Cheikhrouhou, O., & Javed, Y. (2018). Introduction to mobile robot path planning. *Robot Path Planning and Cooperation* (pp. 3–12). Springer, Cham.

Kuwata, Y., Teo, J., Fiore, G., Karaman, S., Frazzoli, E., & How, J. P. (2009). Real-time motion planning with applications to autonomous urban driving. *IEEE Transactions on Control Systems Technology*, 17 (5), 1105–1118.

LaValle, S. M. (1998). *Rapidly-Exploring Random Trees: A New Tool for Path Planning*, , Ames, IA, USA.

Lemaignan, S., Warnier, M., Sisbot, E. A., Clodic, A., & Alami, R. (2017). Artificial cognition for social human–robot interaction: an implementation. *Artificial Intelligence*, 247, 45–69.

Lisetti, C. L., Brown, S. M., Alvarez, K., & Marpaung, A. H. (2004). A social informatics approach to human-robot interaction with a service social robot. *IEEE Transactions on Systems, Man, and Cybernetics, Part C (Applications and Reviews)*, 34 (2), 195–209.

Luber, M., Spinello, L., Silva, J., & Arras, K. O. (2012). Socially-aware robot navigation: A learning approach. *2012 IEEE/RSJ International Conference on Intelligent Robots and Systems*, IEEE, Vilamoura-Algarve, Portugal, 902–907.

Orozco-Rosas, U., Montiel, O., & Sepu´lveda, R. (2019). Mobile robot path planning using membrane evolutionary artificial potential field. *Applied Soft Computing*, 77, 236–251.

Wang, C., Meng, L., She, S., Mitchell, I. M., Li, T., Tung, F., Wan, W., Meng, M. Q.-H., & de Silva, C. W. (2017). Autonomous mobile robot navigation in uneven and unstructured indoor environments. *2017 IEEE/RSJ International Conference on Intelligent Robots and Systems (IROS)*, IEEE, Vancouver, BC, Canada, 109–116.

Zaman, S., Slany, W., & Steinbauer, G. (2011). ROS-based mapping, localization and autonomous navigation using a pioneer 3-dx robot and their relevant issues. *2011 Saudi International Electronics, Communications and Photonics Conference (SIECPC)*, Riyadh, Saudi Arabia, 1–5.

Zhang, L., Lin, Z., Wang, J., & He, B. (2020). Rapidly-exploring Random Trees multi-robot map exploration under optimization framework. *Robotics and Autonomous Systems*, 131, 1–11.

7 Innovations in Healthcare Using Smart Systems Equipped with Evolutionary Computation

Gurpreet Singh, Raman Shergill, and Pradeep Kumar Gaur

CONTENTS

7.1 INTRODUCTION

Human activity recognition (HAR) is considered one of the research fields related to evolutionary computing where small but meaningful object detection plays an important role. This small object-detection activity is considered as one of the emerging areas of image processing and machine vision too. The result of this field of study helps to improve society by keeping a watch over the health of persons (De Jong, 2019; Kicinger, Arciszewski, & De Jong, 2005; Xue, Zhang, Browne, & Yao, 2016). Various applications like tracking of different objects (Crawford & Pineau, 2019; Khoreva, Benenson, Ilg, Brox, & Schiele, 2017; Lee, Liew, Cheah, & Wang, 2014; Y. Liu et al., 2019; Teutsch & Krüger, 2012), segmentation at various time slices (Chen et al., 2019; Leibe, Vi, & Hutchison, 2016; Li, Xie, Lin, & Yu, 2017; S. Liu, Qi, Qin, Shi, & Jia, 2018; Ren, Mengye; Zemel, Ren, & Zemel, 2016), recognition of various actions performed (Cheng, Wan, Saudagar, Namuduri, & Buckles, 2015; DasDawn & Shaikh, 2016; Guo & Lai, 2014; Herath, Harandi, & Porikli, 2017; Kong & Fu,

DOI: 10.1201/9781003158165-7

2018), providing captioning according to the different items existing in the image (Bai & An, 2018; Kalra & Leekha, 2020; Kougia, Pavlopoulos, & Androutsopoulos, 2019; Kumar & Goel, 2017; Srivastava & Srivastava, 2018), to get an idea about the different scenes etc. (Gupta, Arbeláez, Girshick, & Malik, 2015; Hoiem, Efros, & Hebert, 2008; Qiu, Zhuang, Yan, Hu, & Wang, 2019; Sakaridis, Dai, & Van Gool, 2018; Zitnick, Vedantam, & Parikh, 2016) are considered under the huge area of small objects detection. This field of HAR is dealing with the concept of finding and extracting minute information about the activities performed by humans during their different movements. The basic use of this technique helps in different types of applications such as in the development of health and care appliances, human fitness tracking activities like sleeping, running, walking details, etc., and in virtual reality gaming applications. The study in the field of HAR proves that when compared to the well-known statistical and mathematical techniques, machine learning and data science approaches provided very much considerable results. The main idea behind HAR systems is to capture the details about the various physical activities performed by humans, automatically from the raw input. On the basis of this feedback about the different activities performed by any person, some kind of suggestive measures about health can be suggested for the improvement of health. Under the Activity recognition model, the main concern is to track the different regular activities performed by the person. These common activities can be considered as jogging, walking, movement upstairs and downstairs, etc. The above said are fitness activities. By keeping the record of these basic activities, further analysis can be done. The awareness among youth about fitness these days is the main attraction factor of this field of research. In HAR systems, the various daily routine activities discussed can be captured mostly with the help of different sensors or sometimes with the help of recorded videos, etc. As far as the different types of sensors are concerned, mostly infrared motion sensors and magnetic sensors are used to record the abovesaid different activities. The main drawback in the case of activity recognition from video surveillance is the privacy factor. A lot of research has included the development of HAR. The fundamental objective of diverse elements of such systems is to extract the attributes, learn and classify the action, and also to identify and segment the action. A simple process is carried out in three phases that emphasize detecting and tracking the human body. The actions can be recognized using tracking outcomes. To illustrate, the shaking hand activities are identified by detecting and tracking the arms and hands of two persons initially for creating a spatial-temporal description of their action. The existing patterns are employed to compare this description in the training data in order to identify the kind of action.

In general, a HAR framework consists of three main stages: segmentation, feature selection & extraction, and feature classification. Segmentation is a simple approach for attaining this objective works and is defined as: The background is estimated as the preceding frame first of all. Subsequently, this foundation is deducted from the information outline. A limit is used to the total contrast for gaining the frontal area cover. The item shape, speed, outline rate, and so on are considered to verify the efficiency of this technique. To illustrate, even though this kind of method is adaptable for dynamic environments and there is not any necessity of knowledge regarding background, this technique is affected by foreground aperture issue because of the homogeneous color of moving objects. Thus, this technique is unable to detect all the moving pixels. A background subtraction algorithm that has reliability and robustness is capable of dealing with unexpected or progressive enlightenment changes, high recurrence behind the scenes, and long-haul scene changes. Feature selection and extraction method is executed to attain the features of action and to analyze these attributes afterward. At first, the video sequence provides the representative descriptors of attributes at every time interval. A significant design decision is concentrated on developing the feature vector for which the attributes are selected in the resulting highlight characterization requirement of a huge amount of storage and battery backup. For example, if for recognizing or detecting human activities of one persona video camera is used, then the camera may capture the different minute movements of other persons also, who for whatsoever reasons appeared in front of the camera placed. To avoid the issue of security and privacy, instead of video cameras to capture the activity details of a single person, some wearable sensors have been proven to be effective. These wearable sensors can be worn by the person, whom you want to treat as an object, on the body. The energy consumption requirement of these sensors is very less. So these sensors can easily be placed inside smartphones or smartwatches, which can easily be carried by the personas one of the body accessories. The Accelerometer is one of these wearable sensors which is used to capture the detail about the acceleration of a person around the three dimensions. Similarly, the sensor Gyroscope falls under the same category and is used to capture the information about the orientation details around all the three axes. The use of these wearable devices solved the problem of security and surveillance but also faced a number of other challenges like interclass similarity of devices and their compatibility, intraclass variability, the issue of class balancing and imbalance, the issue of picking the exact start and close of activities, different nature of sensors meant for different purposes but need to collaborate with one another to capture various unique factors, etc. During the experimental setup in these HAR applications, the other major factor is to capture the data of the same device for the purpose of analysis and testing but from different persons or participants. Many of the times, the captured data reflects totally different observations. This leads to the major challenge of intraclass similarity. On the other hand, sometimes the data recorded for different activities may reflect similar kind of observations; this leads to the interclass ambiguity. The data of human activity jogging and human activity running sometimes reflect the same kind of information. These two activities are the example of interclass ambiguity or similarity. The factor of class balancing can be affected by the situation in which one activity is extended for a very long duration and can provide a similar kind of result as produced by the other activity in quicker time duration. In this case, the activities walking and jogging can be considered as per the slow nature and the fast nature, respectively. In the beginning, HAR problem was considered as a problem under the category of pattern recognition problems and a number of researchers applied the classification techniques like SVM, HMM, MLP (Singh& Sachan, 2015a, b, 2019, n.d.) etc. to convert the various observations into meaningful results. But in these days, after observing the effectiveness of deep learning approaches in various fields, the concept of data science and machine learning (ML) has been proved to be effective in the field of HAR. The major dependency in the case of classification models was the collection and the selection of feature vectors and early human knowledge. But, due to the involvement of a number of micro-activities, the extraction of these feature vectors is not a simple task. This problem of selection and extraction of various features is not a barrier in case of deep net technique as in the case of data science and machine learning systems, the systems are fetching the information about various features from the hierarchy of the given data itself. The different properties of deep learning models as scale invariance and local dependencies help in the automatic extraction of required feature sets. The other major positive factor associated with ML and data science fields is the existence of a transfer learning approach. With the help of transfer learning, the knowledge gained after learning from one model can be shifted as input to any other model to provide some kind of strength and power to understand various features without prior selection of any feature set. The next part of this chapter shows the state-of-the-art-work already done in this field of research with the help of the literature review section. This section is followed by the experimental section, which shows the model proposed to handle the work of HAR. After that, the results and discussion section shows the effectiveness of the proposed model based on different observations. In the end, the conclusion and future scope sections will sum up the pros and cons of the model used for HAR.

7.2 LITERATURE REVIEW

Ramamurthy and Roy (2018) presented a survey on the use or the impact of machine learning algorithms during HAR process. The authors presented a comparison among the different machine learning alternatives available for the field of HAR. Along with the concepts of machine learning, they also discussed the power of data science in this field of research. They mainly focused on the concept of state-of-the-art techniques or solutions available for HAR. The authors implemented the concept of data mining to get meaningful and more refined information from the raw data. They checked the main scope of the field of HAR under medical and fitness applications or wearable devices. The other observation recorded by the authors is that the ML algorithms worked on the concept of interpreting mined information by the data science techniques to get an idea about the activity involved with the situation. They worked on the concept to observe the effects of transfer learning. Due to the small volume of the dataset used, authors suggested the use of synthetic data generation to check the effectiveness of the model proposed for HAR.

Ke et al. (2013) proposed a method based on the input as video stream rather than the use of more refined sensor data for the purpose of HAR. They worked on three important concerns of the HAR system. The first one is the segmentation of different frames from the video input. The second important step they performed was the extraction of various features. Later on, these features were also represented effectively to identify various activities involved. The third step considered the classification method to categorize the activity performed by the user in the current video frame. Authors concluded the various challenges faced after taking video as input for HAR applications. One of the challenges they considered was the interpretation of different cameras from different angles for the same view. Here, they suggested constrained camera angles as the solution for the abovesaid issue. The authors also concluded that the background subtraction technique only was not that beneficial as the objects were moving. They concluded that the surface where the person is moving also made an impact on the overall observation. Along with this, the clothes and other body accessories also interfere in the detection and identification part. In the end, the authors suggested the 3D viewpoint invariant modeling as one of the best solution to start with HAR applications.

Wang, Chen, Hao, Peng and Hu (2019) conducted a survey on the effective use of the deep learning technique over HAR applications. The authors considered sensor-based input to identify the various human activities. They gave stress on the issue of using deep learning concepts as the previous techniques require a proper selection of feature sets. The other point they mentioned was that the selection of a shallow feature set was always based on the experience level of users and may provide different results for the same activity performed by two different users. They also gave a fact that the traditional classification techniques always worked on the part of well-trained systems, which may not be completely possible in the case of HAR applications. Their other finding was that the success achieved in the case of famous classifiers is only after the consideration of static data, not the dynamic one that is there in the case of HAR applications. The authors discussed the Convolution Neural Network (CNN) based model to explain the concept of deep learning and the role of various hidden layers used to provide the accurate results.

Kim, Helal and Cook (2010) worked on a project of HAR. They applied the concept of pattern recognition techniques over the stored data of various activity recognition sensors. The data under consideration was based on the home activity dataset. Authors put the effort to represent the disjoined data in a single shape to reflect some common features in the form of various human activities categories.

Zeng et al. (2015) proposed a deep learning approach by using CNN to extract a specific feature set from the mobile sensing data. Various sensors were used in the mobile phones, capturing different human activity movements. The data collected by these sensors were used by the authors to form a feature set for further identification of different human activities. They tested their system against three different datasets of HAR. They considered the first public dataset "SKODA". This dataset considered the assembly of activities. The next public dataset was "OPPORTUNITY". This dataset was based on human activities performed in the kitchen. The last dataset considered by the authors was "ACTITRACKER". This dataset considered the jogging and running samples of various users to track these different activities. The authors used five different feature extractors for the purpose of further classification.

Xu, Chai, He, Zhang and Duan (2019) proposed a method for the detection of some complex activities belonging to HAR systems. They discussed the drawbacks of various machine learning methods used for the same purpose but heavily relied on the extraction of different feature components. The authors considered the property of deep neural networks to deal with the situation of automatic detection of features. They proposed the hybrid combination of RNN and Inception NN (INN). Under INN, they used a number of kernel-based convolution layers. For the purpose of tracking the different activities, they considered the concept of Body Area Network (BAN) and Wearable sensors.

Wan, Qi, Xu, Tong and Gu (2020) worked on the concept of mobile edge computing. This leads to the creation of an intelligent system that serves society by regularly monitoring the health of old age persons and other adults and children. For creating the BAN, they took the help of smartphone devices with inbuilt sensors. CNN was used by the authors to apply the concept of deep learning. The commonly used sensors during their research observations included accelerometer, gyroscope, light, proximity, barometer, and GPS. They worked over 24 different human activities. These activities included lying, sitting, standing, walking, running, cycling, Nordic walking, watching TV, computer work, car driving, ascending stairs, descending stairs, vacuum cleaning, ironing, folding laundry, house cleaning, playing soccer, and rope jumping. They observed the better performance of CNN as compared to other classifiers such as SVM (Support Vector Machine), MLP (Multi-Layered Perceptron Nerual Netweor), and LSTM.

Liu, Cai, Ju and Liu (2019) presented the comparison between both manually selected set of features and the generated feature set by learning techniques for human activity detection. For a study deeper in this field of research, the authors consider the concept of RGB-D sensors. Authors observed that the main challenges faced in HAR systems are the adaptability with the online environment, variations in the interpretation level in various scenes, sensing, and execution of action time variations, etc. They concluded that the use of depth-based, skeleton-based, and hybrid-feature-based deep net approaches is best for HAR systems.

Liu et al. (2019) focused on the challenge of providing different annotations within lesser time. Semisupervised learning approach was used by them to produce

TABLE 7.1
Summary of the Selected Work Already Done in the Field of HAR

Authors	Considered Work
Ramamurthy and Roy (2018)	Survey on the Impact of machine learning algorithms for human activity recognition
Ke et al. (2013)	Worked of Video input for Human Activity Recognition
Wang et al. (2019)	Worked on sensor-based input and deep learning technique for human activity recognition
Kim et al. (2010)	Worked on sensors dataset for human activity recognition
Zeng et al. (2015)	Considered mobile sensing data using Deep learning for human activity recognition
Xu, Chai, He, Zhang and Duan (2019)	Worked on the concept of Human activity recognition by considering BAN (Body Area Network) and Wearable Sensors
Wan, Qi, Xu, Tong and Gu (2020)	Worked with the concept of Mobile Edge Computing (MEC)
Liu, Cai, Ju and Liu (2019)	Worked over RGB-D technique for human activity recognition and Depth based, Skeleton Based and Hybrid feature based techniques.
Liu et al. (2019)	Reduction of providing annotation without disturbing the overall performance of human activity recognition system

an active learning system. The HAR module was developed by the authors to self-train the system. This module was capable of giving annotations without using any kind of previous learning. They observed that the task of annotations can be reduced drastically by about 90% without disturbing the overall performance of the system.

Table 7.1 shows the summary of the literature discussed above. This table shows the potential in the field of HAR. Various computational models have been proposed by different researchers. The input method of all these models falls in the categories like BAN, sensor datasets, or activity videos. A solution using deep learning approach using CNN was proposed by a number of researchers and shows good results.

7.3 EVOLUTIONARY COMPUTATION AND SOCIETY 5.0

The idea about the smart society under the concept of Society 5.0 brings lots of challenges to explore and also provided the opportunity to serve the society by using the technical knowledge and computational strength of the technology (Kundu, 2016). A number of problems are there to address under this platform. These problems are related to different parts and the needs of the society. Figure 7.1 throws some light over these different aspects. As shown in Figure 7.1, the various parts considered under the platform of Society 5.0 fall under the broader categories: manufacturing, food-chain systems, hospitality systems, health-care systems, infrastructure maintenance etc. The basic purpose is the creation of a smart society in which there must be some provision to access data easily and the work over that data and based on the observations must be able to perform some actions again for the betterment of society. The computational power of computer systems and the advent of technology

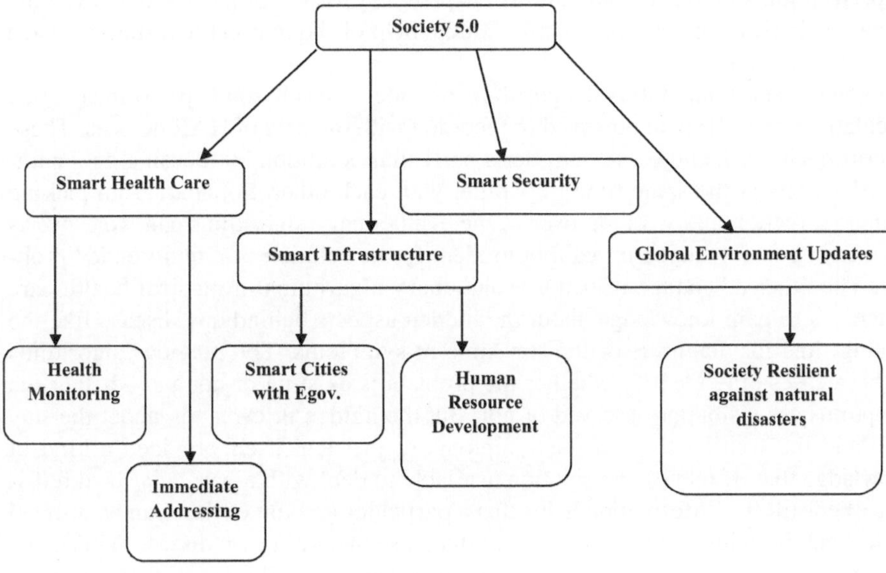

FIGURE 7.1 Platforms considered under Society 5.0.

make these things easy (Taneja & Taneja, 2012; Taneja, 2012; Tyagi, Nigam, & Singh, 2020). The main requirement to build all these smart systems is to channelize the energy in a proper direction. The main tools under consideration are the computational powers of the existing systems to deal with the complex data and the high storage capacity of the current devices to handle data generated by all these smart systems. The starting for storage devices begin from basic floppy drives having storage capacities in kilobytes or megabytes only to hard drives, small pen drives with large capacities, and now, SSDs which have storage capacity along with high-speed data transfer.

The system proposed in the current setup is dealing with an innovative solution for health-related problems by providing a wakeup call to avoid the circumstances in advance. To achieve this target, continuous monitoring of health related activities is required ("Health care system and evolutionary approach," n.d.). To record these monitoring details and analyzing the same with the required thresholds provides a kind of eye opener for different persons. The advent of IoT devices in the current time provides more flexibility to perform different actions based on the health status observed by activity detection analysis. The hybridization of the field of human activity detection and IoT together have the ability to produce real-time systems for the immediate cure of the health of patients. The observation about the timings of any person to complete a specific distance while walking these days is used as a litmus test by the doctors to complete the initial investigation of the human body. Based on these kinds of analysis, the practitioners are able to start the initial medicare process for the patients (Pelikan, Krajic & Conrad, 1997). Evolutionary computation over the data collected by the IoT technologies can provide the possibility to brief the current situation of the human body and in the case of the symptoms of a serious emergency, the system can manage the things in advance by passing the information only or even by performing suitable actions. In this chapter, a system is proposed to analyze the human activities and pass the record with the help of the internet to a suitable place for further consideration.

Evolutionary computational algorithms provided the solution to perform complex calculations over the data observed or recorded with the help of HAR devices. These algorithms have the power to consider a particular condition by focusing on a number of aspects at the same time in parallel with each other. So, apart from picking a single direction and working over it, the evolutionary algorithms can work over a number of considerations in parallel to identify the solution of a multi-ended problem. The other advantage of using evolutionary algorithms in medical health-care systems is to gain knowledge about the hidden aspects behind any disease like the waiting time for diagnosis or the stay time of symptoms. The previous algorithms in these cases only identify whether the problem is existing or not and whether the symptoms are visible or observed or not. But the hidden information about the stay time and the waiting time of the symptoms or problem itself provided a kind of knowledge that, if there is no solution available to deal with a particular health disease, then still this information helps that a particular amount of time can be utilized for finding the solution of the problem or diagnose that unknown disease. This is the main essence of monitoring the different health parameters in a continuous manner. HAR systems provide strength to the monitoring of these parameters. The other role of evolutionary algorithms in the case of analysis of data related to the medical field and health-related issues is to take care of the heterogeneity of the data. Different kinds of data and formats add some power to the complexity of the data analysis part. The evolutionary algorithms in this case helped during analysis and dealing with the complex nature of the problems due to heterogeneity (Gaur, 2017).

Multilevel data items are used to express the feature sets related to individual patients. It requires specificity at a number of different levels. It shows that the collective features of different tests under consideration can conclude a given level of complexity. The best part of these evolutionary approaches is that, under health-care environment, the acceptance level of specificity can vary with the different outcomes of the experimentations or at the times when the outcomes are not different too. The main criteria of these evolution-based techniques is the cancellation of more specific outcomes with the more irrelevant one and focusing on the desired one only. The next part of the chapter will highlight the experimental setup for the things discussed above and for the problem of the analysis of HAR data.

The most important part of the HAR systems is to provide knowledge about the current location of the patient or the person requiring immediate medical attention. In the case of evolutionary algorithms, a number of algorithms can be used for this purpose of finding the location of the person from the data collected by monitoring the various sensor devices carried by the body of the person (Fourie & Groenwold, 2002). Most of the times, the evolutionary algorithms particle swarm optimization (PSO) from the image map covered by the sensors which shows the track route of the person while walking, running, cycling, etc. is used (Wang, Tan, & Liu, 2017). The working phases for this algorithm again fall under the category of training and testing. During the training phase, the learning is involved to study the concept of how to reach the exact location. The next phase of the algorithms is to apply the learning gain from the training phase to get the location of the person. This phase is categorized as the testing phase. The training phase of PSO evolutionary algorithm is always started with random population size (Trelea, 2003). This population can be treated as any sample from the data under the radar. Each population can be treated as a kind of candidate solution. The particles present inside this randomly selected candidate solution are the main objects under consideration. The particles are considered with a random and unique velocity or speed. With this speed factor, all the particles are moving in the desired problem space. After running and observing the different iterations, local and global solutions are to be considered from this kind of movement. The local solution is the best factor achieved by a single particle within its own iteration set. Whereas the global solution has to be considered as the best movement of any particle in different iterations as compared to the iteration values of all the particles.

7.4 EXPERIMENTAL SETUP

An expert system for the purpose of detecting human activities is proposed by keeping the concept and the performance of deep learning models in this field of interest. The main purpose is to serve society by providing a kind of software solution to deal with the serious health issues of elderly persons and provide adequate medical facility to these persons at the earliest. To capture the movements in different time slots, the concept of RNN is used. For the purpose of implementing and reflecting the behavior of the working of deep network, Long Short Term Memory (LSTM) model is proposed. This model is able to record the different movements according to the different time slices. This timely capture of the information is necessary to observe the behavior against different panic situations. The different activities under consideration are standing, walking, sitting, upstairs, downstairs etc. Following are the various phases of research methodology, and Figure 7.2 also represents the exact sequence of these phases.

- **Dataset Collection**: The dataset is collected from different sources for human activity detection. HAR using smartphones ("Human Activity Recognition Dataset," n.d.) dataset is selected to consider various inputs. The main focus is to record time-series output for further use. The other property of this data is that it is multi-variant in nature. A total of 11,299 samples were collected during the implementation step. The entire dataset contains 521 attributes which help in the identification of different activities under consideration. The main use of the dataset is to test and train the proposed system by using the concept of classification and clustering techniques. The dataset consists of the samples for six different activities. These activities include walking, walking up and down the stairs, standing, sitting, and lying. The sensor gyroscope and accelerometer were used to capture the different movements. The device that carried these sensors was a smartphone. With the help of abovesaid sensors, the directional movement of the human body can be captured. To maintain the quality of the signal, a butter-worth low pass filter was used. For the purpose of training and testing of the data from the dataset, 30–70 ratio was considered respectively.
- Training of System: In this phase, the 3D Skelton is created for the feature extraction to identify human actions. The data about three different axes is recorded as "XYZ" information. 70% of the dataset is used for the purpose of training the system. Google collaborative IDE is used to test the proposed system. LSTM model is used to extract the different patterns based on some similarity index.
- Classification: The technique of classification is applied in this phase, which takes the input of the test set. The output of the classification defined the human actions based on the features. Figure 7.3 shows the output after the uploading of dataset content in the system and extraction of different features against the input data. It shows the output in the form of the activity captured.
- Performance Analysis: In this phase, the various parameters like accuracy, precision, and recall are calculated for the performance analysis.

During the selection process of calling the deep learning model, the input required to set to call various functions included the idea about the different parameters present in the hidden layer and different parameters of the output layer. The following function is used at the hidden layer:

$$\text{Normal\_Random}\left(\text{Feature\_Size, Layer\_Size}\right). \tag{7.1}$$

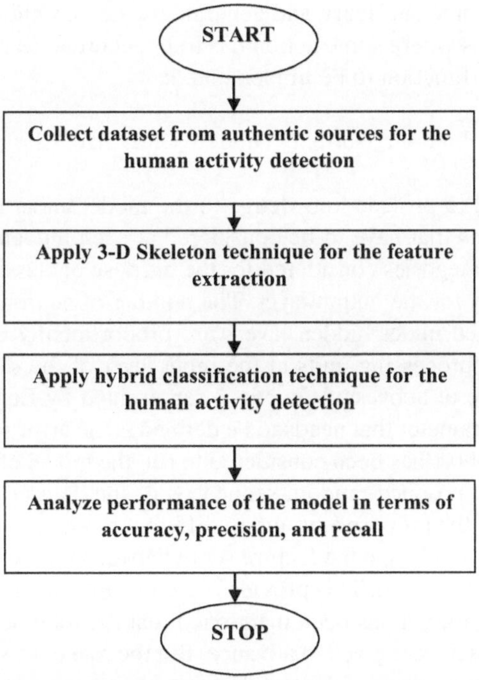

FIGURE 7.2 Proposed flowchart.

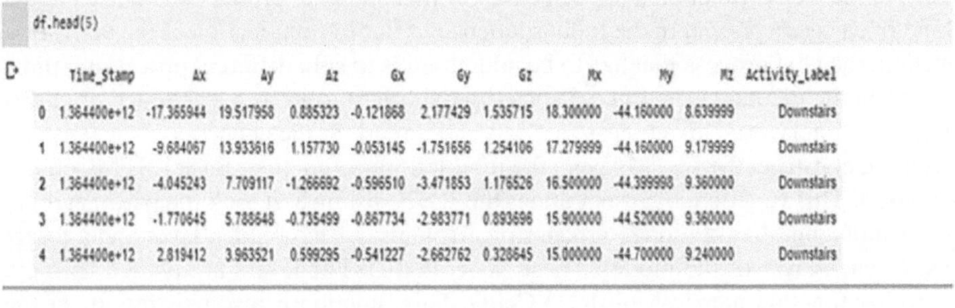

```
df.head(5)
```

	Time_Stamp	Ax	Ay	Az	Gx	Gy	Gz	Mx	My	Mz	Activity_Label
0	1.364400e+12	-17.365944	19.517958	0.885323	-0.121868	2.177429	1.535715	18.300000	-44.160000	8.639999	Downstairs
1	1.364400e+12	-9.684067	13.933616	1.157730	-0.053145	-1.751656	1.254106	17.279999	-44.160000	9.179999	Downstairs
2	1.364400e+12	-4.045243	7.709117	-1.266692	-0.596510	-3.471853	1.176526	16.500000	-44.399998	9.360000	Downstairs
3	1.364400e+12	-1.770645	5.788648	-0.735499	-0.867734	-2.983771	0.893696	15.900000	-44.520000	9.360000	Downstairs
4	1.364400e+12	2.819412	3.963521	0.599295	-0.541227	-2.662762	0.328645	15.000000	-44.700000	9.240000	Downstairs

FIGURE 7.3 labeling of classification results.

The Normal_Random function required input in the form of two different parameters. The first parameter is the idea about the number of features required to build a model. These features are randomly selected from the feature set uploaded in the definition of dataset earlier uploaded and refined. The second parameter to this function is the size of hidden layers used. It means the number of hidden layers. As per the performance of Deep nets in other applications, it has been observed that more number of hidden layers refines more accurate performance of any system. This process is just like providing more and more examples to a child for the understanding with the name of a single object. This concept of the hidden layer is providing the power to the proposed system to self-learn and generate the hidden knowledge to deal with the different situations where some action has to be recorded as per its performance. Following is the next function to be implemented:

$$\text{Output\_Random\_Normal}\left(\text{Hidden\_Unit\_Size, Class\_Size}\right) \qquad (7.2)$$

This function is used to provide knowledge to the model under consideration about the number of neurons that have to be considered at each hidden layer and the total number of different categories considered for the purpose of classification. These credentials are necessary for the output layer. The number of neurons or processing elements to be considered under hidden layers are proportionally dependent upon the previous selection of processing units at the input layer of the same network. Apart from the performance of abovesaid functions represented by Equations 7.1. and 7.2, another important parameter that needs to be defined is the error factor or bias. In this model, a bias factor of 0.1 has been considered to run the model efficiently. Transpose function is applied to produce random sequences of the dataset for getting the idea about the accuracy of the proposed system in different random situations with the use of the same input set. To change the feature set in between the execution of a process, reshape function has been used. This provides better results in the selection of random features. In the beginning, it has been mentioned that the total number of features to be used from the dataset were given in advance. But the role of this reshape function is to select different features under different iteration so that most of the sequences have to be addressed. Rectified Linear activation function (ReLU) is used for activating the model under consideration at different instances when it is required. The binary nature of the ReLU function is the necessity of this model to produce the output about a kind of category present in the input sequence. After the application of the activation function, the next process that has to be addressed is to split different processing units present on the hidden layers into various time-slices to work as a unit together at different time intervals. This process leads to an increase in the accuracy of the system. Further to construct the LSTM layers the inbuilt library function, BasicLSTMCell has been used. This function is present under the RNN instance. It took the parameters in the form of a number of hidden processing elements at each hidden layer and a forget bias for each layer of the model. These generated LSTM layers are passed as input to another function named MultiRNNCell(). This function is also existing under the RNN instance. This is used to produce the more refined LSTM layers. From these LSTM layers, the output can be generated in the form of identification of any human activity with the help of a basic static RNN function. This function accepts LSTM layers as one parameter and the number of hidden layers as another parameter.

7.5 RESULTS AND DISCUSSION

After the uploading of the dataset, many experimental observations are considered. Under one of the observations, the number of epoch or iterations is recorded as 50, and a batch size of 1,024 has been considered. During this experimental setup, the accuracy at the time of training and testing along with the false detection rates is represented by Figure 7.4.

```
epoch: 2 test_accuracy:0.6971962451934814 loss:1.4320471286773682
epoch: 3 test_accuracy:0.8006230592727661 loss:1.2557303905487060
epoch: 4 test_accuracy:0.8467289805412292 loss:1.1295710802078247
epoch: 5 test_accuracy:0.8822429776191711 loss:1.0519989728927612
epoch: 6 test_accuracy:0.8809968829154968 loss:1.0060634613037110
epoch: 7 test_accuracy:0.8785046935081482 loss:1.0059698820114136
epoch: 8 test_accuracy:0.9040498733520508 loss:0.9541240930557251
epoch: 9 test_accuracy:0.9127725958824158 loss:0.9249656200408936
epoch: 11 test_accuracy:0.9202492237091064 loss:0.8723455667495728
epoch: 12 test_accuracy:0.9345794320106506 loss:0.8389042615890503
epoch: 13 test_accuracy:0.9283488988876343 loss:0.8365824222564697
epoch: 14 test_accuracy:0.9289719462394714 loss:0.8442765474319458
epoch: 15 test_accuracy:0.9376947283744812 loss:0.825635552406311
epoch: 16 test_accuracy:0.9308411478996277 loss:0.8527410626411438
epoch: 17 test_accuracy:0.940809965133667 loss:0.7720271348953247
epoch: 18 test_accuracy:0.9644860029220581 loss:0.772343099117279
epoch: 19 test_accuracy:0.9239875674247742 loss:0.8235347270965576
epoch: 21 test_accuracy:0.9595015645027161 loss:0.7210665941238403
epoch: 22 test_accuracy:0.9688473343849182 loss:0.7046180963516235
epoch: 23 test_accuracy:0.9514018893241882 loss:0.7708120346069336
epoch: 24 test_accuracy:0.9532710313796997 loss:0.7053214311599731
epoch: 25 test_accuracy:0.9732087254524231 loss:0.7038005590438843
epoch: 26 test_accuracy:0.9713395833969116 loss:0.6641620397567749
epoch: 27 test_accuracy:0.9763239622116089 loss:0.6519589424133301
epoch: 28 test_accuracy:0.9750778675079346 loss:0.6372754573822021
epoch: 29 test_accuracy:0.9700934290885925 loss:0.6410614252090454
epoch: 31 test_accuracy:0.9744548201560974 loss:0.6219320893287659
epoch: 32 test_accuracy:0.9788162112236023 loss:0.6144272089004517
epoch: 33 test_accuracy:0.9806853532791138 loss:0.6052818298339844
epoch: 34 test_accuracy:0.9769470691680908 loss:0.6073766946792603
epoch: 35 test_accuracy:0.9813084006309509 loss:0.5920457243919373
epoch: 36 test_accuracy:0.9831775426864624 loss:0.5833461284637451
epoch: 37 test_accuracy:0.9769470691680908 loss:0.5925641059875488
```

FIGURE 7.4 Accuracy of the model at training and testing phases.

The highest test accuracy of 98.31% reflects the effectiveness of the proposed model for activity recognition. Figure 7.4 also reflects the nature of deep learning models. It can easily be observed from these results that as the number of iterations increased, it reflected in the increase in the detection rate also.

Figure 7.5 represents the overall performance of the proposed model by reflecting the accuracy achieved at the level of training and testing. The loss function is also calculated for the considered input. Approximately 98% accuracy has been achieved for extracting the information about the exact activity performed by the human. During the initial iterations, it can clearly be checked that the system is reflecting about 70% accuracy. But as the number of epoch are increasing, it reflects the increase in accuracy level.

Figure 7.6 shows the overall results of thc implementation process. This represents the average accuracy of 96% for all categories of human activities under consideration. During the recognition at the testing phase, it has been observed that the system shows the confusion factor while detecting "lying" activity with "sitting" activity. This is the reason for the fluctuation of recognition accuracies of "sitting" and "lying" activities. To resolve this issue, more parameters have to be considered. Similar kind of mismatch has been observed during the identification of "standing" and "sitting" activities. This again happened due to the negligible movement observed during both these activities and no such change was also observed in the background.

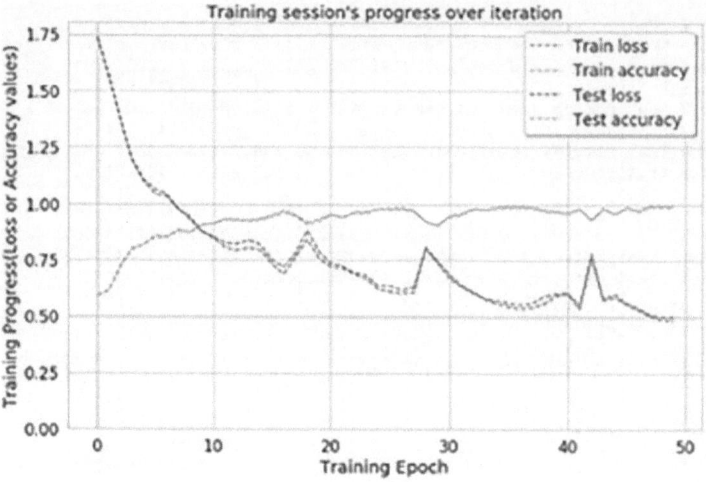

FIGURE 7.5 Performance accuracy of LSTM Model.

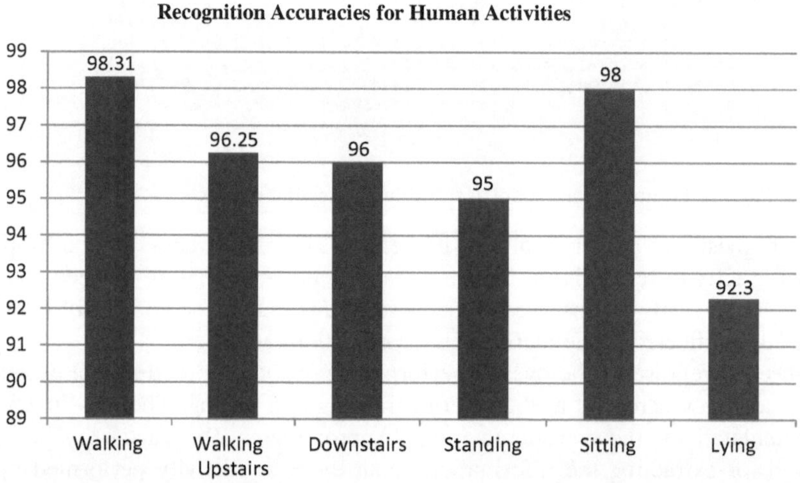

FIGURE 7.6 Recognition accuracies for human activities.

7.6 CONCLUSION

HAR field of research opens a huge amount of probability to track the different activities performed by a person. This field of research is treated as one of the major application domains of evolutionary computing. This provides a kind of application to the society which helps in monitoring the health issues of the elderly persons and other members of the family too. With the help of HAR applications, the dataset can be generated to monitor the various health issues of the person. Apart from monitoring the different activities, suggestive measures can also be presented to the users after proper analysis of various activities. These applications fall under the category of Small Objects Identification (SOI). In this research, the HAR system is proposed to track the different activities like sitting, lying, walking, walking upstairs, and walking downstairs. The dataset used contains the information recorded with the help of an accelerometer and gyroscope. A total of 11,299 samples were collected. The main focus while data collection was on 521 different attributes. Here, a deep learning technique is proposed by using the concept of LSTM technique. RNN model is used to form multiple hidden layers model to refine the data in a more accurate manner by using the concepts of deep nets. The proposed system is reflecting 98% overall accuracy for different activities. For training and testing purposes, 70% and 30% data is used respectively.

7.7 FUTURE SCOPE

As HAR falls under the field of small object identification, the work can be extended by considering more activities rather than just walking, sitting, etc. The availability of other different kinds of sensors like temperature tracking sensors, humidity sensors, proximity sensors, level sensors, etc. attracted the use of this proposed model in other fields related to the said sensors. The other part of the research in which the effectiveness of deep learning model is observed motivated to use other combinations of different classifiers for the purpose of identification and tracking of small movements of different objects. The combination of the RNN and CNN model in deep nets can prove to be effective for HAR analysis. Some other complex activities related to human activity detection have been considered for future work. These activities included the part of recognizing concurrent tasks to be performed by a human. All these kinds of concurrent activities require a switch of the moment to be performed. Most of the times activities like this look parallel to each other. The second major part of the concern is the recognition of the different activities which look interleaved in nature. These activities can be considered parallel or simultaneous in nature. One more area of concern can be considered as the interpretations of the recognized activities. At the time of interpretation, a number of conclusions can be drawn which may lead to ambiguity. All these reasons show the different opportunities in this field for various researchers.

REFERENCES

Bai, S., & An, S. (2018). A survey on automatic image caption generation. *Neurocomputing*, 311, 291–304. Doi: 10.1016/j.neucom.2018.05.080.

Chen, K., Pang, J., Wang, J., Xiong, Y., Li, X., Sun, S., Lin, D. (2019). Hybrid task cascade for instance segmentation. *ArXiv*, 4974–4983.

Cheng, G., Wan, Y., Saudagar, A. N., Namuduri, K., & Buckles, B. P. (2015). Advances in human action recognition: a survey, 1–30. Retrieved from http://arxiv.org/abs/1501.05964.

Crawford, E., & Pineau, J. (2019). Exploiting spatial invariance for scalable unsupervised object tracking. *ArXiv*. Doi: 10.1609/aaai.v34i04.5777.

Das Dawn, D., & Shaikh, S. H. (2016). A comprehensive survey of human action recognition with spatio-temporal interest point (STIP) detector. *Visual Computer*, 32(3), 289–306. Doi: 10.1007/s00371-015-1066-2.

De Jong, K. (2019). Evolutionary computation: a unified approach. *GECCO 2019 Companion - Proceedings of the 2019 Genetic and Evolutionary Computation Conference Companion,* (January 2006), 507–522. Doi: 10.1145/3319619.3323379.

Fourie, P. C., & Groenwold, A. A. (2002). The particle swarm optimization algorithm in size and shape optimization. *Structural and Multidisciplinary Optimization*, 23, 259–267.

Gaur, P. K. (2017). Scalability analysis using auto-configuration process for coalescing of heterogeneous wireless adhoc networks II related work. *International Journal of Applied Engineering Research*, 12(14), 4427–4432.

Guo, G., & Lai, A. (2014). A survey on still image based human action recognition. *Pattern Recognition*, 47(10), 3343–3361. Doi: 10.1016/j.patcog.2014.04.018.

Gupta, S., Arbeláez, P., Girshick, R., & Malik, J. (2015). Indoor scene understanding with RGB-D images: bottom-up segmentation, object detection and semantic segmentation. *International Journal of Computer Vision*, 112(2), 133–149. Doi: 10.1007/s11263-014-0777-6.

Health Care System and Evolutionary Approach. (n.d.). Retrieved February 20, 2021, from https://www.emerald.com/insight/content/doi/10.1108/14777260510600059/full/html

Herath, S., Harandi, M., & Porikli, F. (2017). Going deeper into action recognition: a survey. *Image and Vision Computing*, 60, 4–21. Doi: 10.1016/j.imavis.2017.01.010.

Hoiem, D., Efros, A. A., & Hebert, M. (2008). Closing the loop in scene interpretation. *26th IEEE Conference on Computer Vision and Pattern Recognition*, CVPR. Doi: 10.1109/CVPR.2008.4587587.

Human Activity Recognition Dataset. (n.d.). Retrieved from https://archive.ics.uci.edu/ml/datasets/human+activity+recognition+using+smartphones.

Kalra, S., & Leekha, A. (2020). Survey of convolutional neural networks for image captioning. *Journal of Information and Optimization Sciences*, 41(1), 239–260. Doi: 10.1080/02522667.2020.1715602.

Ke, S. R., Thuc, H. L. U., Lee, Y. J., Hwang, J. N., Yoo, J. H., & Choi, K. H. (2013). A review on video-based human activity recognition. Computers (Vol. 2). Doi: 10.3390/computers2020088.

Khoreva, A., Benenson, R., Ilg, E., Brox, T., & Schiele, B. (2017). [2017 davis-2] Lucid data dreaming for object tracking. Cvprw. Retrieved from https://davischallenge.org/challenge2017/papers/DAVIS-Challenge-2nd-Team.pdf.

Kicinger, R., Arciszewski, T., & De Jong, K. (2005). Evolutionary computation and structural design: asurvey of the state-of-the-art. *Computers and Structures*, 83(23–24), 1943–1978. https://doi.org/10.1016/j.compstruc.2005.03.002

Kim, E., Helal, S., & Cook, D. (2010). Human activity recognition and pattern discovery. *IEEE Pervasive Computing*, 9(1), 48–53. Doi: 10.1109/MPRV.2010.7.

Kong, Y., & Fu, Y. (2018). Human action recognition and prediction: a survey. *ArXiv*, 13(9).

Kougia, V., Pavlopoulos, J., & Androutsopoulos, I. (2019). A survey on biomedical image captioning. *ArXiv*, 26–36. Doi: 10.18653/v1/w19-1803.

Kumar, A., & Goel, S. (2017). A survey of evolution of image captioning techniques. *International Journal of Hybrid Intelligent Systems*, 14(3), 123–139. Doi: 10.3233/his-170246.

Kundu, J. (2016, April). Development of computerized maintenance management system development of computerized maintenance management system and root cause analysis for PTPS Panipat. *International Journal of Applied Engineering Research*, 6, 2183–2186.

Lee, B. Y., Liew, L. H., Cheah, W. S., & Wang, Y. C. (2014). Occlusion handling in videos object tracking: a survey. *IOP Conference Series: Earth and Environmental Science*, 18(1). Doi: 10.1088/1755-1315/18/1/012020.

Leibe, B., Vi, P., & Hutchison, D. (2016). Computer vision – ECCV 2016, Part VI (echte Seite 249: deep learning 3D shape surfaces using geometry images) (Vol. 9905). Retrieved from http://link.springer.com/10.1007/978-3-319-46448-0.

Li, G., Xie, Y., Lin, L., & Yu, Y. (2017). Instance-level salient object segmentation. *ArXiv*, 2386–2395.

Liu, B., Cai, H., Ju, Z., & Liu, H. (2019). RGB-D sensing based human action and interaction analysis: a survey. *Pattern Recognition*, 94, 1–12. Doi: 10.1016/j.patcog.2019.05.020.

Liu, S., Qi, L., Qin, H., Shi, J., & Jia, J. (2018). Path aggregation network for instance segmentation. (arXiv:1803.01534v3 [cs.CV] UPDATED). Cvpr, 8759–8768. Retrieved from http://arxiv.org/abs/1803.01534.

Liu, Y., Jing, X. Y., Nie, J., Gao, H., Liu, J., & Jiang, G. P. (2019). Context-aware three-dimensional mean-shift with occlusion handling for robust object tracking in RGB-D videos. *IEEE Transactions on Multimedia*, 21(3), 664–677. Doi: 10.1109/TMM.2018.2863604.

Pelikan, J. M., Krajic, K., & Conrad, G. (1997). Quality and sustainability of health promoting. *Proceedings of the 5th International Conference on Health Promoting Hospitals Vienna, Austria Edited by Health Promotion Publications Hamburg* Germany. World Health.

Qiu, Z., Zhuang, Y., Yan, F., Hu, H., & Wang, W. (2019). RGB-DI images and full convolution neural network-based outdoor scene understanding for mobile robots. *IEEE Transactions on Instrumentation and Measurement*, 68(1), 27–37. Doi: 10.1109/TIM.2018.2834085.

Ramasamy Ramamurthy, S., & Roy, N. (2018). Recent trends in machine learning for human activity recognition—a survey. *Wiley Interdisciplinary Reviews: Data Mining and Knowledge Discovery*, 8(4), 1–11. Doi: 10.1002/widm.1254.

Ren, M., Zemel, R. S.., Ren, M., & Zemel, R. S. (2016). End-to-end instance segmentation and counting with recurrent attention. *ArXiv Preprint*, 1–17. Retrieved from http://arxiv.org/abs/1605.09410%5Cnhttps://arxiv.org/abs/1605.09410

Sakaridis, C., Dai, D., & Van Gool, L. (2018). Semantic foggy scene understanding with synthetic data. *International Journal of Computer Vision*, 126(9), 973–992. Doi: 10.1007/s11263-018-1072-8.

Singh, G., & Sachan, M. K. (2015a). Multi-layer perceptron (MLP) neural network technique for offline handwritten Gurmukhi character recognition. In *2014 IEEE International Conference on Computational Intelligence and Computing Research*, IEEE ICCIC 2014. Doi: 10.1109/ICCIC.2014.7238334.

Singh, G., & Sachan, M. K. (2015b). Data capturing process for online Gurmukhi script recognition system. In *IEEE International Conference on Computational Intelligence and Computing Research (ICCIC)* (pp. 518–521). Doi: 10.1109/ICCIC.2015.7435778.

Singh, G., & Sachan, M. K. (2019). A Bilingual (Gurmukhi-Roman) online handwriting identification and recognition system. *International Journal of Recent Technology and Engineering (IJRTE)*, 8, 2936–2952.

Singh, G., & Sachan, M. K. (n.d.). An unconstrained and effective approach of script identification for online bilingual handwritten text. *National Academy Science Letters*, Doi: 10.1007/s40009-020-00889-0.

Srivastava, G., & Srivastava, R. (2018). A survey on automatic image captioning. *Communications in Computer and Information Science* (Vol. 834). Springer Singapore. Doi: 10.1007/978-981-13-0023-3_8.

Taneja, K. (2012). Resource management schemes for QoS in hybrid wireless networks. *International Journal of Computer Applications*, 60(12), 32–35.

Taneja, H., & Taneja, K. (2012). Trends for web information processing over world wide web. *International Journal of New Innovation in Engineering and Technology*, 1(2), 11–15.

Teutsch, M., & Krüger, W. (2012). Detection, segmentation, and tracking of moving objects in UAV videos. *Proceedings -2012 IEEE 9th International Conference on Advanced Video and Signal-Based Surveillance*, AVSS 2012, 313–318. Doi: 10.1109/AVSS.2012.36.

Trelea, I. C. (2003). The particle swarm optimization algorithm : convergence analysis and parameter selection. *Information Processing Letters*, 85, 317–325.

Tyagi, A., Nigam, S., & Singh, R. (2020). A concise review of baseline facts of SARS-CoV-2 for interdisciplinary research. *Chemistry Select*, 10897–10923. Doi: 10.1002/slct.202002420.

Wan, S., Qi, L., Xu, X., Tong, C., & Gu, Z. (2020). Deep learning models for real-time human activity recognition with smartphones. *Mobile Networks and Applications*, 25(2), 743–755. Doi: 10.1007/s11036-019-01445-x.

Wang, D., Tan, D., & Liu, L. (2017). Particle swarm optimization algorithm : an overview. *Soft Computing*. Doi: 10.1007/s00500-016-2474-6.

Wang, J., Chen, Y., Hao, S., Peng, X., & Hu, L. (2019). Deep learning for sensor-based activity recognition: a survey. *Pattern Recognition Letters*, 119, 3–11. Doi: 10.1016/j.patrec.2018.02.010.

Xu, C., Chai, D., He, J., Zhang, X., & Duan, S. (2019). InnoHAR: a deep neural network for complex human activity recognition. *IEEE Access*, 7, 9893–9902. Doi: 10.1109/ACCESS.2018.2890675.

Xue, B., Zhang, M., Browne, W. N., & Yao, X. (2016). A survey on evolutionary computation approaches to feature selection. *IEEE Transactions on Evolutionary Computation*, 20(4), 606–626. Doi: 10.1109/TEVC.2015.2504420.

Zeng, M., Nguyen, L. T., Yu, B., Mengshoel, O. J., Zhu, J., Wu, P., & Zhang, J. (2015). Convolutional neural networks for human activity recognition using mobile sensors. *Proceedings of the 2014 6th International Conference on Mobile Computing, Applications and Services*, MobiCASE 2014, 6, 197–205. Doi: 10.4108/icst.mobicase.2014.257786.

Zitnick, C. L., Vedantam, R., & Parikh, D. (2016). Adopting abstract images for semantic scene understanding. *IEEE Transactions on Pattern Analysis and Machine Intelligence*, 38(4), 627–638. Doi: 10.1109/TPAMI.2014.2366143.

8 Exploiting Evolutionary Computation Techniques for Service Industries

Alejandro Rodríguez-Molina, José Solís-Romero, Miguel Ángel Paredes-Rueda, Carlos Alberto Guerrero-León, Manuel Eduardo Mora-Soto, and Axel Herroz Herrera

CONTENTS

DOI: 10.1201/9781003158165-8

8.1 INTRODUCTION

Since the beginning of humanity, humans have sought to simplify tasks that require significant effort, a large number of resources, are dangerous, or demand great precision. Over the centuries, the performance of these tasks has been perfected through tools or mechanisms. These mechanisms are arrangements of rigid bodies coupled to each other to transfer motion or energy to a determined action point (Singh, 2005).

In early civilizations, mechanisms were manually operated by people, and the task performance was highly dependent on the operator's skills and experience (Putt et al., 2019). During the first industrial revolution, the introduction of mass production promoted the development of steam mechanisms with greater autonomy (Iordache, 2017), i.e., with less human intervention. After the subsequent two revolutions, the evolution of electrical and computational sciences allowed the construction of automated machines (Zemrane et al., 2021). The above marked the birth of modern automated mechanical systems such as robotic manipulators (Spong, Hutchinson, & Vidyasagar, 2020), autonomous vehicles (Siegwart et al., 2011), androids (Matsui et al., 2005), and more sophisticated tools (Son et al., 2009) that nowadays we can find practically anywhere.

The development of automated machines has been such that today they bring a wide variety of services, many of them provided only by humans in the service industries until just a few years ago. Then, one can find those systems providing healthcare (Liu & Cavusoglu, 2016), rescue (Tesen et al., 2013), cooking (Bollini et al., 2013), childcare (Tanaka & Ghosh, 2011), and defense services (Abate et al., 2020), to mention a few examples.

In Society 5.0, there is a strong relationship between those automated machines and the society that will empower humans to solve the most challenging problems of the current age. So, humans benefit from the knowledge and intelligence generated by machines through digital technologies (Granrath, 2019). The above is translated in the form of services and benefits used when and how humans need them. In the basic schema of Society 5.0, all the information generated by machines in the physical space (the real world) is acquired and processed by computer systems in the cyberspace (simulated world) to make decisions and take proper actions to satisfy human needs (Deguchi et al., 2020). Consequently, the above requirements need modern mechanical systems that become more complex every day, and governing their dynamic behaviors to the desired ones is challenging and concerned to control engineering.

Control engineering is responsible for designing and tuning control strategies that govern complex systems (Ogata & Yang, 2002). The above two tasks require a model that successfully describes the system's behavior to be governed or controlled (Brosilow & Joseph, 1995). This model is used in cyberspace to make predictions about the actual system, so various control schemes or multiple configurations (controller parameter tuning) can be tested to obtain the desired performances before deploying them in the physical space, preventing damage, reducing costs, and allowing correct decision-making that satisfies human needs.

For this, the system must be studied in depth to gain enough insights for the model building. Fortunately, the behavior of mechanical systems can be described using Newton's laws (also known as classical mechanics' axioms) (Ogata, 2004). So, the models of such systems can be derived systematically. Nevertheless, these models have a set of parameters that determine their accuracy, i.e., the extent to which these resemble reality.

A system identification process performs the optimization of the model parameters (Ljung, 1998). Since the model establishes a relationship between the system's inputs and outputs, the identification uses these actual signal measurements to obtain the best model parameters. When a new set of inputs is provided to the best-identified model, it can predict the minimum error outputs. System identification is also a difficult task, and its complexity depends on the model structure, the amount of available input/output data, and the number of model parameters (Rojas et al., 2011).

Once an optimized model is obtained, it can be deployed to design or adjust a suitable control scheme. There are currently many control schemes in the specialized literature that have been successfully applied to mechanical systems under diverse operating conditions (Fedor & Perdukova, 2016; Romero et al., 2016; Wang & Hou, 2018; Xiao et al., 2018). No matter which scheme is adopted, all of them have parameters determining their performance when governing a system. The adjustment of these parameters is known as the controller tuning problem and is a crucial task in control engineering (O'Dwyer, 2009). Overlapping the basic scheme of Society 5.0 with the aims of control engineering in the model identification and controller tuning, a significant amount of information must be acquired from the actual mechanical system in the physical space through sensing elements and a communication interface; the data is processed by a computer in the cyberspace, where simulations of the mechanism and its controller take place; and, depending on the human necessities (e.g., the need for the mechanism that develops a desired trajectory with a certain performance level and low energy consumption to provide a given service), a result that includes the suitable controller parameters is computed and then implanted into the actual system in the physical space to satisfy them.

In this way, Artificial Intelligence (AI) is one of the main tools to perform data processing in cyberspace considered by Society 5.0. The AI can be incrementally classified into three types depending on the extent that the computer incorporates, uses, and generates knowledge (Salgues, 2018): (i) Computer programming based on algorithms, (ii) Expert systems, and (iii) Machine learning.

As the most extensive type of AI today, Machine Learning (ML) is a subfield of computer science that builds a model for a given phenomenon or system from data (Kitchin, 2018). In this way, ML includes various techniques to make a model learn from data. Depending on the level of human supervision, ML can be supervised, semi-supervised, unsupervised, and reinforced (Chinnamgari, 2019).

In the context of mechanical systems, the identification process can be encompassed in supervised learning, where a predictor model is adjusted based on a set of known features (actual inputs) and targets (actual outputs). Then, this predictor can be deployed in the controller tuning process.

Therefore, the system identification process can be developed using an ML methodology. Among available methodologies, the Cross-Industry Standard Process for Data Mining (CRISP-DM) (Wirth, 2000) is one of the most successful alternatives that have been exploited in many study areas such as biology (Liu et al., 2020), accounting (Moro et al., 2011), engineering (Wiemer et al., 2019), among others.

On the other hand, the learning process in ML has been successfully supported by meta-heuristics for a wide variety of applications where classical learning algorithms cannot provide good enough results (Mirjalili et al., 2020). Meta-heuristics are stochastic computational methods that can find good solutions to very complex problems using an accessible number of resources (Talbi, 2009). Most of these methods belong to the evolutionary computation (Bäck et al., 1997) and swarm intelligence (Kennedy, 2006) computing science fields. The success of meta-heuristics in the system identification problem is highlighted in the recent specialized literature (Garg et al., 2017; Gotmare et al., 2017; Huang et al., 2018; Worden et al., 2018).

The ML problems related to the identification and controller tuning are very complex. The above can be attributed to the highly non-linear behavior of mechanical systems and control schemes, the inherent presence of uncertainties and disturbances in the physical space, and the limited resources of the current computer systems. Then, the use of meta-heuristics can be necessary as in many other engineering applications (Reynoso-Meza et al., 2014; Rodríguez-Molina et al., 2020).

This chapter presents an Evolutionary Machine Learning Framework for Identification and Control (EMLF-IC) of mechanical systems to integrate them into Society 5.0. This framework aims to describe the process of implementing highly effective automated machines that can help meet the needs of the service industries. In this way, EMLF-IC encompasses the different activities involved in system identification and controller tuning discussed in this section and aligns them with the CRISP-DM methodology, from the business understanding to the final deployment of the optimized controller.

The rest of this chapter is organized as follows. Section 8.2 presents in detail the steps of the proposed evolutionary machine learning framework. Section 8.3 describes two study cases, the position regulation of the simple pendulum and the trajectory tracking with a fully-actuated inverted-pendulum, and each framework step's development. Conclusions are included in Section 8.4.

8.2 FRAMEWORK

The proposed Evolutionary Machine Learning Framework for Identification and Control (EMLF-IC) of mechanical systems is based on the Cross-Industry Standard Process for Data Mining (CRISP-DM), a well-defined and widely-used methodology to properly establish and develop the sequential activities in a machine learning project (Wirth, 2000).

Due to the similarities of the identification and controller tuning processes with the flow of a machine learning project, some of the CRISP-DM ideas can be landed in developing well-controlled mechanical systems.

Then, EMLF-IC aims to provide a well-organized batch of sequential activities to successfully govern the behavior of mechanisms in the context of Society 5.0.

The activities, aligned to those found in CRISP-DM, are detailed next.

8.2.1 Business Understanding

In mechanics, the business is around a mechanical system. Then, this stage is related to understanding the goal of this system in a particular application. According to the above, it is necessary to know or study several aspects of the machine as the first step to design a successful control strategy. The most relevant factors are described next:

- **Arrangement of the Mechanical Parts**: This is related to knowing where the rigid bodies of the mechanism and their joints are located and how they interact with each other. A schematic diagram of the system is helpful to obtain this knowledge.
- **Actuation Elements**: These are elements located in specific joints of the mechanism to produce a rotary or linear motion between two or more rigid bodies. Actuators generate a force or torque when fed by an energy signal determined by the controller. A typical example of an actuator for mechanical systems is the electric motor. All actuators have limitations regarding the magnitude of the energy they can receive and the maximum force or torque they can grant; these are important aspects to understand.
- **Generalized Coordinates**: This is the minimum set of variables that can describe the mechanism's motion. For a wide variety of mechanical systems (fully-actuated systems (Bullo & Murray, 1999), these variables are related to the configuration of the actuation elements (e.g., the angular position of a DC motor shaft).
- **Kinematics**: This study helps to understand the motion of the mechanism concerning its geometrical properties and the generalized coordinates. In this way, it is possible to determine the location of any point in the system if the values of the generalized coordinates are provided and vice versa. Analogously, differential kinematics can obtain the velocities and accelerations.
- **Location of the End-Effector**: The main goal of a mechanical system is to perform a task with a tool so-called end-effector. Because of the above, it is important to know this device's location to determine how the mechanism has to move to locate and orientate it.
- **Desired Trajectory**: Depending on the final application, the end-effector is required to change its location and orientation over time. These end-effector movements are specified in the Cartesian or the generalized coordinate spaces and parametrized concerning time. The above conforms to the desired trajectory, which defines exactly what the mechanism must do for a given application.
- **Available Transducers or Observers**: These measures or estimate system variables of interest (e.g., a rotary encoder to determine a motor's position). These elements are crucial for identification since they are the source of valuable input and output information from physical space.

In addition to the above business aspects, understanding the user necessities such as the acceptable performance levels when performing a task with the mechanism is valuable. Moreover, the knowledge of disturbances or uncertainties affecting the mechanism is also desirable, but, in practice, they are frequently unpredictable.

8.2.2 Data Understanding

This step implies a deep analysis of the available information to perform the system identification. The variables that make up the sets of features (acquired input information) and targets (acquired output information)from physical space are determined with the above.

For every instant, the internal behavior of all dynamic systems (mechanical systems included) can be fully described by a reduced set of variables, so-called state variables (Friedland, 2005).

Therefore, the actual system behavior can be expressed with the state equation in (8.1), where t is the time, $\boldsymbol{x}(t)$ is the vector of state variables, $\boldsymbol{u}(t)$is the vector of control inputs, and $\boldsymbol{p}$ is the vector of the actual dynamic system parameters (these parameters are unknown or difficult to measure).

$$\dot{x} = f\left(t(x), \boldsymbol{u}(t), \boldsymbol{p}\right) \tag{8.1}$$

For the EMLF-IC, it is assumed that the state variables include the values of the generalized coordinates and their derivatives (i.e., the position and velocity of each coordinate) since they are typically measurable or observable (Tian et al., 2006). Moreover, these state variables are used as system outputs.

Hence, if the state of the mechanical system is $\boldsymbol{x}(t)$ for an instant t, and the control action $\boldsymbol{u}(t)$is applied to the system actuators at that moment, the state of the system after a short interval dt(known as sampling interval) will be $\boldsymbol{x}(t+dt)$, which is implicitly given by the solution to the initial value problem associated to the differential equation in (8.1).

Since the states of the system are intimately related to its dynamic behavior, the information $\boldsymbol{x}(t)$and $\boldsymbol{u}(t)$can be used as features, while the targets can be $\boldsymbol{y}(t) = \boldsymbol{x}(t+dt)$.

8.2.3 Data Preparation

At this point, the variables of the set of features and targets are already defined. So, it is necessary to extract the values of these variables experimentally from the actual system in physical space.

The way to perform the above is by exciting the system dynamics with a strategic control signal during a predefined time window (the more extensive the window, the greater the information generated). It is important to mention that not all the control signals are suitable for this purpose. An appropriate control signal induces a wide variety of behaviors in the system (Shardt et al., 2015).

While the system is being excited, its state variables must be measured or observed in short sampling intervals dt. The generated data is then archived in the sets of features and targets.

It must be highlighted that the acquired data is homogeneous (all the variables are real-valued), does not have missing values, and its normalization is not required for system identification. So, additional cleaning or pre-processing of data is not necessary.

Once the set of targets and features contains the generated system data, it is divided into two subsets, one to train the model and the other to evaluate it. Typically, 80% data is used for training and the remaining 20% for evaluation.

Since the generated data is sorted concerning time, each subset's samples can be chosen sequentially by as small as possible batches. The percentage of data used for training is picked first from each batch, and the remaining is picked after for evaluation. For example, if the aforementioned rates are selected (i.e., 80% for training and 20% for evaluation), the smallest possible batch would have five samples. Then, the first four samples would be used for training and the last one for evaluation.

8.2.4 Modeling

In this step, a suitable model of the system is found, and its parameters are adjusted using the data in the training set. As discussed before, an adequate model for a mechanical system can be obtained by using the laws of classical mechanics, unlike other systems where different model structures have to be tested to discover a suitable alternative (Ahmed et al., 2010).

For the EMLF-IC, the system model is derived from Lagrangian mechanics, a reformulation of classical mechanics that simplifies the analysis of mechanisms by considering the system energies (Gignoux & Silvestre-Brac, 2009). Therefore, the model corresponds to the Euler-Lagrange equations of motion in (8.2), where t is the time, $\mathcal{L}$ is the system Lagrangian, $\mathcal{D}$ is the Rayleigh dissipation function, q_i is the i-th generalized coordinate, Q_i is the i-th generalized force/torque, and n is the number of generalized coordinates. In this equation, the Lagrangian $\mathcal{L} = \mathcal{K} - \mathcal{U}$ is the difference between the total kinetic energy $\mathcal{K}$ and the total potential energy $\mathcal{U}$ of the system. At the same time, the Rayleigh dissipation function $\mathcal{D}$ involves the non-conservative effects in the system. Additional details on Lagrangian mechanics can be consulted in (Ogata, 2004).

$$\frac{d}{dt}\left(\frac{\partial \mathcal{L}}{\partial \dot{q}_i}\right) - \frac{\partial \mathcal{L}}{\partial q_i} + \frac{\partial \mathcal{D}}{\partial \dot{q}_i} = Q_i,\ i = 1,2,\ldots,\ n \tag{8.2}$$

Based on (8.2), the corresponding state equation can be stated in (8.3), where t is the time, $\hat{\boldsymbol{x}}(t)$ is the vector of state variables, $\hat{\boldsymbol{u}}(t)$is the vector of control inputs, and $\hat{\boldsymbol{p}}(t)$ is the vector of the model parameters.

$$\dot{\hat{\boldsymbol{x}}} = \hat{f}\left(t, \hat{\boldsymbol{x}}(t), \hat{\boldsymbol{u}}(t), \hat{\boldsymbol{p}}\right) \tag{8.3}$$

The vector $\hat{\boldsymbol{p}}$ in (8.3) includes the model parameters that must be correctly adjusted to minimize the differences between the actual system behavior, described in (8.1) and the model in (8.3).

The above can be formally stated in the optimization problem in (8.4) using the sum of the squared error as the objective function, where $\boldsymbol{y}_i$ is the actual system output in the i-th target vector of the training set, N_T is the number of samples in the same set, and $\hat{\boldsymbol{y}}_i$ is the predicted system output. The vector $\hat{\boldsymbol{y}}_i$ is the solution of the initial value problem given by $\dot{\hat{x}} = \hat{f}(t, \boldsymbol{x}_i, \boldsymbol{u}_i, \hat{\boldsymbol{p}})$ and the initial condition $\hat{\boldsymbol{x}}_0 = \boldsymbol{x}_i$, where $\boldsymbol{x}_i$ and $\boldsymbol{u}_i$ are respectively the actual system state and control action in the i-th feature vector of the training set. The solution to this initial value problem can be obtained using a numerical integration method such as Euler's.

$$\min_{\hat{\boldsymbol{p}}} J_{\mathrm{MI}}(\hat{\boldsymbol{p}}) = \sum_{i=1}^{N_T} (y_i - \hat{y}_i)^2 \tag{8.4}$$

Due to the complexity of the problem in (8.4), it is a candidate to be solved by meta-heuristics from evolutionary computation and swarm intelligence.

As mentioned before, meta-heuristics are stochastic computational techniques that can find good solutions to hard optimization problems using an affordable number of computational resources. Despite this, there is no meta-heuristic with the ability to solve all kinds of problems (Wolpert & Macready, 1997). Due to the above, using a diverse enough variety of these meta-heuristic to solve the problem in (8.4) is highly recommendable in the EMLF-IC.

In many studies on meta-heuristics, around 30 independent runs are necessary to get relevant insights about their performance (Veček et al., 2017). For this reason, the same number of runs are performed in EMLF-IC to obtain potentially good sets of model parameters.

8.2.5 Evaluation

After the modeling step, several sets of model parameters are obtained from all meta-heuristic independent runs. In the evaluation step, the performance of each set of model parameters is determined using the data in the evaluation set. The above allows to determine the extent to which each set of model parameters allows generalizing the dynamic behavior of the actual mechanism.

The evaluation criteria adopted in the EMLF-IC is focused on the model accuracy and is the same as the objective function J_{MI} in (8.4), but this time, using the data in the evaluation set. Once all the parameter sets are evaluated, it is possible to choose the best alternative and deploy it in the next step.

8.2.6 Deployment

The deployment step refers properly to the use of the best-identified model for a specific purpose.

In the EMLF-IC, the best-identified model is used to tune the system controller. Among the available controller alternatives for mechanical systems, the model-based ones can achieve outstanding performances and are robust to uncertainties and disturbances (Farahmand et al., 2019; Müller & Hufnagel, 2012). Model-based controllers are control schemes that incorporate the system model in their operation (Brosilow & Joseph, 1995). The main drawback of these controllers is the necessity of an accurate model. However, the model identified in the previous steps of the EMLF-IC can be exploited in one of these control schemes.

Although the previously identified parameters are part of the model-based controller, some additional control variables are focused on tracking the desired trajectory with the mechanism and have to be tuned. The correct tuning of these controller parameters is crucial since the controller performance in the final application strongly depends on it. Hence, a dynamic simulation of the identified model in a defined time window is necessary to evaluate different controller parameters in cyberspace without damaging the actual system with possible dangerous alternatives. The dynamic simulation is referred to as the iterative integration of the identified state equation in (8.3), which can be performed using the Euler's integration method starting from a known initial state condition (typically the rest configuration).

The searching for the best controller parameters is established in the EMLF-IC as the optimization problem in (8.5), wheret_f is the simulation end time, z is the vector of controller parameters, and $\boldsymbol{e}(t) = \boldsymbol{y}_d(t) - \boldsymbol{y}(t)$ is the error between the desired trajectory $\boldsymbol{y}_d(t)$ and the system output $\boldsymbol{y}(t)$ in the instant t. The problem is also subject to the minimum $\boldsymbol{u}_{\min}$ and maximum $\boldsymbol{u}_{\max}$values of the control action $\boldsymbol{u}(t)$generated by the controller; both depend on the actuator limits of the system. In this problem, the Integral of the Time-weighted Squared Error (ITSE) is adopted as the objective function. Still, other performance indicators can be adopted to achieve diverse controller behaviors (Rodríguez-Molina et al., 2020). The ITSE metric is commonly used in the controller tuning task (Verma & Padhy, 2018), and its value is more affected by the error as time progresses. The value of the ITSE can be obtained by the Euler's integration method using the initial condition $\boldsymbol{e}_0 = 0$.

$$\min_{z} J_{\text{CT}}(z) = \text{ITSE}(\boldsymbol{e}(t)) = \int_0^{t_f} t\boldsymbol{e}^T(t)\boldsymbol{e}(t)dt$$

subject to:

$$\boldsymbol{u}_{\min} \leq \boldsymbol{u}(t) \leq \boldsymbol{u}_{\max} \tag{8.5}$$

It is noticeable that the tuning problem in (8.5) is complex. Nevertheless, it can be successfully solved by meta-heuristic methods as performed nowadays for a wide variety of controllers (Rodríguez-Molina et al., 2020).

The same meta-heuristics used in the modeling step are used to solve the problem in (8.5). Again, 30 independent runs of the meta-heuristic are required in the EMLF-IC to find potentially good sets of controller parameters. After these 30 independent runs, the best controller parameters can be selected from the generated alternatives. Finally, the best controller parameters and the best model ones are implanted in the controller for the actual mechanical system.

Additional activity in the deployment step is related to the evaluation of the actual system behavior. This can be performed using the objective function in (8.5), but this time, considering the mechanical system's actual outputs. The above evaluation helps decide if the adjusted controller is adequate enough for the particular application or if a complete redesign is required with the EMLF-IC.

8.3 STUDY CASES

In this section, the steps of the EMLF-IC are described for two study cases: the position regulation of the simple pendulum and the trajectory tracking with the fully-actuated inverted pendulum.

8.3.1 Position Regulation with the Simple Pendulum

The simple pendulum is a well-known mechanism adopted in the test-beds of many works in the study of control systems. Each step of the EMLF-IC is detailed in developing the control scheme for this mechanical system in the position regulation task.

8.3.1.1 Business Understanding

The simple pendulum can be conceptualized with the schematic diagram in Figure 8.1 to enhance the understanding of this system. The simple pendulum includes a bar of length l coupled to the ground with a rotary joint. A point mass m is located at the end of the bar. The system is vertically oriented, so it is affected by the gravity acceleration g. Moreover, the pendulum includes non-conservative forces such as the air resistance associated with the friction coefficient b.

The simple pendulum is a fully-actuated system. In this way, a single rotary actuator is located in the joint between the bar and the ground to variate the angle θ. The actuator limits are assumed as known and are $u_{\min} = -200\ (\mathrm{N \cdot m})$ and $u_{\max} = -200\ (\mathrm{N \cdot m})$. Also, $\theta$is the single generalized coordinate of the system.

The end-effector of the pendulum shares the location of the point mass, and the desired trajectory for the regulation task is proposed as$y_d = \left[\theta_d, \dot{\theta}_d\right]^T$, with $\theta_d = \pi\ (\mathrm{rad})$ as the desired angle and $\dot{\theta}_d = 0\ (\mathrm{rad/s})$as the desired angular speed. Then, the task is to bring and hold the pendulum to its unstable equilibrium configuration.

Finally, it is assumed that both $\theta$and $\dot{\theta}$ are measurable or observable variables.

8.3.1.2 Data Understanding

The most relevant information about the simple pendulum behavior is contained in its state variables. The state variables are given by the generalized coordinate θ and its time derivative $\dot{\theta}$ for this system. Then, the state vector of the pendulum is $x(t) = \left[\theta(t), \dot{\theta}(t)\right]^T$.

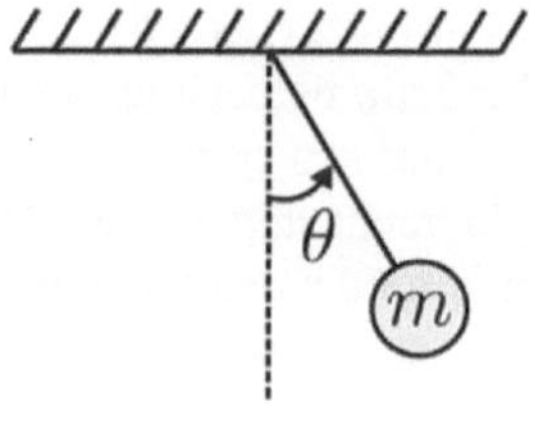

FIGURE 8.1 Schematic diagram of the simple pendulum.

After a short interval of time established $dt = 5$ (ms), the state change in the pendulum produced by the control action $u(t)$ is considered as the system output $y(t) = \boldsymbol{x}(t+dt)$.

Then, the information of $u(t)$ and $\boldsymbol{x}(t)$ is taken as the features, and the variables $\boldsymbol{x}(t+dt)$ are selected as the targets.

8.3.1.3 Data Preparation

The actual system is assumed to behave according to the model (8.6). Based on the above, the actual pendulum's state equation can be defined in (8.7). The actual parameters of the state equation (8.7) are given in Table 8.1.

$$\ddot{\theta} = \frac{1}{ml^2}u - \frac{g}{l}\sin(\theta) - \frac{b}{m} \tag{8.6}$$

$$\dot{\boldsymbol{x}} = \left[\begin{array}{c} \theta \\ \hline \frac{1}{ml^2}u - \frac{g}{1}\sin(\theta) - \frac{b}{m}\dot{\theta} \end{array}\right] \tag{8.7}$$

Then, a dynamic simulation using Euler's integration method over (8.7) and the initial condition $\boldsymbol{x}(0) = 0$ is performed to generate the feature and target information in the time window $t \in [0,10](s)$. The above simulation is not necessary for practice since the feature and target information is acquired directly from the actual mechanism in physical space through the available transducers or observers.

A control signal $u(t)$ is used to excite as many as possible behaviors in the actual pendulum. The damped sine wave in (8.8) is adopted with this purpose, with an amplitude $A = 5$, a decay constant $\lambda = 0.5$, an angular frequency $\omega = 2\ \pi$, and a phase angle $\phi = 0$.

$$u(t) = Ae^{-\lambda t}\left(\cos(\omega t + \phi) + \sin(\omega t + \phi)\right) \tag{8.8}$$

Additionally, random noise signals up to ±5% are included in all the feature and target variables to emulate the data acquisition difficulties from the actual system in the physical space. The acquired feature and target information from the actual simple pendulum can be observed in Figure 8.2.

TABLE 8.1
Actual Parameters of the Simple Pendulum

Parameter	Actual Value
m	1.000 (kg)
l	0.500 (m)
b	0.500 (kg/s)
g	9.810 (m/s^2)

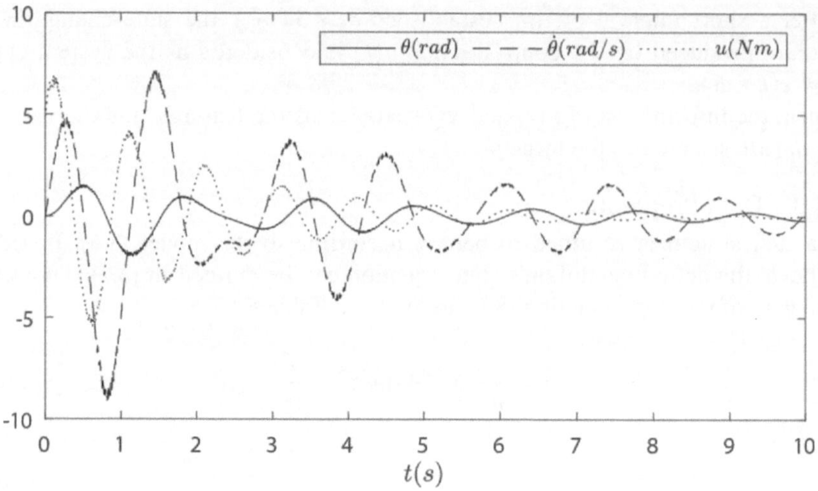

FIGURE 8.2 Feature and target information acquired from the actual simple pendulum.

TABLE 8.2
Input and Output Information Acquired from the Simple Pendulum

Time	Feature Data			Target Data	
T	$\theta(t)$	$\dot{\theta}(t)$	$u(t)$	$\theta(t+dt)$	$\dot{\theta}(t+dt)$
0.0000	0.0000	0.0000	4.8820	0.0000	0.0965
0.0050	0.0000	0.0991	5.1445	0.0004	0.1945
⋮	⋮	⋮	⋮	⋮	⋮
9.9900	−0.1867	0.1164	0.0305	−0.1802	0.1273
9.9950	−0.1792	0.1351	0.0333	−0.1944	0.1587

Since the sampling time $dt = 5$ (ms) is adopted for the time window $t \in [0,10](s)$, a series of $N = 2000$ target and feature vectors are generated in the form depicted in Table 8.2.

As discussed before, 80% of the rows in Table 8.2 are used to train the model, while the remaining 20% are considered for its evaluation. The selection of each row is based on the strategy described at the end of Section 8.2.3.

8.3.1.4 Modeling

The simple pendulum equation can be derived using Lagrangian mechanics and is observed in (8.9). In a state-space representation, the above equation takes the form in (8.10) and serves as the pendulum model.

$$\ddot{\hat{\theta}} = \frac{1}{\hat{m}\hat{l}^2}\hat{u} - \frac{g}{\hat{l}}\sin\left(\hat{\theta}\right) - \frac{\hat{b}}{\hat{m}}\dot{\hat{\theta}} \tag{8.9}$$

$$\dot{\hat{\boldsymbol{x}}} = \begin{bmatrix} \dot{\hat{\theta}} \\ \hline \frac{1}{\hat{m}\hat{l}^2}\hat{u} - \frac{g}{\hat{l}}\sin\left(\hat{\theta}\right) - \frac{\hat{b}}{\hat{m}}\dot{\hat{\theta}} \end{bmatrix} \tag{8.10}$$

The model's adjustable parameters are chosen as the mass $\hat{m}$, the bar length $\hat{l}$, and the friction coefficient $\hat{b}$. Gravity $g = 9.81\ \left(\text{m/s}^2\right)$ is considered as a fixed parameter. In practice, parameters such as the mass and the bar length can be easily measured with a scale and a ruler. Nevertheless, both parameters are set as adjustable to study the behavior of the EMLF-IC. The pendulum identification problem has the form in (8.4), with $\boldsymbol{p} = \left[\hat{m}, \hat{l}, \hat{b}\right]^T$. Then, this problem is to find $\hat{m}$, $\hat{l}$, and $\hat{b}$ that minimize the error between the actual system output (target values) and the output predicted by the model (obtained as the solution of the initial value problem associated with the state equation (8.10) by Euler's method for a future period dt) when they both start from the same initial state and are fed by the same control action (feature values) for all samples in the training set.

The pendulum identification problem in (8.4) is complex since it depends on the solution of several initial value problems, which are in turn related to the nonlinear differential equation (8.10). So, classical optimization techniques may fail in finding a suitable set of model parameters. Then, meta-heuristic techniques from evolutionary computation and swarm intelligence are preferred to solve this problem.

Three well-known meta-heuristics are selected with the above purpose. The first one is the rand/1/bin variant of the Differential Evolution (DE) (Mezura-Montes et al., 2006). The second one is the Genetic Algorithm (GA) version found in (Rodríguez-Molina et al., 2019). The last one is the Particle Swarm Optimization (PSO), with a fully-connected topology and linear decreasing inertia (Arasomwan & Adewumi, 2013). The hyper-parameters of these meta-heuristics are proposed empirically and are shown in Table 8.3. A fine-tuning of these hyper-parameters may require specialized tools such as *i-race* (López-Ibáñez et al., 2016).

In order to determine the solution that better fits the identification problem, 30 independent runs of each algorithm are carried out. Then, the same number of candidate solutions (sets of model parameters) are obtained. These alternatives are evaluated next.

TABLE 8.3
Hyper-Parameters of the Selected Meta-Heuristics

Method	Hyper-Parameters			
DE	CR = 0.5	$F = 0.5$		
GA	$P_c = 1.0$	$P_m = 0.5$	$\eta_c = 10.0$	$\eta_m = 10.0$
PSO	$C_1 = 2.0$	$C_2 = 2.0$	$w_1 = 1.0$	$w_2 = 0.0$

TABLE 8.4
Evaluation of the Obtained Model Alternatives for the Simple Pendulum

Optimizer	$\hat{m}$	$\hat{l}$	$\hat{b}$	J_{MI} (Training)	J_{MI} (Evaluation)
DE	0.7699	0.5662	0.2741	17.8334	4.2522
GA	0.7554	0.5683	0.2742	17.8329	4.2500
PSO	0.7232	0.5787	0.2639	17.8322	4.2569

8.3.1.5 Evaluation

The model parameters optimized by DE, GA, and PSO in the previous section are evaluated by the criterion established as the objective function J_{MI} in (8.4) using the samples in the evaluation set. The best alternative found by each meta-heuristic is displayed in Table 8.4. The first columns of this table show the meta-heuristic technique and the obtained model parameters. The last two columns indicate the value of J_{MI} obtained for training and evaluation.

From Table 8.4, it is observed that the best parameters found by all meta-heuristics are similar. Moreover, these parameters are relatively close but not equal to the actual ones presented in Table 8.1. This is due to the noise included in the input and output signals that emulate the uncertainties in the data acquisition from the actual pendulum in physical space.

It is important to observe that the parameters found by all meta-heuristics in Table 8.4 have different performances. If J_{MI} of the training stage is considered, PSO finds the best model parameters. On the other hand, if J_{MI} of the evaluation phase is taken into account, GA finds a better alternative. Since the J_{MI} used for evaluation denotes the extent to which the model can generalize the behavior of the actual system, the model parameters found by GA are selected to be deployed in the controller tuning task.

8.3.1.6 Deployment

At this point, the best-identified model is used to adjust a model-based control scheme. The scheme selected to control the pendulum is the Computed-Torque Controller (CTC) (Spong, Hutchinson, & Vidyasagar, 2020). This controller is observed in (8.11) and (8.12), with $\hat{m}$, $\hat{l}$, and $\hat{b}$ as the model parameters optimized in the previous step by GA, $e = \theta_d - \theta$ as the angular error, and k_p, k_i, and k_d, as the proportional, integral, and derivative gains of the Proportional Integral Derivative (PID) compensator.

$$u = \hat{m}\hat{l}^2 v + \hat{m} g \hat{l} \sin(\theta) + \hat{b}\hat{l}^2 \ \dot{\theta} \tag{8.11}$$

$$v = k_p e + k_i \int e dt + k_d \ \dot{e} \tag{8.12}$$

With the selected control scheme, the tuning problem is established as in (8.5), where the vector $z = \left[k_p, k_i, k_d\right]^T$ contains the gains of the PID compensator in (8.12). Therefore, the tuning problem is to find the control parameters k_p, k_i, and k_d that minimize the ITSE in the execution of the position regulation task with the pendulum during the dynamic simulation of (8.10) by the Euler's method. The end time of the simulation is proposed as $t_f = 10\ (s)$.

The tuning problem is also complex, and advanced optimization techniques such as meta-heuristics can find suitable results. In this way, the same optimizers used before, i.e., DE, GA, and PSO, are selected to search for the best controller parameters. Again, 30 independent runs of each optimizer are required to find the most promising solution.

Table 8.5 describes the best performances achieved with the controller parameters obtained by each meta-heuristic concerning the J_{CT} index in (8.5) for tuning. The first columns of this table indicate the meta-heuristic method and the calculated controller parameters. The last two columns show the value J_{CT} obtained for tuning and during the final implementation in the actual system. In this work, the J_{CT} for implementation is calculated over the dynamic simulation of (8.7) with the CTC controller in (8.11) and (8.12).

According to Table 8.5, all meta-heuristics develop similar performances in the tuning task, but GA can find the best set of controller parameters. The three best controller configurations also have similar performances in the final implementation, close to those obtained for tuning, which describes the effectiveness of the EMLF-IC in the identification and adjustment of controllers.

Finally, the simple pendulum's behavior that uses the best CTC controller, obtained with the EMLF-IC and GA, is depicted in Figure 8.3. This figure indicates that the CTC successfully takes the pendulum from the rest configuration to the reference position in about 2 (s) without overshoot.

8.3.2 Trajectory Tracking with the Fully-Actuated Inverted Pendulum

Compared to the simple pendulum, the fully-actuated inverted pendulum is a notably more complex system regarding its number of inputs and outputs and higher nonlinear behavior. For this system, the steps of the EMLF-IC are detailed on the development of a control scheme for the trajectory tracking task.

8.3.2.1 Business Understanding

The schematic diagram of the fully-actuated inverted pendulum is depicted in Figure 8.4 to better understand its details. This mechanism has a car of mass M that moves linearly. The car is coupled to a pendulum with a rotary joint. This pendulum has

TABLE 8.5
Evaluation of the Tuned Controller for the Simple Pendulum

Optimizer	k_p	k_i	k_d	J_{CT} (Tuning)	J_{CT} (Implementation)
DE	257.9240	4.2170E-14	72.3814	2.7697	2.8635
GA	260.9240	1.8563E-14	72.9313	2.7668	2.8632
PSO	262.5900	1.0906	70.8834	2.7717	2.8722

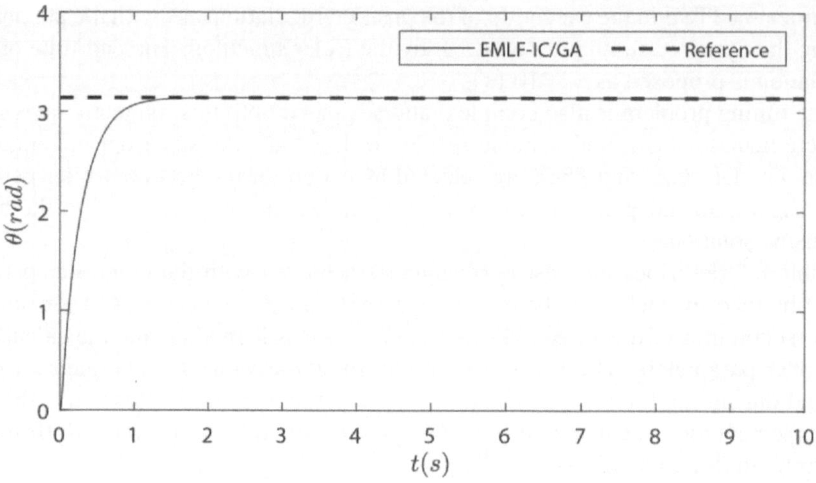

FIGURE 8.3 Regulated output of the actual simple pendulum using the EMLF-IC with GA.

a bar of length l with a point mass m located at the end. The system is vertically oriented and is affected by the gravity acceleration g. The car and the pendulum are also affected by non-conservative forces generated by the friction with the ground and the air resistance. These two forces are associated with the friction coefficients of b_1 and b_2.

The inverted pendulum is typically an under-actuated system since only one element, the car or the pendulum, is actuated. The fully-actuated variant of the inverted pendulum has two actuators, one that generates the linear movement $d$of the car and the other, located in the joint of the bar with the car, that varies the angle θ. The limits of the actuators are assumed as known and are $u_{\text{min},1} = -200\ (N)$ and $u_{\text{max},2} = 200\ (N)$ for the car and $u_{\text{min},2} = -200\ (\text{N}\cdot\text{m})$ and $u_{\text{max},2} = 200(\text{N}\cdot\text{m})$for the pendulum. The generalized coordinates of the system are given by d and θ.

The end-effector of the fully-actuated inverted pendulum is located in the same place as the point mass. On the other hand, the desired trajectory for the tracking task is proposed as $y_d = \left[d_d, \theta_d, \dot{d}_d, \dot{\theta}_d\right]^T$, with d_d and $\dot{d}_d$ as the desired linear displacement and speed of the car and θ_d and $\dot{\theta}_d$ as the desired angular position and speed of the pendulum given in (8.13).

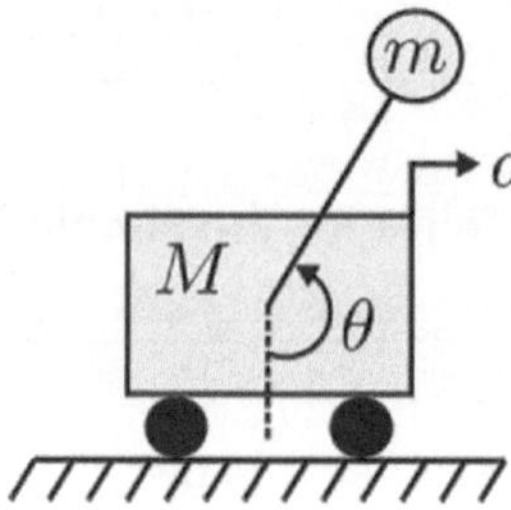

FIGURE 8.4 Schematic diagram of the fully-actuated inverted pendulum.

$$d_d = \cos\left(\frac{\pi}{5}t\right)$$

$$\dot{d}_d = -\frac{\pi}{5}\sin\left(\frac{\pi}{5}t\right)$$

$$\theta_d = \pi\sin\left(\frac{\pi}{5}t\right) \tag{8.13}$$

$$\dot{\theta}_d = \frac{\pi^2}{5}$$

The variables $d, \dot{d}, \theta$, and $\dot{\theta}$ are assumed to be measurable or observable.

8.3.2.2 Data Understanding

The dynamic information of the fully-actuated inverted pendulum is related to its state variables. These state variables are given by the values of the generalized coordinates d, θ and their time derivatives $\dot{d}, \dot{\theta}$. In this way, the state vector of the inverted pendulum is $\boldsymbol{x}(t) = \left[d(t),\ \theta(t),\ \dot{d}(t),\ \dot{\theta}(t)\right]^T$

The state change in the inverted pendulum induced by a control action $\boldsymbol{u}(t)$ is considered as the system output $\boldsymbol{y} = \boldsymbol{x}(t + dt)$, with a sampling interval established as$dt = 5$ (ms).

Hence, the feature data contains $\boldsymbol{u}(t)$ and $\boldsymbol{x}(t)$, while the target information includes $\boldsymbol{x}(t + dt)$.

8.3.2.3 Data Preparation

The actual inverted pendulum behavior is assumed to be equivalent to the model in (8.14) to (8.17). Therefore, the corresponding state equation can be defined in (8.7). The values of actual parameters included in the state equation (8.18) are described in Table 8.6.

$$\begin{bmatrix} \ddot{\boldsymbol{d}} \\ \ddot{\theta} \end{bmatrix} = \mathcal{M}^{-1}\left(\boldsymbol{u} - c\begin{bmatrix} \dot{d} \\ \dot{\theta} \end{bmatrix} - g\right) \tag{8.14}$$

$$\mathcal{M} = \begin{bmatrix} M + m & ml\cos(\theta) \\ ml\cos(\theta) & ml^2 \end{bmatrix} \tag{8.15}$$

$$\mathcal{C} = \begin{bmatrix} b_1 & -ml\dot{\theta}\sin(\theta) \\ 0 & b^2 \end{bmatrix} \tag{8.16}$$

$$g = \begin{bmatrix} 0 \\ mgl\sin(\theta) \end{bmatrix} \tag{8.17}$$

TABLE 8.6
Actual Parameters of the Simple Pendulum

Parameter	Actual Value
M	1.000 (kg)
m	1.000 (kg)
l	0.500 (m)
b_1	0.500 (N)
b_2	0.500 (kg/s)
g	9.810 (m/s^2)

$$\dot{x} = \begin{bmatrix} \dot{d} \\ \dot{\theta} \\ \mathcal{M}^{-1}\left(u - c\begin{bmatrix}\dot{d} \\ \theta\end{bmatrix} - g\right) \end{bmatrix} \tag{8.18}$$

The feature and target information can be then obtained by simulating the system dynamics with the Euler's integration method over (8.18) and the initial condition $\boldsymbol{x}(0) = 0$ within the time window $t \in [0,10](s)$. Again, this dynamic simulation is unnecessary in practice since the available transducers and observers can obtain the feature and target data in the actual inverted pendulum in physical space.

Two damped sine waves are used as the control signal $\boldsymbol{u}(t) = [u_1(t),\, u_2(t)]^T$ to excite different behaviors of the actual inverted pendulum. These waves are in (8.19), where the amplitudes are $A_1 = 5$ and $A_2 = 5$, the decay constants are $\lambda_1 = 0.5$ and $\lambda_2 = 0.5$, the angular frequencies are $\omega_1 = 2\pi$ and $\omega_2 = \pi$, and the phase angles are $\phi_1 = 0$ and $\phi_2 = 0$.

$$\begin{aligned} u_1(t) &= A_1 e^{-\lambda_1 t}\left(\cos(\omega_1 t + \phi_1) + \sin(\omega_1 t + \phi_1)\right) \\ u_2(t) &= A_2 e^{-\lambda_2 t}\left(\cos(\omega_1 t + \phi_1) + \sin(\omega_1 t + \phi_1)\right) \end{aligned} \tag{8.19}$$

Random noise signals up to ±5% are also included in all the feature and target variables to emulate the uncertainties in the data acquisition from the actual mechanism in physical space. Figure 8.5 shows the acquired feature and target data from the actual inverted pendulum.

Since the sampling time $dt = 5$ (ms) and the time window $t \in [0,10](s)$ are used for the data acquisition, a number of $N = 2000$ samples of target and feature vectors are obtained from the actual system, and some of them are shown in Table 8.7.

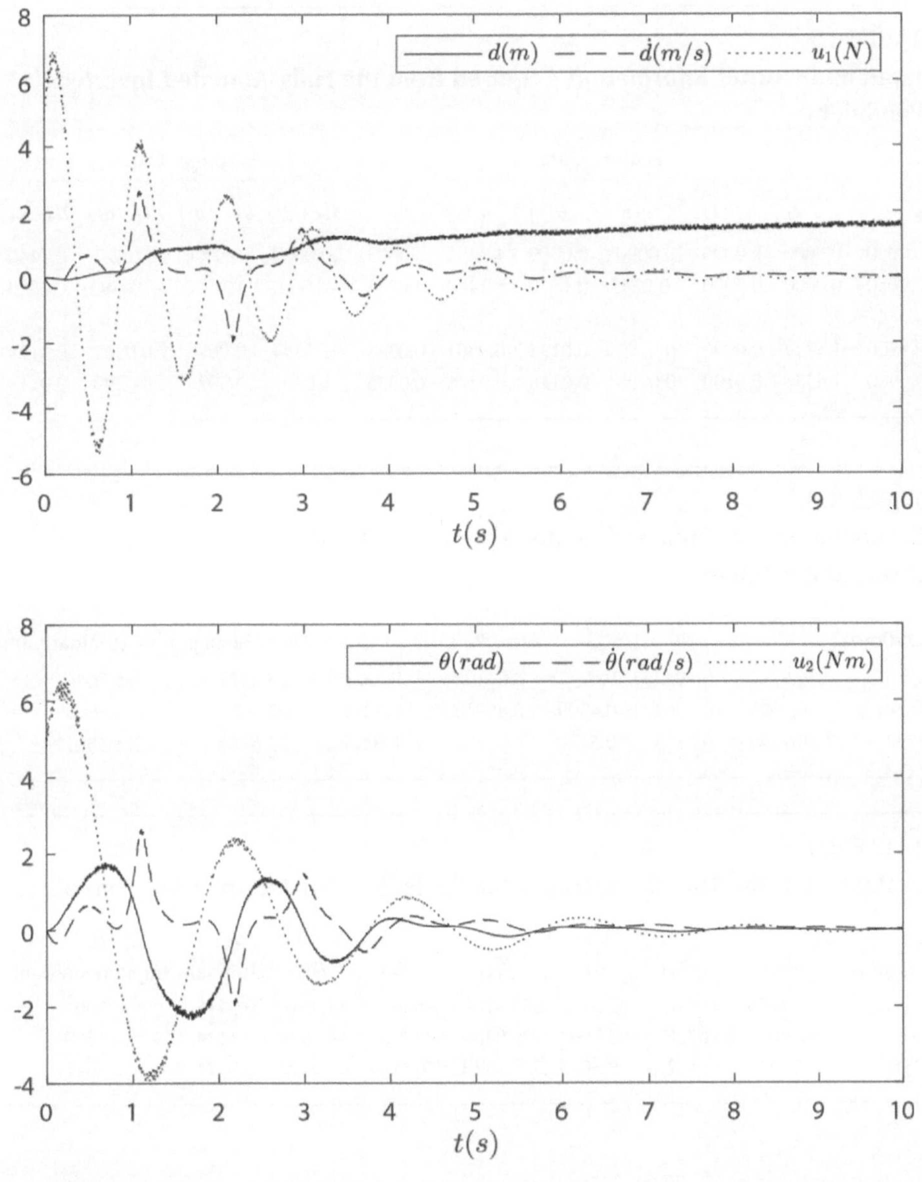

FIGURE 8.5 Feature and target information acquired from the actual fully-actuated inverted pendulum.

Eighty percent of the samples in Table 8.7 are used for training, and the remaining 20% for evaluation. Each sample is selected based on the strategy described at the end of Section 8.2.3.

TABLE 8.7
Input and Output Information Acquired from the Fully-Actuated Inverted Pendulum

Time	Feature Data						Target Data			
t	$d(t)$	$\theta(t)$	$\dot{d}(t)$	$\dot{\theta}(t)$	$u_1(t)$	$u_2(t)$	$d(t+dt)$	$\theta(t+dt)$	$\dot{d}(t+dt)$	$\dot{\theta}(t+dt)$
0.0000	0.0000	0.0000	0.0000	0.0000	4.8258	5.0683	0.0000	0.0000	−0.0248	0.1440
0.0050	0.0000	0.0000	−0.0243	0.1570	5.2910	4.8243	−0.0001	0.0007	−0.0492	0.2880
⋮	⋮	⋮	⋮	⋮	⋮	⋮	⋮	⋮	⋮	⋮
9.9900	1.5739	0.0096	0.0272	0.0135	0.0301	0.0338	1.6544	0.0098	0.0297	0.0126
9.9950	1.5354	0.0100	0.0298	0.0128	0.0326	0.0335	1.5307	0.0099	0.0299	0.0113

TABLE 8.8
Evaluation of the Obtained Model Alternatives for the Fully-Actuated Inverted Pendulum

Optimizer	$\hat{M}$	$\hat{m}$	$\hat{l}$	$\hat{b}_1$	$\hat{b}$	J_{MI} (Training)	J_{MI} (Evaluation)
DE	0.9222	0.9042	0.4777	3.6180E^{-13}	0.5222	23.6127	5.5636
GA	0.9095	0.8963	0.4809	2.4591E^{-14}	0.5212	23.6124	5.5633
PSO	0.7696	0.8165	0.5170	0.0012	0.5271	23.6308	5.5501

TABLE 8.9
Evaluation of the Tuned Controller for the Fully-Actuated Inverted Pendulum

Optimizer	$k_{p,1}$	$k_{i,1}$	$k_{d,1}$	$k_{p,2}$	$k_{i,2}$	$k_{d,2}$	J_{CT} (Tuning)	J_{CT} (Implementation)
DE	80.7960	5.1514E^{-09}	32.9435	635.4180	2.3660E^{-05}	57.4610	0.3436	0.3604
GA	82.0041	3.1672E^{-14}	32.9994	706.9300	7.2409E^{-13}	56.9929	0.3424	0.3600
PSO	92.2993	0.1892	39.3045	518.2870	107.3390	48.3722	0.3397	0.3647

8.3.2.4 Modeling

Lagrangian mechanics can be used to determine the equation of motion of the fully-actuated inverted pendulum. This equation is given in (8.20–8.23). In a state-space representation, the equation of motion is rewritten in (8.24).

$$\begin{bmatrix} \ddot{\hat{d}} \\ \ddot{\hat{\theta}} \end{bmatrix} = \mathcal{M}^{-1}\left(u - c\begin{bmatrix} \dot{\hat{d}} \\ \dot{\hat{\theta}} \end{bmatrix} - \hat{g} \right) \tag{8.20}$$

$$\hat{\mathcal{M}} = \begin{bmatrix} \hat{M} + \hat{m} & \hat{m}\hat{l}\cos(\hat{\theta}) \\ \hat{m}\hat{l}\cos(\theta) & \hat{m}\hat{l}^2 \end{bmatrix} \tag{8.21}$$

$$\hat{\mathcal{C}} = \begin{bmatrix} \hat{b}_1 & -\hat{m}\hat{l}\dot{\hat{\theta}}\sin(\hat{\theta}) \\ 0 & \hat{b}^2 \end{bmatrix} \tag{8.22}$$

$$\hat{g} = \begin{bmatrix} 0 \\ \hat{m}\hat{g}\hat{l}\sin(\hat{\theta}) \end{bmatrix} \tag{8.23}$$

$$\dot{\hat{x}} = \begin{bmatrix} \dot{\hat{d}} \\ \dot{\hat{\theta}} \\ \hat{\mathcal{M}}^{-1}\left(\hat{u} - \hat{\mathcal{C}}\begin{bmatrix} \dot{\hat{d}} \\ \dot{\hat{\theta}} \end{bmatrix} - \hat{g}\right) \end{bmatrix} \tag{8.24}$$

The adjustable parameters of the model are selected as the car mass $\hat{M}$, the pendulum mass $\hat{m}$, the bar length $\hat{l}$, and the friction coefficients $\hat{b}_1$ and $\hat{b}_2$. On the other hand, the gravity acceleration $g = 9.81\ (m / s^2)$ remains a fixed parameter. Therefore, the vector of the model parameters is $\boldsymbol{p} = \left[\hat{M}, \hat{m}, \hat{l}, \hat{b}_1, \hat{b}_2\right]^T$ and is obtained as the solution to the identification problem in (8.4). Therefore, the solution to the above problem contains the values of $\hat{M}$, $\hat{m}$, $\hat{l}$, $\hat{b}_1$, and $\hat{b}_2$ that minimize the error between the actual system output (target values) and the output predicted by the model (calculated as the solution of the initial value problem that includes the state equation (8.24) by Euler's method for a future period *dt*) when both are provided with the same inputs (feature values) for all samples in the training set.

Due to the noticeable complexity of the identification problem, the meta-heuristics chosen for the study case in Section 8.3.1 (i.e., DE, GA, and PSO) are also adopted to solve it through 30 independent runs using the hyper-parameters in Table 8.3. After all meta-heuristic runs, several sets of model parameters are calculated, and the performance of these alternatives is evaluated next.

8.3.2.5 Evaluation

The model parameters, generated in the previous step after each independent run with all meta-heuristics, are evaluated by the objective function J_{MI} in (8.4), this time, using the features and targets in the evaluation set. Table 8.8 shows the best results obtained by each meta-heuristic. This table describes the used meta-heuristic, the best set of parameters found by it, and the value J_{MI}computed for training and evaluation.

The best solutions found by each optimizer have noticeable differences as observed in Table 8.8. Nevertheless, they deploy very similar performances according to J_{MI} for training and evaluation. The above suggests that the identification problem for the inverted pendulum could be multimodal, i.e., the problem has more than one optimal solution. Also, the best model parameters in Table 8.8 are somewhat different from the actual ones in Table 8.6. However, they are still handy for the consequent application in the controller tuning task since they satisfactorily make the model generalize the behavior of the actual inverted pendulum.

Table 8.8 indicates that PSO found the best model parameters concerning the value of J_{MI} for evaluation. Although this alternative has the worst performance according to the J_{MI} for training, its outstanding value of J_{MI} for evaluation indicates that this model configuration can better generalize the actual system behavior. For this reason, the model parameters calculated by GA are used in the next stage for the controller tuning.

8.3.2.6 Deployment

In this stage, the best model parameters detected in the previous evaluation are used in a Computed-Torque Controller (CTC) (Spong, Hutchinson, & Vidyasagar, 2020) for the trajectory tracking task with the inverted pendulum. The CTC is given in (8.25) and (8.26), with $\hat{\mathcal{M}}$, $\hat{\mathcal{C}}$, and $\hat{g}$ as the terms in (8.20–8.23) that include the model parameters previously optimized with PSO, $\boldsymbol{e} = [d_d - d,\ \theta_d - \theta]^T$ as the error vector, and $K_p = \mathrm{diag}(k_{p,1},\ k_{p,2})$, $K_i = \mathrm{diag}(k_{i,1},\ k_{i,2})$, and $K_d = \mathrm{diag}(k_{d,1},\ k_{d,2})$, as the proportional, integral, and derivative gain matrices of the PID compensator.

$$\boldsymbol{u} = \hat{\mathcal{M}}\boldsymbol{v} + \hat{\mathcal{C}}\begin{bmatrix} \dot{d} \\ \dot{\theta} \end{bmatrix} + \hat{g} \tag{8.25}$$

$$\boldsymbol{v} = K_p\boldsymbol{e} + K_i \int \boldsymbol{e}dt + K_d\ \dot{\boldsymbol{e}} \tag{8.26}$$

Based on the above CTC structure, the parameters in the vector $z = [k_{p,1},\ k_{i,1},\ k_{d,1},\ k_{p,2},\ k_{i,2},\ k_{d,2}]^T$ must be found as the solution of the controller tuning problem in (8.5).The tuning problem is to find the gain matrices K_p, K_i, and K_d that achieve the minimum ITSE for the trajectory tracking task during the dynamic simulation of (8.24) by Euler's method. The dynamic simulation stops when the time reaches $t_f = 10\ (s)$.

Since the controller tuning problem is complex, the DE, GA, and PSO optimizers are used to find promising solutions. In this way, 30 independent runs of each optimizer are performed to find several controller parameter alternatives. The alternative with the best performance can be implanted in the actual system.

Table 8.9 shows the performances of the best controller parameters found by each meta-heuristic. The first columns show the used meta-heuristic and the obtained controller parameters. The last two columns describe the value of J_{CT} for the tuning process and the final implementation in the actual inverted pendulum. In this work, the J_{CT} for implementation is computed over the dynamic simulation of (8.18) with the CTC controller in (8.25) and (8.26).

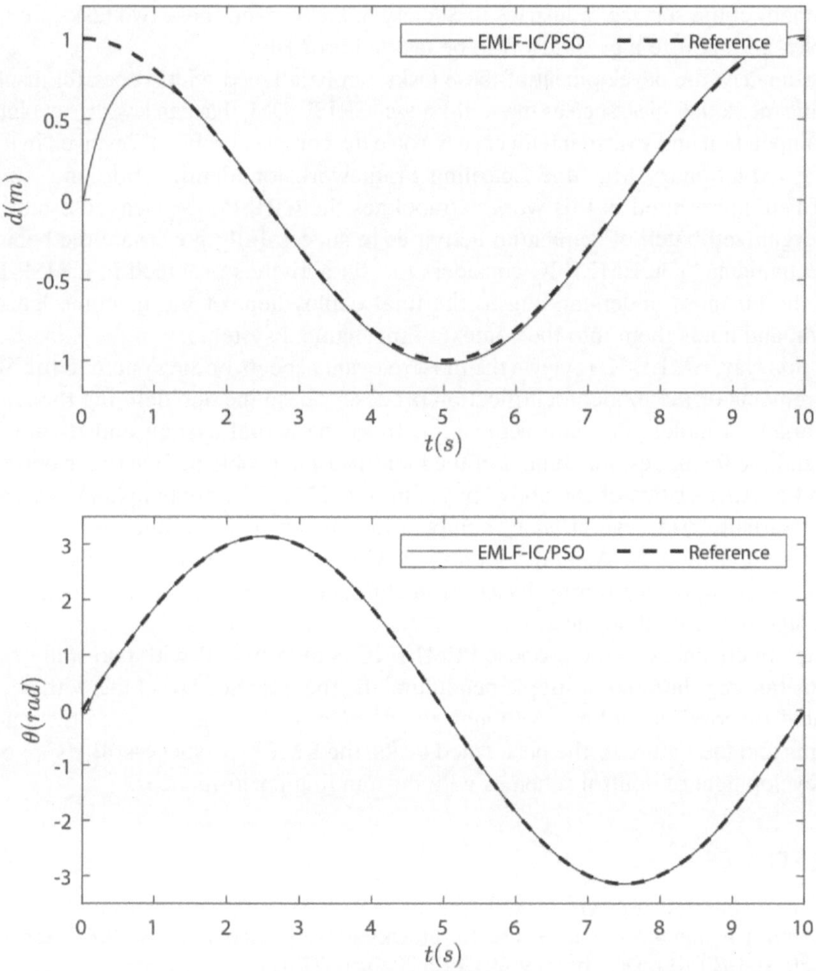

FIGURE 8.6 Trajectory tracked with the actual fully-actuated inverted pendulum using the EMLF-IC with PSO.

Similar performances are observed in Table 8.9 concerning J_{CT} for tuning and implementation, despite the visible differences among the reported controller parameters. The solution found by PSO is implemented in the actual system since it shows the best performance when considering the value of J_{CT} for tuning.

Finally, the fully-actuated inverted pendulum's behavior using the best CTC controller found by PSO in the EMLF-IC is displayed in Figure 8.6. This figure shows that the fully-actuated inverted pendulum can accurately track the proposed reference trajectory.

8.4 CONCLUSION

The system identification and controller tuning tasks are crucial in designing well-controlled mechanical systems that can be used in a wide variety of applications for the benefit of the service industries in Society 5.0. However, these two tasks are quite complex and require a great effort to be tackled properly.

Fortunately, the development of these tasks can be aligned with successful machine learning methodologies, such as the well-known CRISP-DM, that can leverage evolutionary computation and swarm intelligence to solve the complex problems involved in them.

The Evolutionary Machine Learning Framework for Identification and Control (EMLF-IC) presented in this work extrapolates the CRISP-DM ideas to establish a well-organized batch of sequential activities to successfully governate the behavior of mechanisms. The EMLF-IC considers the six activities provided in CRISP-DM, from the business understanding to the final deployment of the machine learning system, and lands them into the context of mechanical systems.

In this way, EMLF-IC reviews the most relevant aspects of the system to fulfill the requirements of the model identification process. These include defining the feature and target variables, the data acquisition from the actual system and its preparation, and the formal establishment of the identification problem. Due to the complexity and characteristics of the above problem, the EMLF-IC contemplates the use of meta-heuristics from evolutionary computation and swarm intelligence to find optimized system models. A model identified with EMLF-IC is deployed to perform a successful controller tuning based on meta-heuristic optimization, which can be implanted in the final application.

The effectiveness of the proposed EMLF-IC is determined with two study cases: the position regulation of a simple pendulum and the trajectory tracking with a fully-actuated inverted pendulum. Although the notable differences between the studied systems and the nature of the performed tasks, the EMLF-IC successfully supported the development of control schemes with outstanding performances.

REFERENCES

Abate, A. F., Barra, P., Bisogni, C., Cascone, L., & Passero, I. (2020). Contextual trust model with a humanoid robot defense for attacks to smart eco-systems. *IEEE Access*, *8*, 207404–207414. Doi: 10.1109/ACCESS.2020.3037701.

Ahmed, N. K., Atiya, A. F., El Gayar, N., & El-Shishiny, H. (2010). An empirical comparison of machine learning models for time series forecasting. *Econometric Reviews*, *29*(5), 594–621. Doi: 10.1080/07474938.2010.481556.

Arasomwan, M. A., & Adewumi, A. O. (2013). On the performance of linear decreasing inertia weight particle swarm optimization for global optimization. *The Scientific World Journal*, *2013*, 1–12. Doi: 10.1155/2013/860289.

Bäck, T., Fogel, D. B., & Michalewicz, Z. (1997). Handbook of evolutionary computation. In T. Back, D. B. Fogel, & Z. Michalewicz (Eds.), *Handbook of Evolutionary Computation* (Vol. 97, Issue 1). IOP Publishing Ltd. Doi: 10.1887/0750308958.

Bollini, M., Tellex, S., Thompson, T., Roy, N., & Rus, D. (2013). Interpreting and executing recipes with a cooking robot. In J. P. Desai, G. Dudek, O. Khatib, & V. Kumar (Eds.), *Experimental Robotics: The 13th International Symposium on Experimental Robotics* (pp. 481–495). Springer International Publishing. Doi: 10.1007/978-3-319-00065-7_33.

Brosilow, C., & Joseph, B. (1995). *Methods of Model Based Process Control*, R. Berber (Ed.), Springer Netherlands. Doi: 10.1007/978-94-011-0135-6.

Bullo, F., & Murray, R. M. (1999). Tracking for fully actuated mechanical systems: a geometric framework. *Automatica*, *35*(1), 17–34. Doi: 10.1016/S0005-1098(98)00119-8.

Chinnamgari, S. K. (2019). *R Machine Learning Projects: Implement Supervised, Unsupervised, and Reinforcement Learning Techniques using R 3.5*. Birmingham, UK: Packt Publishing Ltd. https://www.packtpub.com/product/r-machine-learning-projects/9781789807943.

Deguchi, A., Hirai, C., Matsuoka, H., Nakano, T., Oshima, K., Tai, M., & Tani, S. (2020). What is society 5.0? In *Society 5.0* (pp. 1–23). Springer Singapore. Doi: 10.1007/978-981-15-2989-4_1.

Farahmand, B., Dehghani, M., & Vafamand, N. (2019). Fuzzy model-based controller for blood glucose control in type 1 diabetes: an LMI approach. *Biomedical Signal Processing and Control*, *54*, 101627. Doi: 10.1016/j.bspc.2019.101627.

Fedor, P., & Perdukova, D. (2016). Model-based fuzzy control applied to a real nonlinear mechanical system. *Iranian Journal of Science and Technology - Transactions of Mechanical Engineering*, *40*(2), 113–124. Doi: 10.1007/s40997-016-0005-9.

Friedland, B. (2005). *Control System Design: An Introduction to Space-State Methods*. Mineola, NY: Dover Publications.

Garg, A., Tai, K., & Panda, B. N. (2017). System identification: survey on modeling methods and models. In *Advances in Intelligent Systems and Computing* (Vol. 517, pp. 607–615). Springer. Doi: 10.1007/978-981-10-3174-8_51.

Gignoux, C., & Silvestre-Brac, B. (2009). *Solved Problems in Lagrangian and Hamiltonian Mechanics*. Springer Netherlands. Doi: 10.1007/978-90-481-2393-3.

Gotmare, A., Bhattacharjee, S. S., Patidar, R., & George, N. V. (2017). Swarm and evolutionary computing algorithms for system identification and filter design: a comprehensive review. *Swarm and Evolutionary Computation*, *32*, 68–84. Doi: 10.1016/j.swevo.2016.06.007.

Granrath, L. (2019). Large scale optimization is needed for industry 4.0 and society 5.0. In M. Fathi, M. Khakifirooz, & P. M. Pardalos (Eds.), *Springer Optimization and Its Applications* (Vol. 152, pp. 3–6). Springer International Publishing. Doi: 10.1007/978-3-030-28565-4_1.

Huang, Y., Gao, L., Yi, Z., Tai, K., Kalita, P., Prapainainar, P., & Garg, A. (2018). An application of evolutionary system identification algorithm in modelling of energy production system. *Measurement: Journal of the International Measurement Confederation*, 114, 122–131. Doi: 10.1016/j.measurement.2017.09.009.

Iordache, O. (2017). *Implementing Polytope Projects for Smart Systems* (Vol. 92). Springer International Publishing. Doi: 10.1007/978-3-319-52551-8.

Kennedy, J. (2006). Swarm intelligence. In *Handbook of Nature-Inspired and Innovative Computing* (pp. 187–219). Kluwer Academic Publishers. Doi: 10.1007/0-387-27705-6_6.

Kitchin, J. R. (2018). Machine learning in catalysis. *Nature Catalysis*, *1*(4), 230–232. Doi: 10.1038/s41929-018-0056-y.

Liu, T., & Cavusoglu, M. C. (2016). Needle grasp and entry port selection for automatic execution of suturing tasks in robotic minimally invasive surgery. *IEEE Transactions on Automation Science and Engineering*, *13*(2), 552–563. Doi: 10.1109/TASE.2016.2515161.

Liu, Z. L., Hu, J. H., Jiang, F., & Wu, Y. D. (2020). CRiSP: accurate structure prediction of disulfide-rich peptides with cystine-specific sequence alignment and machine learning. *Bioinformatics*, *36*(11), 3385–3392. Doi: 10.1093/bioinformatics/btaa193.

Ljung, L. (1998). System identification. In Procházka A., Uhlíř J., Rayner P.W.J. & Kingsbury N. G. (Eds.), *Signal Analysis and Prediction. Applied and Numerical Harmonic Analysis* (pp. 163–173). Birkhäuser. Doi: 10.1007/978-1-4612-1768-8_11.

López-Ibáñez, M., Dubois-Lacoste, J., Pérez Cáceres, L., Birattari, M., & Stützle, T. (2016). The irace package: iterated racing for automatic algorithm configuration. *Operations Research Perspectives*, *3*, 43–58. Doi: 10.1016/j.orp.2016.09.002.

Matsui, D., Minato, T., MacDorman, K. F., & Ishiguro, H. (2005). Generating natural motion in an android by mapping human motion. *2005 IEEE/RSJ International Conference on Intelligent Robots and Systems*, 3301–3308.

Mezura-Montes, E., Velázquez-Reyes, J., & Coello Coello, C. A. (2006). A comparative study of differential evolution variants for global optimization. *Proceedings of the 8th Annual Conference on Genetic and Evolutionary Computation - GECCO '06*, 485. Doi: 10.1145/1143997.1144086.

Mirjalili, S., Faris, H., & Aljarah, I. (2020). Introduction to evolutionary machine learning techniques. In S. Mirjalili, H. Faris, & I. Aljarah (Eds.), *Evolutionary Machine Learning Techniques: Algorithms and Applications* (pp. 1–7). Springer Singapore. Doi: 10.1007/978-981-32-9990-0_1.

Moro, S., Laureano, R. M. S., & Cortez, P. (2011). Using data mining for bank direct marketing: an application of the CRISP-DM methodology. *ESM 2011–2011 European Simulation and Modelling Conference: Modelling and Simulation 2011*, 117–121.

Müller, A., & Hufnagel, T. (2012). Model-based control of redundantly actuated parallel manipulators in redundant coordinates. *Robotics and Autonomous Systems*, *60*(4), 563–571. Doi: 10.1016/j.robot.2011.11.014.

O'Dwyer, A. (2009). *Handbook of PI and PID Controller Tuning Rules*. London, UK: Imperial College Press.

Ogata, K. (2004). *System Dynamics* (4th ed.). Upper Saddle River, NJ: Pearson/Prentice Hall.

Ogata, K., & Yang, Y. (2002). *Modern Control Engineering* (Vol. 4). India: Prentice Hall.

Putt, S. S. J., Wijeakumar, S., & Spencer, J. P. (2019). Prefrontal cortex activation supports the emergence of early stone age toolmaking skill. *NeuroImage*, *199*, 57–69. Doi: 10.1016/j.neuroimage.2019.05.056.

Reynoso-Meza, G., Blasco, X., Sanchis, J., & Martínez, M. (2014). Controller tuning using evolutionary multi-objective optimisation: current trends and applications. *Control Engineering Practice*, *28*(1), 58–73. Doi: 10.1016/j.conengprac.2014.03.003.

Rodríguez-Molina, A., Mezura-Montes, E., Villarreal-Cervantes, M. G., & Aldape-Pérez, M. (2020). Multi-objective meta-heuristic optimization in intelligent control: a survey on the controller tuning problem. *Applied Soft Computing Journal*, *93*, 106342. Doi: 10.1016/j.asoc.2020.106342

Rodríguez-Molina, A., Villarreal-Cervantes, M. G., Álvarez-Gallegos, J., & Aldape-Pérez, M. (2019). Bio-inspired adaptive control strategy for the highly efficient speed regulation of the DC motor under parametric uncertainty. *Applied Soft Computing Journal*, *75*, 29–45. Doi: 10.1016/j.asoc.2018.11.002.

Rojas, C. R., Barenthin, M., Welsh, J. S., & Hjalmarsson, H. (2011). The cost of complexity in system identification: the output error case. *Automatica*, *47*(9), 1938–1948. Doi: 10.1016/j.automatica.2011.06.021.

Romero, J. G., Ortega, R., & Donaire, A. (2016). Energy shaping of mechanical systems via PID control and extension to constant speed tracking. *IEEE Transactions on Automatic Control*, *61*(11), 3551–3556. Doi: 10.1109/TAC.2016.2521725.

Salgues, B. (2018). Information technology 2.0, the foundation of society 5.0. In *Society 5.0* (pp. 75–90). John Wiley & Sons, Inc. Doi: 10.1002/9781119507314.ch5.

Shardt, Y. A. W., Huang, B., & Ding, S. X. (2015). Minimal required excitation for closed-loop identification: some implications for data-driven, system identification. *Journal of Process Control*, *27*, 22–35. Doi: 10.1016/j.jprocont.2015.01.009.

Siegwart, R., Nourbakhsh, I. R., & Scaramuzza, D. (2011). Introduction to autonomous mobile robots. *Choice Reviews Online*, *49*(03), 49–1492-49–1492. Doi: 10.5860/CHOICE.49–1492.

Singh, S. (2005). *Theory of Machines*. India: Pearson Education.

Spong, M.W., Hutchinson, S. and Vidyasagar, M. (2020). *Robot Dynamics and Control* (2nd ed.). John Wiley & Sons. https://www.wiley.com/en-us/Robot+Modeling+and+Control%2C+2nd+Edition-p-9781119524045.

Son, S., Kim, T., Sarma, S. E., & Slocum, A. (2009). A hybrid 5-axis CNC milling machine. *Precision Engineering*, *33*(4), 430–446. Doi: 10.1016/j.precisioneng.2008.12.001.

Talbi, E. G. (2009). Metaheuristics: from design to implementation. In *Metaheuristics: From Design to Implementation* (Vol. 74). John Wiley & Sons. Doi: 10.1002/9780470496916.

Tanaka, F., & Ghosh, M. (2011). The implementation of care-receiving robot at an English learning school for children. *Proceedings of the 6th International Conference on Human-Robot Interaction*, 265–266. Doi: 10.1145/1957656.1957763.

Tesen, S., Saga, N., Satoh, T., & Nagase, J. (2013). Peristaltic crawling robot for use on the ground and in plumbing pipes. In V. Padois, P. Bidaud, & O. Khatib (Eds.), *Romansy 19-- Robot Design, Dynamics and Control* (pp. 267–274). Springer Vienna.

Tian, Y. L., Zou, H. J., & Guo, W. Z. (2006). An integrated knowledge representation model for the computer-aided conceptual design of mechanisms. *International Journal of Advanced Manufacturing Technology*, *28*(5–6), 435–444. Doi: 10.1007/s00170-004-2399-6.

Veček, N., Črepinšek, M., & Mernik, M. (2017). On the influence of the number of algorithms, problems, and independent runs in the comparison of evolutionary algorithms. *Applied Soft Computing Journal*, *54*, 23–45. Doi: 10.1016/j.asoc.2017.01.011.

Verma, B., & Padhy, P. K. (2018). Optimal PID controller design with adjustable maximum sensitivity. *IET Control Theory and Applications*, *12*(8), 1156–1165. Doi: 10.1049/iet-cta.2017.1078.

Wang, X., & Hou, B. (2018). Trajectory tracking control of a 2-DOF manipulator using computed torque control combined with an implicit lyapunov function method. *Journal of Mechanical Science and Technology*, *32*(6), 2803–2816. Doi: 10.1007/s12206-018-0537-6.

Wiemer, H., Drowatzky, L., & Ihlenfeldt, S. (2019). Data mining methodology for engineering applications (DMME)-A holistic extension to the CRISP-DM model. *Applied Sciences*, *9*(12), 403–408. Doi: 10.3390/app9122407.

Wirth, R. (2000). CRISP-DM : towards a standard process model for data mining. *Proceedings of the Fourth International Conference on the Practical Application of Knowledge Discovery and Data Mining*, *24959*, 29–39.

Wolpert, D. H., & Macready, W. G. (1997). No free lunch theorems for optimization. *IEEE Transactions on Evolutionary Computation*, *1*(1), 67–82. Doi: 10.1109/4235.585893.

Worden, K., Barthorpe, R. J., Cross, E. J., Dervilis, N., Holmes, G. R., Manson, G., & Rogers, T. J. (2018). On evolutionary system identification with applications to nonlinear benchmarks. *Mechanical Systems and Signal Processing*, *112*, 194–232. Doi: 10.1016/j.ymssp.2018.04.001.

Xiao, B., Yin, S., & Gao, H. (2018). Reconfigurable tolerant control of uncertain mechanical systems with actuator faults: a sliding mode observer-based approach. *IEEE Transactions on Control Systems Technology*, *26*(4), 1249–1258. Doi: 10.1109/TCST.2017.2707333.

Zemrane, H., Baddi, Y., & Hasbi, A. (2021). Internet of Things industry 4.0 ecosystem based on zigbee protocol. In *Advances in Intelligent Systems and Computing* (Vol. 1188, pp. 249–260). Springer. Doi: 10.1007/978-981-15-6048-4_22.h

9 Evolutionary Computation Techniques for Strengthening Performance of Commercial MANETs in Society 5.0

Ramanpreet Kaur, Kavita Taneja, and Harmunish Taneja

CONTENTS

9.1 INTRODUCTION

Wireless networks have revolutionized interconnectivity approach in recent decades by allowing devices to link and share the information stored on them wirelessly (Reina et al., 2013a). As the number of wireless handheld devices is increasing, the number of installed wireless communication devices is constantly increasing. With this huge increase in the use of ubiquitous computing, there is a need for strengthening

DOI: 10.1201/9781003158165-9

the performance of commercial mobile ad hoc networks (MANETs) in society 5.0. Evolutionary computing (EC) is a significant field of study in adaptation and optimization. The technique is based on Darwin's natural selection concept, commonly known as survival of the fittest. Knowledge gained by an individual cannot be transmitted into its genome and then passed on to the next generation, according to the Darwinian hypothesis (Zhang et al., 2011). In recent decades, Ad hoc networks have fascinated the whole world due to their dynamic characteristics. They reflect a modern communication model in which decentralized wireless nodes work together to achieve a common goal. When the nodes in the network are considered mobile, the network is termed as mobile ad hoc network (MANET). Today, we are entering the age of ubiquitous computing, in which a user may get information from many electronic platforms whenever and wherever it is needed. Commercial MANETs are being utilized in a variety of applications such as military operations, weather monitoring, disaster recovery, emergency operations, commercial applications, home automation, education, and entertainment. In the context of society 5.0, MANETs may reflect situations like individuals using cell phones, a search and rescue squad in an emergency mission, warriors in military applications, etc. Due to the diverse range of commercial MANET application areas, there is a need for effective routing techniques to provide quality service to users. In order to tackle these challenges and to strengthen the performance of commercial MANETs, EC techniques are particularly effective for optimizing problems as these techniques are bio-inspired and have self-healing properties. EC methods are particularly used in domains where the calculus is difficult to apply or is inapplicable. These approaches are a type of directed stochastic global search that may be used for a variety of purposes. It can solve difficult problems consistently and quickly by searching across a vast, complicated, non-continuous, non-differentiable, and multimodal surface. It can always achieve the near-optimal or global optimum solution. Figure 9.1 shows the configuration of mobile ad hoc network. The mobile devices are connected through wireless links.

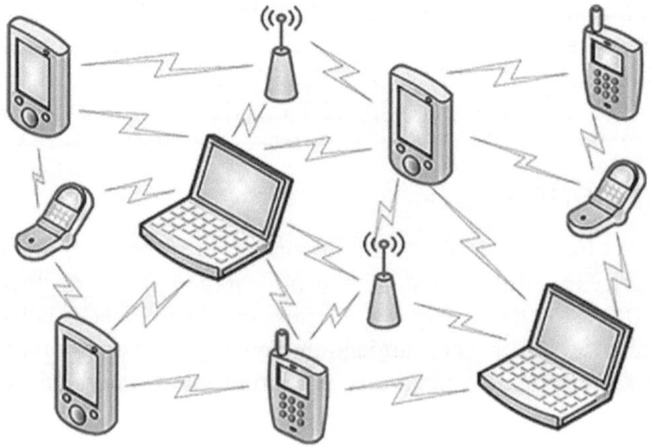

FIGURE 9.1 Ad hoc network of large number of connected devices per user.

The Internet of Things (IoT) environment has resulted in a significant rise in the number of devices connected to wireless networks. These gadgets have become a part of our everyday life, evolving into a powerful tool with wireless networking capabilities giving rise to commercial MANETs. Due to this drastic increase in the number of connected devices, there is a need to design smart, adaptive, and effective routing protocols to provide better services to a huge set of mobile users. In this chapter, the evolutionary computation techniques in designing effective routing protocols in commercial MANETs to enhance the network performance in society 5.0 is discussed.

9.2 COMMERCIAL MANET: ISSUES AND CHALLENGES IN SOCIETY 5.0

MANET is a dynamic system that is self-adjustable and is made up of wireless mobile nodes that form a temporary wireless network without the use of a centralized system and communicate through several hops. It seeks to deliver networking over cellular connections rather than relying on fixed network topologies. It is leading in a wide range of smart environment and application areas to address challenges that people face in their everyday lives. This type of network provides greater mobility to users while further lowering network implementation costs. However, it also poses new daunting issues in terms of networking (Taneja & Patel, 2003) Figure 9.2 shows various application areas of mobile ad hoc networks such as military areas, wireless LANs, emergency areas, rescue areas in smart environment.

The following points elaborate the issues of commercial MANET in society 5.0:

1. **Dynamic Topology**: Since majority nodes in the network are mobile in commercial MANET, they are free to travel in every direction at any moment. The nodes can enter and exit the network as and when desired. The topology will shift promptly and arbitrarily.

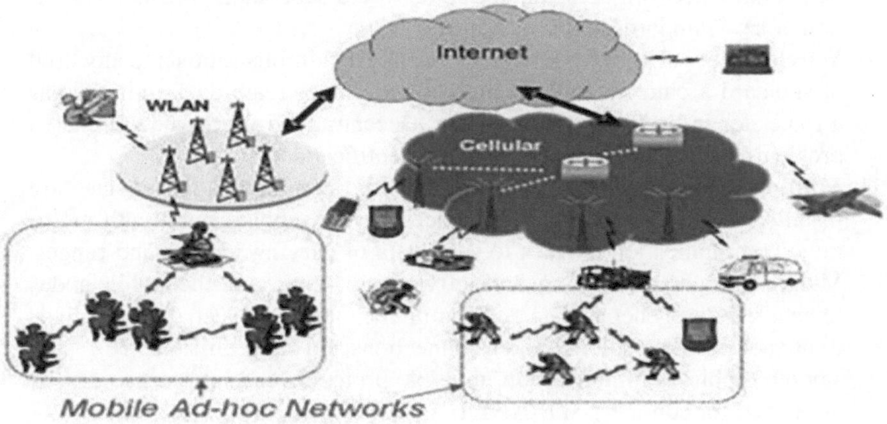

FIGURE 9.2 Evolving applications of commercial MANET in society 5.0.

2. **Heterogeneous Network**: Commercial MANET nodes have different radio reception (upstream) and radio transmitting (downstream) frequencies.
3. **Resource Constrained**: Nodes have limited processing power, storage, and battery life. The transmission and forwarding capabilities of wireless devices are limited by power constraints. As a result, the mobile device can only support a limited range of utilities and applications.
4. **Security**: Commercial MANETs are more vulnerable to protection and privacy threats than fixed networks. Access protection, database integrity, and encryption are all security issues that necessitate increased security requirements and procedures. Several key management mechanisms for providing security have been developed, but wireless network security still remains an open challenge.
5. **Fast Deployment**: Since MANET does not need any prior infrastructure or installation, the versatility level for setting up networks is even higher than their commercial counterparts, and they can be installed and dismantled in a short period of time.
6. **Fault Tolerance**: The rate of repair/link failure is high when nodes move. Commercial MANETs require immediate support for connection failures, and intelligent routing protocols need to be designed to maintain these situations.
7. **Ad hoc Routing**: There is no single node responsible for routing because of wireless connections. In order to transmit incoming packets to the destined node, each node may act as a router.
8. **Network Partition**: In Commercial MANETs, mobile nodes often alter the locations, which result in network partitions. Network management is essential to maintain these network partition issues.
9. **Quality of Service (QoS)**: Meeting the QoS standards for commercial MANET is a daunting challenge. Since the network facilitates versatility and wireless communication, achieving QoS metrics such as packet distribution ratio, network lifespan, and so on would necessitate the implementation of new standards and procedures.
10. **Wireless Network**: MANETs lack centralized administration and any kind of standard architecture. Each mobile node may create data and serve as a router for information transmission. Decentralized infrastructure causes problems such as failure detection and identification.
11. **Multiple Communication Channels**: Radio communication devices are installed in wireless nodes. The devices can be supplemented with multi-radio communication devices to send data of varying lengths and ranges. Moreover, separate hardware and software are required inside mobile nodes to handle heterogeneous data and communication (Taneja et al., 2018). This might potentially result in broken connections and data collisions. To ensure smooth mobile communication, multiple protocols and interfaces must be developed for commercial MANET.
12. **Scalability**: As infrastructure is wireless, the network size may be scaled to cover a large region for application areas in society 5.0. The needs for this

include the adoption of additional mobile nodes capable of connecting to the network, the implementation of new security measures to manage scalability, and so on.

The implementation of commercial MANET in society 5.0 necessitates the creation of new effective routing protocols using evolutionary computation techniques to strengthen the performance of the network. Routing protocols are typically necessary to maintain efficient and well-organized transmission from source to destination, as well as to maintain routes among mobile nodes and design paths for forwarding packets (Devi & Gill, 2019a, b). These smart routing protocols must ensure fairness, quality of service, and communication across the network.

Despite the high quality of today's wireless networks, meeting the needs of the evolving smart infrastructure poses significant challenges, particularly in terms of the number of connections, bandwidth requirements, reach, latency, support for low-energy devices, anonymity, protection, and resilience (Weitnauer et al., 2017).

- **Support for Large Number of Devices**: The number of Internet-connected computers will exceed immensely in the coming decades. This system will generate massive amounts of data which goes beyond the limits of current wireless networks.
- **Support for Real Time Applications**: To support real-time control applications such as Tactile Internet, multi-player gaming, and virtual reality in commercial MANET network latency, or round-trip delay, must be reduced by more than a factor of ten.
- **Support for Energy Efficient Devices**: Commercial MANET must accommodate a large number of devices with high bandwidth, energy-constrained smart devices such as sensors whose batteries must last for several years to handle the data in smart environment, and such network must have evolved framework to combat selfish behavior among the connected nodes. Figure 9.3 shows the increase in connected devices according to growing population. This chart indicates there is a huge spike in connected devices which goes from 8 billion devices to 52 billion devices towards society 5.0.

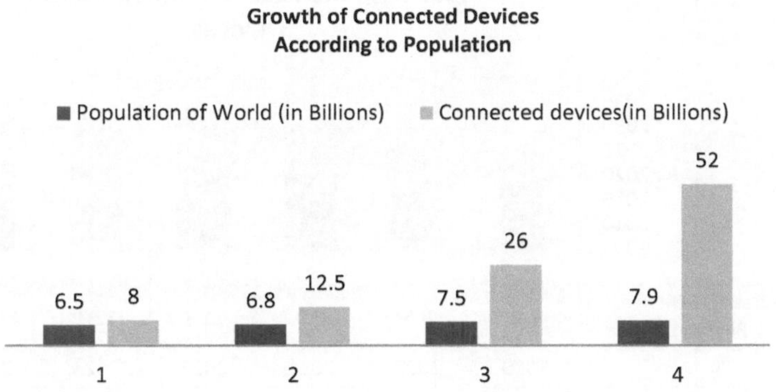

FIGURE 9.3 Increase in the number of connected devices towards society 5.0.

- **Support for Privacy & Security**: The sensor-based devices in a smart world can track human position and movement that pose critical privacy concerns. As the number of linked devices increases day by day, wireless networks must be secured from cyber-attacks as well.
- **Support for Quality of Service (QoS)**: As smart device users are increasing, the wireless networks must have protocols and algorithms to deliver quality of service to these users as per their diverse requirements.

9.3 VULNERABILITIES AND THREATS TO ROUTING IN COMMERCIAL MANET IN SOCIETY 5.0

Communication networks have had a significant influence on modern societies. No one could have predicted the extraordinary popularity and involvement of mobile phones in our daily lives. Networking technology is advancing at a quicker rate these days. Even small gadgets are already equipped with communication capabilities. Most users were pleased by the recent improvements in wireless sensor networks, which extended their service to "mobility," i.e., the usage of mobile nodes.

With technology enhancements and huge spike in mobile users, the total number of connected devices would be around 13.7 billion in the near future, as shown in Figure 9.4. In order to tackle this gigantic number of devices in the network, the smart environment has to face some routing issues in wireless networks as follows:

- **Mobility**: Traditional networks previously had stable topologies, but it is not viable to maintain the constant and stable topology in MANETs, the key constituent of IoT.
- **Extensive Range of Devices**: The devices will change depending on the network standard utilized, the kind of resources, and the category of application. Some devices will experience resource limitations, while others will not.
- **Fault Tolerance**: Energy constraints and device positioning are some of the external variables that cause network performance to deteriorate. To address such external variables, an effective routing protocol should be suggested.

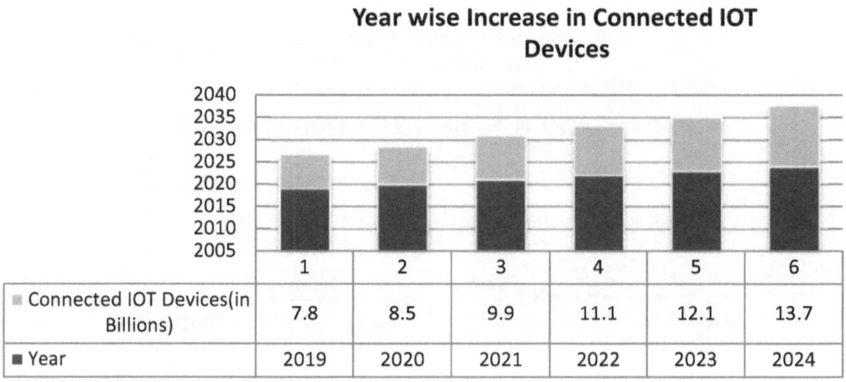

FIGURE 9.4 Total number of connected IoT devices till 2024 towards society 5.0.

- **Networking Principles and Standards**: The IoT includes different techniques such as wireless sensor networks (WSNs), Wi-Fi networks, conventional networks, radio frequency identification (RFID), MANETs, vehicular ad hoc networks (VANETs), and so on. Each of these networks exploits a distinct protocol stack. By integrating different technologies with EC, the performance of such wireless networks can be strengthened.

9.4 COMPUTATIONAL INTELLIGENCE FOR EFFECTIVE ROUTING IN COMMERCIAL MANET

Routing plays a significant role in designing MANET. The function of routing protocol in an ad hoc network is to enable the source to find routes to the destination with the help of other intermediate nodes. Routing protocols can be classified into the following types:

- **Proactive Routing Protocols**: Table-driven routing protocols are another name for these protocols. In these protocols, each node stores routing information of every other node in the network, i.e., a path is available before the need arises. Routing tables are used to store routing data (Mbarushimana & Shahrabi, 2007). Routing tables must be updated on a regular basis as the network's topology changes. Furthermore, these protocols are inefficient for big networks since they need all nodes to upgrade their routing table entries. Table-driven protocols include optimized link state routing (OLSR), destination sequenced distance vector routing (DSDV).
- **Reactive Routing Protocols**: Reactive routing protocols are also known as on-demand routing protocols since they find the route on demand, which means that the route is only accessible when it is required (Patel et al., 2015). Examples are on-demand distance vector (AODV), dynamic source routing (DSR)
- **Hybrid Routing Protocols**: Hybrid protocols use the characteristics of both reactive and proactive protocols to obtain superior results. The network is divided into zones, each of which uses a distinct protocol. One protocol is used within the zone, while the other is used between them. Zone routing protocol (ZRP) is an example of a hybrid routing protocol.

The main challenge in developing routing protocols for commercial MANET is to ensure that protocols should be adaptive in nature to accommodate according to dynamic features of the network. Due to the stochastic behavior of nodes in commercial MANETs, designing intelligent and dynamic routing protocols is the need of the hour. For developing dynamic and effective routing protocols computational intelligence techniques are best suited as these techniques are inspired by natural creatures having self-healing properties. Computational intelligence (CI) is an emerging concept in the field of machine learning. CI is an innovative concept that consists of a highly multidisciplinary framework for assisting in the design and development of cognitive systems (Jain et al., n.d.). It is the investigation of adaptive systems, practical adaptability, self-organization principles, paradigms, algorithms, and

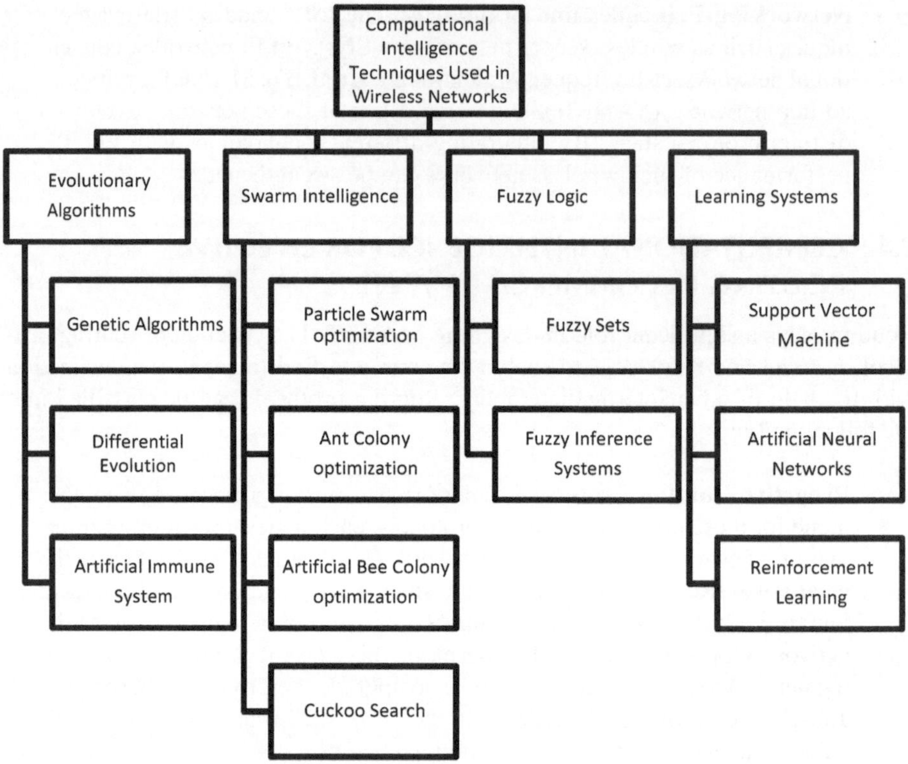

FIGURE 9.5 Computational intelligence techniques used in society 5.0.

implementations that enable or assist intelligent behavior in complex and dynamic settings (Pedrycz et al., 2016). In commercial MANETs, finding routes that may enable efficient data transmission is quite tough due to the wireless channel characteristics and various transmission landscapes. Due to the mobility of nodes and the nomadic nature of MANET, connection failure and route breaking happen very frequently. In order to handle these issues, majorly used CI techniques are neural networks, fuzzy systems, EC, artificial immune systems, and many more (Fogel, 2000).

Figure 9.5 describes various types of computational intelligence techniques used in wireless networks such as evolutionary algorithms, swarm intelligence, fuzzy logic, learning systems, genetic algorithms, differential evolution.

9.5 EVOLUTIONARY COMPUTATION TECHNIQUES FOR COMMERCIAL MANETs IN SOCIETY 5.0: A COMPARATIVE STUDY

Numerous nature-inspired computer models have developed through time such as ant colony optimization (ACO), binary particle swarm optimization (BPSO), firefly optimization, earthworm optimization, lion optimization, etc. (Siddique & Adeli, 2015). These techniques also include computer paradigms such as cultural learning, ambient intelligence, artificial life, artificial endocrine networks, social reasoning, and artificial hormone networks (Kulkarni et al., 2011). It plays a vital role in the creation of effective intelligent systems, such as games and cognitive developmental systems. CI incorporates novel models, many of which have a high machine learning quotient (Munot & Kulkarni, 2016). CI techniques are best suited for designing dynamic routing protocols for commercial MANETs in society 5.0 to tackle link and connection failure issues.

Krishna and Swapna (2020) proposed power-aware routing protocol for military applications using cross-layer approach. This methodology is based on the trust model, optimal forwarder; selection function, and improved pheromone update model using factors such as distance, energy, and link quality. A friendship-based handshaking scheme is used in cross-layer method between the data link layer and network layer. Experimental setup used for performance evaluation is NS-2.34 with 40 nodes.

Alappatt and Joe Prathap (2020) proposed ACO- BPSO-based hybrid algorithm to enhance the lifetime of the network by recognizing the energy-efficient path. ACO is used to find the optimal energy-efficient path based on residual energy of the nodes, and BPSO is used to optimize the energy consumption among the network nodes. Performance evaluation environment used is the NS-2 simulator with 50 nodes.

Yang et al. (2018) proposed ACO-based routing protocol to balance the energy consumption between network nodes. The algorithm is based on the offset coefficient of transition probability to reduce the number of route discovery packets. The remaining lifetime of node and remaining lifetime of link is calculated and used to design a pheromone generation approach in order to select the best routing path. MATLAB® is used to evaluate performance parameters.

Sethi and Udgata (2010) proposed an ACO-based on-demand routing system that incorporates principles like local retransmissions, and blocking expanding ring search to minimize routing costs and increase packet delivery ratio.

Hussein and Dahnil (2017) proposed a hybrid energy-aware path selection approach based on ACO and lion optimization. The methodology is based on eligible energetic path detection to identify paths between nodes and then adjust the power transmission of the nodes in the network.

Sapienza (2008) presented an evolutionary computation-based approach for improving QoS in MANETs in accordance with IEEE 802.11 specifications. Intelligent approaches such as fuzzy systems in combination with associative memory and evolutionary algorithms are used to improve the network layer and MAC layer implementation details. The suggested model's performance is evaluated using a network simulator.

Hussein and Saadawi (2003) discussed the swarm intelligence-based routing method inspired by ant natural behavior. The basic concept of this method is to focus on optimizing several QoS parameters. The study presented in this paper focused on the equitable distribution of energy in a dynamic network while taking node mobility into account. The algorithm's performance is tested using the OPNET simulator. Figure 9.6 shows operators, various parameters, and application areas of ant colony optimization algorithm.

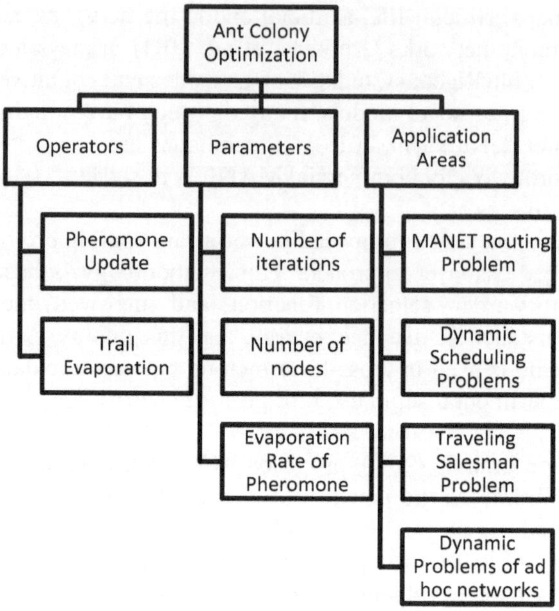

FIGURE 9.6 Working of ant colony optimization.

Devika and Sudha (2020) presented a hybrid algorithm based on earthworm optimization and chronological concept using topology management and clustering. The proposed algorithm works in two phases: in the first phase, a graph is constructed; and in the second phase, clustering of the graph is performed. A fitness function is used based on various parameters like power, energy, distance, mobility, and connectivity for the selection of cluster head. The concept of the Gabriel graph is also used after the selection of optimal clusters to provide a convenient way for the nodes to update the list of neighbors, connectivity, and energy parameters. Experimental setup is done using NS-2.

Reina et al. (2013) showed the topological design of a MANET using a genetic algorithm. The network structure is optimized by taking into account the speed of mobile nodes and the distance between static nodes. The basic concept of this technique is to determine the best position and speed of the node in order to improve network performance.

Wedde et al. (2005) proposed energy-efficient routing method inspired by the foraging behavior of honeybees. The algorithm operated in reactive mode to save energy by using fewer control messages. Using a network simulator, the algorithm's performance is compared to that of existing classical routing methods.

Sharma and Agarwal (2017) proposed to enhance the QoS parameters inside the network, through a hybrid technique integrating the features of automatic neuro-fuzzy inference system (ANFIS) with the idea of Kalman filtration. The aim of this study is to automatically learn and calculate dynamic parametric values.

Vishnu Balan et al. (2015) proposed to identify security assaults in the network with a fuzzy logic-based intrusion detection system. The approach's main focus is identifying black hole and grey hole assaults. To prevent hostile network assaults, this system employs a node blocking technique. The approach's performance is measured using measures like as throughput, packet loss, and jitter.

Revathi (2020), to maximize QoS routing, a hybrid intelligent algorithm was proposed. This method is built on two meta-heuristic techniques inspired by the cuckoo search algorithm and the differential evolution algorithm. The search algorithm handles parameter exploration, and the updating method is developed from differential evolution.

Nandgave-usturge (2020) proposed a security algorithm based on water spider monkey optimization using two phases such as bi-filtering phase to identify secure nodes using trust factors, and in second phase, routing is performed to find the best path depending on the factors like distance and delay. Figure 9.7 shows working of fire fly algorithm by specifying various operators, parameters, and applications areas.

Persis and Robert (2016) proposed a multi-criteria based optimal route path set identification algorithm using firefly optimization. The network performance is enhanced by considering various factors such as load, hop-count, distance, and delay using Pareto principle.

Hassan and Muniyandi (2017) proposed an energy-aware hybrid algorithm based on cellular automata, African buffalo optimization, and genetic algorithm. A set of available routes according to the delay rules are identified using cellular automata; then, the optimum route is selected using two optimization algorithms.

In commercial MANET, designing and developing safe, robust, and efficient routing protocols has become one of the major issues due to link instability, dynamic topologies, and node mobility (Kaur, 2021). The key concern with most routing

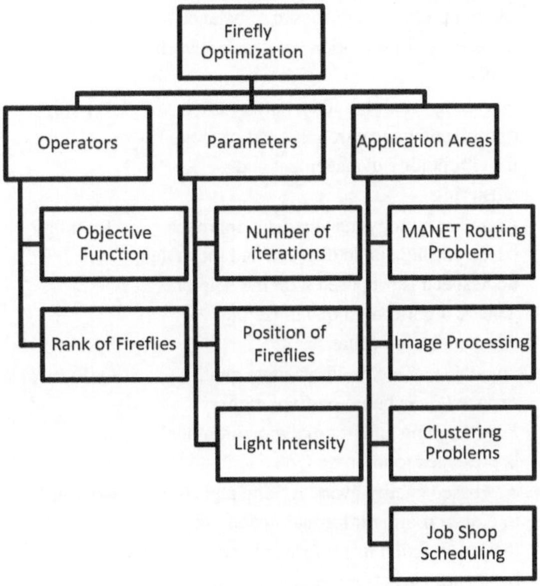

FIGURE 9.7 Working of firefly optimization algorithm.

protocols is that they depend on mobile nodes and are predicated on the presumption that nodes will corporate or behave accordingly, but there may be occasions where a certain group of nodes does not comply appropriately (Devi & Gill, 2018), and these issues need to be addressed properly.

The comparative analysis presented in Table 9.1 shows a variety of bio-inspired techniques such as ant colony, cuckoo search, lion optimization, African buffalo optimization, genetic algorithms, bee colony, water spider monkey, differential evolution,

TABLE 9.1
Comparative Analysis of Evolutionary Computation Techniques for Commercial MANETs in Society 5.0

Evolutionary Technique	Proposed Methodology	Challenge Addressed	Limitations
Ant colony Krishna and Swapna (2020)	Power-aware routing protocol for military applications using cross-layer approach, trust model, function for forward selection and pheromone updation is proposed.	Routing	Evaluation is done with only 40 nodes
Ant colony and binary particle swarm optimization Alappatt and Joe Prathap (2020)	Hybrid algorithm Ant colony Optimization (ACO) and Binary Particle Swarm Optimization (BPSO) based is proposed for increasing the lifetime of the network by recognizing the energy-efficient path	Routing	Performance evaluation is done with only AODV
Ant colony Yang et al. (2018)	Ant colony Optimization and offset coefficient of transition probability-based routing protocol is proposed to balance the energy consumption between network nodes	Routing	Evaluation is done with only 40 nodes
Earthworm optimization Devika and Sudha (2020)	Hybrid algorithm based on earthworm optimization and chronological concept using topology management and clustering	Topology	Congestion control factor not considered
Genetic algorithm Reina et al. (2013)	Dynamic topology management approach based on static node distance and movable node speed is proposed with the help of genetic algorithm to maximize the communication distance	Topology	Only suitability for railway scenarios are considered
Evolutionary programming Sapienza (2008)	A model showing multi-attribute quality of service (QoS) based on Evolutionary Programming and non-polynomial model is presented to improve QoS	QoS	Typical MANET setup testing is not considered
Ant colony Hussein and Saadawi (2003)	ACO based energy-aware routing algorithm to collect routing information and update that information using forward and backward packets respectively to improve QoS is proposed	Routing	Not fit for complex dynamic environments

(Continued)

TABLE 9.1 (*Continued*)
Comparative Analysis of Evolutionary Computation Techniques for Commercial MANETs in Society 5.0

Evolutionary Technique	Proposed Methodology	Challenge Addressed	Limitations
Bee colony Wedde et al. (2005)	Proposed energy-efficient routing method inspired by the foraging behavior of honeybees based on two phases route discovery using scouts and transmit data using foragers while providing quality route path	Routing	Testing is done with only 50 nodes
Fuzzy logic Sharma and Agarwal (2017)	Hybrid technique integrating the features of Automatic Neuro-Fuzzy Inference System (ANFIS) with the idea of Kalman filteration was proposed	QoS	Implementation is done with a very less number of nodes
Fuzzy logic Vishnu balan et al. (2015)	Fuzzy logic-based intrusion detection system is proposed to identify malicious behavior of nodes from grey and black hole attacks using the node blocking method	Security	Route adjustment not considered
Cuckoo search Revathi (2020)	A hybrid algorithm based on the cuckoo search and differential evolution method with a position update mechanism is developed.	Routing	Not suitable for real-time networks
Ant colony Sethi and Udgata (2010)	To enhance the packet delivery ratio, an Ant-based protocol with blocking expanding ring search is proposed.	Routing	Performance comparison is done with AODV and DSR
Water spider monkey optimization Nandgave-usturge (2020)	Water spider monkey optimization based routing protocol using trust factors is proposed to provide a secure route among nodes	Routing	Only three performance metrics are considered
African Buffalo optimization and genetic algorithm Hassan and Muniyandi (2017)	Energy and delay aware routing protocol is proposed based on cellular automata, genetic algorithm, and African buffalo optimization	Routing	Not suitable for scalable complex networks
Ant colony and lion optimization Hussein & Dahnil (2017)	Energy-aware routing approach to improve path selection in the network is proposed based on ant colony and lion optimization techniques	QoS	Performance comparison is done with only two protocols
Fire fly optimization Persis and Robert (2016)	Firefly optimization based routing protocol using Pareto principle and impact factors like hop count, cost, distance, etc. is proposed	Routing	More impact factors can be added like bandwidth and signal strength

fuzzy logic, evolutionary programming, and earthworm optimization implemented in commercial MANETs for enhancing network performance in society 5.0. The table shows there are some issues and challenges that still need to be addressed in order to enhance network lifetime and to provide the best services to a varied range of users.

As per the literature survey, the existing routing protocols need improvement in network with respect to complexity, scalability, mobility, load balancing, and congestion control issues. Due to the continuous use of ubiquitous computing, the number of connected devices is increasing day by day, and it will reach around 13.7 billion by 2024. In order to cater to this gigantic number of devices in commercial MANETs, the routing protocols need to be much adaptive, dynamic, and intelligent, having auto-configuration properties to handle link failures and frequent network partitions. EC or bio-inspired techniques are best suited for handling these challenges, as these approaches are inspired by natural organisms having features of self-configuring and self-healing, which is very helpful in improving performance of commercial MANETs in society 5.0.

9.6 AREAS FOR STRENGTHENING MANET PERFORMANCE WITH EVOLUTIONARY COMPUTATION TECHNIQUES

As the number of users of mobile networks is increasing drastically, there is a need to design and develop dynamic routing protocols having additional features such as energy-saving, topology management, link failure, QoS, and so on. The inclusion of these factors makes the routing protocols adaptive according to the changing environment conditions. However, present communication technologies are insufficient for such diverse networks.

Self-organization methods that can deal with heterogeneity, dynamic nature, resource restrictions, scalability, failures, and other issues are required (Dorronsoro et al., 2014). CI-based methods that are inspired by natural organisms play an important role in handling these challenges, as these natural creatures are equipped with self-organization and self-healing properties.

Following are the areas for strengthening MANET performance with evolutionary computation techniques:

1. **Broadcasting**: In MANETs, broadcasting is the process of transmitting a message from a source node to the rest of the network's nodes (Mkwawa & Kouvatsos, 2011). It is one of the most essential operations since it is used by many applications and protocols. In MANETs, it is impossible to ensure complete coverage because of issues like network partitions, dynamic topology, unexpected properties of medium, mobility, varying channel and signal strength with respect to dynamic environments, and so forth. Because of node mobility and limited system resources, broadcasting in MANETs presents greater hurdles than in wired networks. Due to these factors, there is no one ideal method that applies to all cases. The capacity to properly distribute information across network nodes is a crucial aspect that determines the quality of a MANET (Mkwawa & Kouvatsos, 2011). The comparative

analysis of various EC techniques elaborated in Table 9.1 suggests that EC methods are promising to result in effective message distribution thereby strengthening the performance in commercial MANET.

2. **Topology Management**: Topology Management is one of the most significant strategies used in MANETs for reducing energy consumption and signal interference which is required to increase network lifetime. Given that network nodes can modify their capacity of transmission, topology control is concerned with determining the transmission range that offers a desirable attribute to the network (e.g., connectivity). The primary aim of topology control through bio-inspired methods is to minimize node energy consumption, enhance network capacity, and extend network lifetime.
3. **Multipath Routing**: Multipath routing is a routing approach that uses many alternate pathways through a network at the same time. This can result in a number of advantages, including fault tolerance, greater bandwidth, and enhanced security. For commercial MANETs, multi-path routing is a potential routing approach. Concurrent multipath routing (CMR) is frequently used to increase performance or fault tolerance by managing and utilizing many accessible routes for the transmission of data streams at the same time. Load balancing is achieved via multipath routing, and it is more resilient to route failures.
4. **Clustering**: Clustering indigenously is evolved from evolutionary computation and is the process of organizing the network into groups or re-organizing all nodes into tiny clusters based on their geographic proximity such that some nodes (typically known as cluster heads) have a specific function in governing the nearby devices. The cluster head is in charge of overseeing cluster activities such as cluster process management, routing table updates, and route discovery (Devika & Sudha, 2020). It provides benefits like robust routing tables, making scalable protocols, increasing the overall lifespan of network, and so on.
5. **Protocol** Optimization: Due to the dynamic nature of commercial MANETs, the factors like unpredictability, mobility, and changing topology play an important role in designing and optimization of communication protocols. These protocols often rely on various characteristics that adjust their behavior to the present situation. The protocol's performance is extremely sensitive to slight changes in the collected information of configuration settings in these dynamic environments. As a result, precisely tuning them for optimal communication within the network while bearing in mind these unpredictable impact factors is a daunting task. In order to enhance the network performance in commercial MANETs, there is a need to optimize QoS, available resources, bandwidth, energy consumption etc.
6. **Mobility**: Creating test beds for MANETs is not only expensive but also time-consuming. Reproducible experiments, extensive networks for assessing scalability, are required for creating and testing methods. A test bed may be possible for small static networks, but not for commercial MANETs.

7. **Selfish Behavior**: Commercial MANET nodes function as access points, passing information to other network nodes. All the nodes in ad hoc networks are battery operated, and to handle these communication tasks, nodes consume battery power. In order to save the battery, each and every node particularly in commercial MANET wants to discard any messages that aren't intended for the node itself. It is considered selfish conduct and a serious concern in such MANETs, which needs to be addressed for varied applications in society 5.0.

9.7 CONCLUSION

The constant availability of smart devices and new mobile applications has resulted in a significant increase in mobile network users. Mobile applications are currently one of the most popular choices for businesses such as commerce, academics, industry, government, and entertainment. People want to be able to complete everyday chores using commercial MANETs using ubiquitous computing. This enormous growth in mobile network users offers a significant problem in delivering QoS to a broad variety of users in terms of link connection, higher throughput, and quick processing in society 5.0. To address these difficulties, evolutionary computation techniques for strengthening the performance of commercial MANETs in society 5.0 are required. This chapter discusses the use of evolutionary computation techniques to handle these issues of implementing commercial MANETs in society 5.0. This study investigates computational intelligence approaches for improving commercial MANET performance. The comparative analysis in this chapter provides insight into bio-inspired intelligent methods as well as their limitations in the context of commercial MANETs for society 5.0. The incorporation of evolutionary computing techniques is required for improving performance in terms of network lifetime, security, congestion control, and energy efficiency while bearing in mind the adaptability and scalability factors in commercial MANETs, and this chapter highlights the major areas for the same.

REFERENCES

Alappatt, V., & Joe Prathap, P. M. (2020). A hybrid approach using ant colony optimization and binary particle swarm optimization (ACO: BPSO) for energy efficient multi-path routing in MANET. *Proceedings -2020 Advanced Computing and Communication Technologies for High Performance Applications, ACCTHPA 2020*, 175–178. Doi: 10.1109/ACCTHPA49271.2020.9213196.

Devi, M., & Gill, N. S. (2018). Study of mobile ad hoc network routing protocols in smart environment. *International Journal of Applied Engineering Research, 13*(16), 12968–12975.

Devi, M., & Gill, N. S. (2019a). Mobile ad hoc networks and routing protocols in IoT enabled smart environment: a review. *Journal of Engineering and Applied Sciences, 14*(3), 802–811. Doi: 10.3923/jeasci.2019.802.811.

Devi, M., & Gill, N. S. (2019b). Novel algorithm for enhancing manet protocol in smart environment. *International Journal of Innovative Technology and Exploring Engineering, 8*(10), 1830–1835. Doi: 10.35940/ijitee.J9214.0881019.

Devika, B., & Sudha, P. N. (2020). Power optimization in MANET using topology management. *Engineering Science and Technology, an International Journal, 23*(3), 565–575. Doi: 10.1016/j.jestch.2019.07.008.

Dorronsoro, B., Ruiz, P., Danoy, G., Pigne, Y., & Bouvry, P. (2014). *Evolutionary Algorithms for Mobile Ad Hoc Networks*, Hoboken, New Jersey: John Wiley & Sons, Inc.

Fogel, D. B. (2000). What is evolutionary computation? *IEEE Spectrum*, *37*(2). Doi: 10.1109/6.819926.

Hassan, M. H., & Muniyandi, R. C. (2017). An improved hybrid technique for energy and delay routing in mobile ad-hoc networks. *International Journal of Applied Engineering Research*, *12*(1), 134–139.

Hussein, O., & Saadawi, T. (2003). Ant routing algorithm for mobile ad-hoc networks (ARAMA). *IEEE International Performance, Computing and Communications Conference, Proceedings*, 281–290. https://doi.org/10.1109/pccc.2003.1203709.

Hussein, S. A., & Dahnil, D. P. (2017). A new hybrid technique to improve the path selection in reducing energy consumption in mobil AD-HOC. *International Journal of Applied Engineering Research*, *12*(3), 277–282.

Jain, L. C., Sato-ilic, M., Virvou, M., Tsihrintzis, G. A., & Balas, V. E. Eds. (n.d.). *Studies in Computational Intelligence,* Volume 137. Berlin Heidelberg: Springer-Verlag.

Kaur, R. (2021). Evolutionary computation techniques for intelligent computing in commercial mobile adhoc networks. *International Journal of Next-Generation Computing*, *12*(2).

Krishna, G. S., & Swapna, P. (2020). A new cross layer design for efficient routing protocol using ant colony optimization. *01*, 133–139.

Kulkarni, R. V., Forster, A., & Venayagamoorthy, G. K. (2011). Computational intelligence in wireless sensor networks: a survey. *IEEE Communications Surveys and Tutorials*, *13*(1), 68–96.

Mbarushimana, C., & Shahrabi, A. (2007). Comparative study of reactive and proactive routing protocols performance in mobile ad hoc networks. *Proceedings -21st International Conference on Advanced Information Networking and Applications Workshops/Symposia, AINAW'07, 1*, 679–684. Doi: 10.1109/AINAW.2007.123.

Mkwawa, I. H. M., & Kouvatsos, D. D. (2011). Broadcasting methods in MANETS: an overview. *Lecture Notes in Computer Science (Including Subseries Lecture Notes in Artificial Intelligence and Lecture Notes in Bioinformatics)*, *5233*(NoE 028022), 767–783. Doi: 10.1007/978-3-642-02742-0_32.

Munot, H., & Kulkarni, P. H. (2016). Survey on computational intelligence based routing protocols in WSN. *International Research Journal of Engineering and Technology (IRJET)*, *3*(9), 122–127.

Nandgave-usturge, S. (2020). Water spider monkey optimization algorithm for trust-based MANET secure routing in IoT. *International Journal of Scientific Research and Engineering Trends*, *6*(2), 980–984.

Patel, D. N., Patel, S. B., Kothadiya, H. R., Jethwa, P. D., & Jhaveri, R. H. (2015). A survey of reactive routing protocols in MANET. *2014 International Conference on Information Communication and Embedded Systems, ICICES 2014, May 2016.* Doi: 10.1109/ICICES.2014.7033833.

Pedrycz, W., Sillitti, A., & Succi, G. (2016). Computational intelligence: an introduction. *Studies in Computational Intelligence*, *617*(Ci), 13–31. Doi: 10.1007/978-3-319-25964-2_2.

Persis, D. J., & Robert, T. P. (2016). Reliable mobile ad-hoc network routing using firefly algorithm. *International Journal of Intelligent Systems and Applications*, *8*(5), 10–18. Doi: 10.5815/ijisa.2016.05.02.

Reina, D. G., Toral, S. L., Barrero, F., Bessis, N., & Asimakopoulou, E. (2013b). The role of ad hoc networks in the Internet of Things: a case scenario for smart environments. *Studies in Computational Intelligence*, *460*(January), 89–113. Doi: 10.1007/978-3-642-34952-2_4.

Reina, D. G., Toral Marin, S. L., Bessis, N., Barrero, F., & Asimakopoulou, E. (2013a). An evolutionary computation approach for optimizing connectivity in disaster response scenarios. *Applied Soft Computing Journal*, *13*(2), 833–845. Doi: 10.1016/j.asoc.2012.10.024.

Revathi, P. (2020). Quality of service routing in manet using a hybrid intelligent algorithm inspired by ant colony optimization. *International Journal of Advanced Science and Technology*, *29*(3), 4033–4046.

Sapienza, T. J. (2008). Optimizing quality of service of wireless mobile ad-hoc networks using evolutionary computation. CSIIRW'08-*4th* Annual Cyber Security and Information Intelligence Research Workshop: Developing Strategies to Meet the Cyber Security and Information Intelligence Challenges Ahead, 1–5. Doi: 10.1145/1413140.1413182.

Sethi, S., & Udgata, S. K. (2010). The efficient ant routing protocol for MANET. *International Journal on Computer Science and Engineering*, *2*(7), 2414–2420.

Sharma, S., & Agarwal, R. (2017). Optimizing QoS parameters using computational intelligence in MANETS. *Proceeding - IEEE International Conference on Computing, Communication and Automation, ICCCA 2017*, 2017-January, 708–715. Doi: 10.1109/CCAA.2017.8229893.

Siddique, N. and Adeli, H. (2015). *Computational Intelligence Computational Intelligence Synergies of Fuzzy Logic*. United Kingdom: John Wiley & Sons, Ltd.

Taneja, K., & Patel, R. (2003). Mobile ad hoc networks. *National Conference on Challenges & Opportunities in Information Technology (COIT-2007)*, 133–135. Doi: 10.1201/9780203011690.ch16.

Taneja, K., Taneja, H., & Kumar, R. (2018). Multi-channel medium access control protocols: review and comparison. *Journal of Information and Optimization Sciences*, *39*(1), 239–247. Doi: 10.1080/02522667.2017.1372921.

Vishnu Balan, E., Priyan, M. K., Gokulnath, C., & Usha Devi, G. (2015). Fuzzy based intrusion detection systems in MANET. *Procedia Computer Science*, *50*, 109–114. Doi: 10.1016/j.procs.2015.04.071.

Wedde, H. F., Farooq, M., Pannenbaecker, T., Vogel, B., Mueller, C., Meth, J., & Jeruschkat, R. (2005). Bee AdHoc. *Proceedings of the 7th Annual Conference on Genetic and Evolutionary Computation*, 153–160. Doi: 10.1145/1068009.1068034.

Weitnauer, M. A., Rexford, J., Laneman, N., Bloch, M., Griljava, S., Ross, C., & Chang, G. K. (2017). Smart wireless communication is the cornerstone of smart infrastructures. *ArXiv*, 1–5.

Yang, D., Xia, H., Xu, E., Jing, D., & Zhang, H. (2018). Energy-balanced routing algorithm based on ant colony optimization for mobile ad hoc networks. *Sensors (Basel, Switzerland)*, *18*(11), 1–19. Doi: 10.3390/s18113657.

Zhang, B., Wu, Y., Lu, J., & Du, K.-L. (2011). Evolutionary computation and its applications in neural and fuzzy systems. *Applied Computational Intelligence and Soft Computing*, *2011*, 1–20. Doi: 10.1155/2011/938240.

10 Availability Optimization of a Rice Finishing and Grading System Using Evolutionary Computation Techniques

Parveen Kumar Saini, Vikas Modgil, and Rajiv Khanduja

CONTENTS

10.1 INTRODUCTION

Continuous improvement in the performance of any industrial system is highly desirable. The word reliability originates from failure itself. Frequent failures in the system make it unreliable for use. The cost of an unreliable system is so high in terms of economy and safety. The failures in the system can lead to serious accidents. It is thus required in industries to use the Reliability Availability and Maintainability RAM approach for Continuous improvement in the performance of any industrial system is highly desirable. To have this, the overall reliability and availability of the various systems/subsystems used may be maintained at the highest working level. It is not possible to have fault-free operation because the failure can never be avoided; rather, the efforts can be exerted to minimize it to an extent using various kinds of reliable components, proper preventive maintenance of the components or subsystem, and by

DOI: 10.1201/9781003158165-10

inculcating positive environment in the work area. To analyze the performance of a system in various process industries, a number of performance measures are used regarding reliability and maintainability. The availability (i.e. inherent, operational, or steady state), is one of the indexes to measure the system performance.

Reliability of the system is the ability to perform its desired function over a stated time duration, under various operational and environmental conditions.

$$R(t) = e^{-\lambda t}$$

Where λ is the mean failure rate i.e., failures per hour and "t" is the time period in hours.

The availability of a system can be stated as the operational readiness of the system or its ability to perform its function at an instant time and for a stated time period.

Availability of a system is of different types : inherent availability, achievable availability, and operational availability. Operational availability of a system in process industries includes both the planned and the unplanned maintenance time and also includes the time loss in administration and operational logistics. Achievable availability includes only the planned and unplanned maintenance time. Whereas only unscheduled maintenance time is considered in inherent availability. Thus, the availability of a system is concerned with its reliability and maintainability.

Mathematically availability function $A(t)$ can be expressed as

$$A(t) = \frac{\text{Up time}}{\text{Uptime} + \text{Down time}}$$

The maintainability of a system is its ability to retain or restore into a working state to perform the intended function satisfactorily under stated operational and environmental conditions by using maintenance with prescribed procedures. From the above definitions, it is clear that system availability can be enhanced either by increasing the reliability at the design stage or by improving the maintainability at the operational stage or both.

The steady state availability (SSA) is considered an important factor for the performance measures of the industrial system. The various subsystems of the rice finishing and grading system (RFGS) are connected in a mixed layout. SSA of an RFGS has been modeled and optimized using evolutionary techniques. Due to many reasons, wrong design, inadequate maintenance, and faulty operations, sometimes the industrial systems undergo a failure state. The SSA modeling and optimization is required for a long time working of the system with high-performance level in industries to increase the productivity and reduce the production cost. The SSA modeling and optimization of RFGS of a Rice Milling Plant could do using different modeling techniques like FMEA, Monte Carlo Simulation, RBD, FTA, and Petri Nets. Similarly, the performance of the industrial system has been optimized using various evolutionary techniques such as Ant Bee Colony (ABC), Difference Evaluation (DE), GA, ACO, PSO, Jaya, Teacher Learner Based Optimization (TLBO), and many more.

The various applications of PSO have been described in the literature part, the advantages of PSO over other meta-heuristic search techniques have been studied by (Waintraub et al., 2009). In the current scenario, the mechanical systems are more complex due to advances in technology; these systems became the focus of attraction among researchers in this area. Dhillon and Singh (1981) applied the MA for the SSA analysis, taking failure and repair rates distributed exponentially. Juang et al. (2008) applied an evolutionary technique i.e. GA to optimize the availability of a hybrid system. The reliability evaluation and optimization methods for non-repairable different subsystems which were cold standby have been developed by Azaron et al. (2009). Kumar and Tewari (2011) suggested an availability optimization model for a fertilizer plant using GA to cool down the CO_2 circulating within the system. Bose et al. (2012) on the basis of the Preventive Maintenance (PM) schedule of a power plant examined the RAM. The finding of the paper was that the economizer was the most critical subsystem. Khanduja et al. (2012) developed an availability model for an important system of the paper industry, i.e. stock preparation unit, using MA that also optimized the performance of the said system by applying GA. Arabi et al. (2012) suggested an availability model for performance optimization SA technique. Kachitvichyanukul (2012) studied the differences and similarities between three EA, namely DE, PSO, and GA. Garg (2013) developed a technique for analyzing the reliability of the washing system of a paper plant unit using intuitionistic fuzzy set (IFS) and PSO.

Modgil et al. (2013) dealt with the SSA model using MA applied to the shoe industry and evaluated the time-based system availability. Usubamatov et al. (2013) suggested a mathematical model for automated production lines considering availability machines as an important parameter. Aggarwal et al. (2014) evaluate the availability of a butter oil production (BOP) system in a dairy plant by using MA, and to calculate the MTBF, the Runge-Kutta approach has been applied. Li and Peng, (2014) evaluated the system availability and cost of operation for hybrid industrial systems. Further, GA has been used as an optimizing tool. Aggarwal et al. (2015) dealt with SSA analysis of a fertilizer plant by applying MA considering the exponential distribution for governing parameters and proposed a suitable maintenance plan to improve the system. Kumar and Tewari (2017) analyzed the SSA of a carbonated soft drink glass bottle-filling system of a beverage plant and optimized the SSA using PSO. Malik and Tewari (2018) carried out performance modeling of a water flow system which is a subsystem of coal-operated thermal power plants. Based upon the performance modeling and failure and repair rates (FRR) of different components, the repair priorities were assigned to various components using the probabilistic approach. Kumar et al. (2018) carried out the analysis of real industry-operated cooling towers, both experimentally and by two-dimensional computational fluid dynamics. Influence of temperature of inlet fan, water flow rate, and air-flow rate were assessed on the working of an induced draft cooling tower. Bahl et al. (2018) developed a methodology to carry out the availability analysis of distillery plants using Petri nets (PN). The study shows the effect of various FRR and availability with respect to repair facilities employed.

Ahmadi et al. (2019) performed the reliability evaluation of material hauling system of the earth pressurized balance tunnel boring machine using the failure and repair data, followed by the statistical techniques used for performance modeling. RAM analysis performed by Rabbani et al. (2018) evaluated the availability of the combined heating and power generation unit and optimized the performance by applying GA and PSO. Diyaley and Chakraborty (2019) applied various techniques to find out the optimal value of various parameters like feed, speed, and cut for milling operation. Kumar et al. (2019) developed a Petri net model to carry out the performance analysis of the subsystem of a milk plant, i.e. refrigeration system. An attempt has been made to assign repair priorities to meet operational objectives, taking spare parts and repair facilities into consideration. Malik and Tewari (2020) developed a mathematical model to improve the performance of a coal crushing unit of a thermal power plant using PSO.

In this chapter, the SSA model has been developed for the RFGS of a rice milling plant. After using the PSO and GA, the availability optimization has been performed, and the comparison of both approaches has been presented.

10.2 INDUSTRIAL SYSTEM DESCRIPTION

RFGS is one of the important sections of the Rice Milling Plant. The system comprises six subsystems viz. Abrasive Whitener (CC), Rotary Shifter (DD), Sizer (EE), Polisher (FF), Sortex (GG), and Grader (HH), which are subjected to major as well minor failures and has been consider for analysis. Figure 10.1 shows the flow diagram of RFGS.

Abrasive Whitener: The rice coming out from the paddy separator is fed to the whitener, which removes the yellow film from the rice due to abrasive action.

Rotary Shifter: This subsystem separates the full-length and short-length rice.

Polisher: It is used to remove the yellow film from the rice due to friction and give extra whiteness to the rice.

Sizer: The work of sizer is to separate the rice of various sizes after polisher.

Sortex: It is used to distinguish rice of black or other color that is not desirable. It works on a colour sensor and is pneumatically operated.

Grader: This subsystem is utilized to separate the undersized rice after passing through the Sortex. Finally, rice coming out from the grader is of the same size and is ready for packing.

10.3 NOTATIONS AND ASSUMPTIONS

To develop the States, Transition Diagram (TD) notations and assumptions used are as under:

Subsystem CC: This subsystem has one Abrasive whitener subjected to failures.

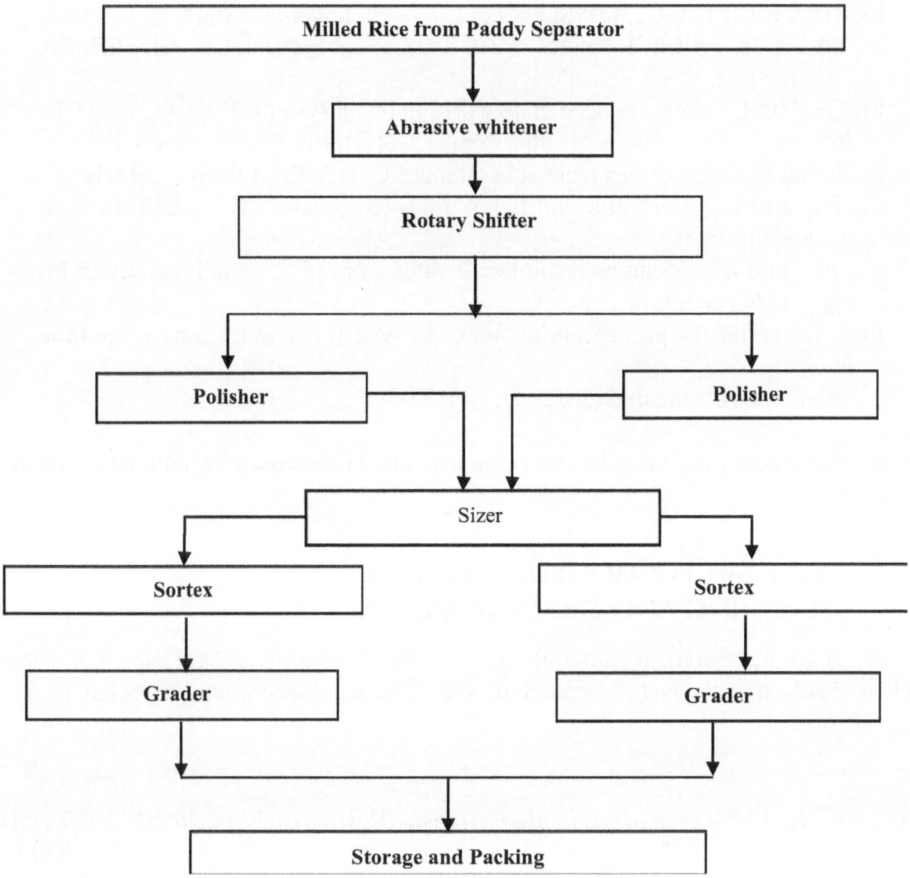

FIGURE 10.1 Process flow diagram of rice finishing and grading system.

Subsystem DD: It consists of one Rotary Shifter subjected to failures.
Subsystem EE: There is one Sizer subjected to failures.
Subsystem FF: This subsystem consists of two Polishers subjected to failure.
Subsystem GG: It consists of two Sortexes subjected to failures arranged in parallel.
Subsystem HH: There are two Graders subjected to failures, also arranged in parallel.

: Represents working state.

: Represents reduced state.

: Represents failed state.

CC, DD, EE, FF, GG, HH: Subsystems in a good working state.
cc, dd, ee, ff, gg, hh: Indicates failed state of subsystems CC, DD, EE, FF, GG, and HH.
Ff, Gg, Hh: Indicates reduced state working of subsystems FF, GG, and HH
λ_i: Represents failure rates of subsystems CC, DD, EE, FF, GG, and HH
μ_i: Represents repair rates from subsystems CC, DD, EE, FF, GG and HH
λ'_{37}, λ'_{38}, **and** λ'_{39}: Mean constant failure rates from states Ff, Gg, and Hh to ff, gg, and hh states.
μ'_{37}, μ'_{38}, **and** μ'_{39}: Mean constant repair rates from ff, gg, and hh states to Ff, Gg, and Hh states.
Pi(t): Describes the probability of state that system lies in ith state when time is 't'.
′: Dash represents the derivatives w.r.t. 't'

Using abovementioned notations and assumptions, TD has been formulated as shown in Figure 10.2.

10.4 AVAILABILITY MODELING OF RICE FINISHING AND GRADING SYSTEM

The various mathematical equations are derived on the basis of MA approach using TD and solved recursively to calculate SSA. The following equations related with the TD are:

$$P_1'(t) + \left(\sum_{i=34}^{39} \lambda_i\right) P_1(t) = \mu_{37} P_{38}(t) + \mu_{38} P_{39}(t) + \mu_{39} P_{40}(t) + \mu_{35} P_2(t) + \mu_{36} P_3(t) + \mu_{37} P_4(t) \tag{10.1}$$

$$P_2'(t) + \mu_{34} P_2(t) = \lambda_{34} P_1(t) \tag{10.2}$$

$$P_3'(t) + \mu_{35} P_3(t) = \lambda_{35} P_1(t) \tag{10.3}$$

$$P_4'(t) + \mu_{36} P_4(t) = \lambda_{36} P_1(t) \tag{10.4}$$

$$P_5'(t) + \mu_{34} P_5(t) = \lambda_{34} P_{38}(t) \tag{10.5}$$

$$P_6'(t) + \mu_{35} P_6(t) = \lambda_{35} P_{38}(t) \tag{10.6}$$

$$P_7'(t) + \mu_{36} P_7(t) = \lambda_{36} P_{38}(t) \tag{10.7}$$

$$P_8'(t) + \mu'_{37} P_8(t) = \lambda'_{37} P_{38}(t) \tag{10.8}$$

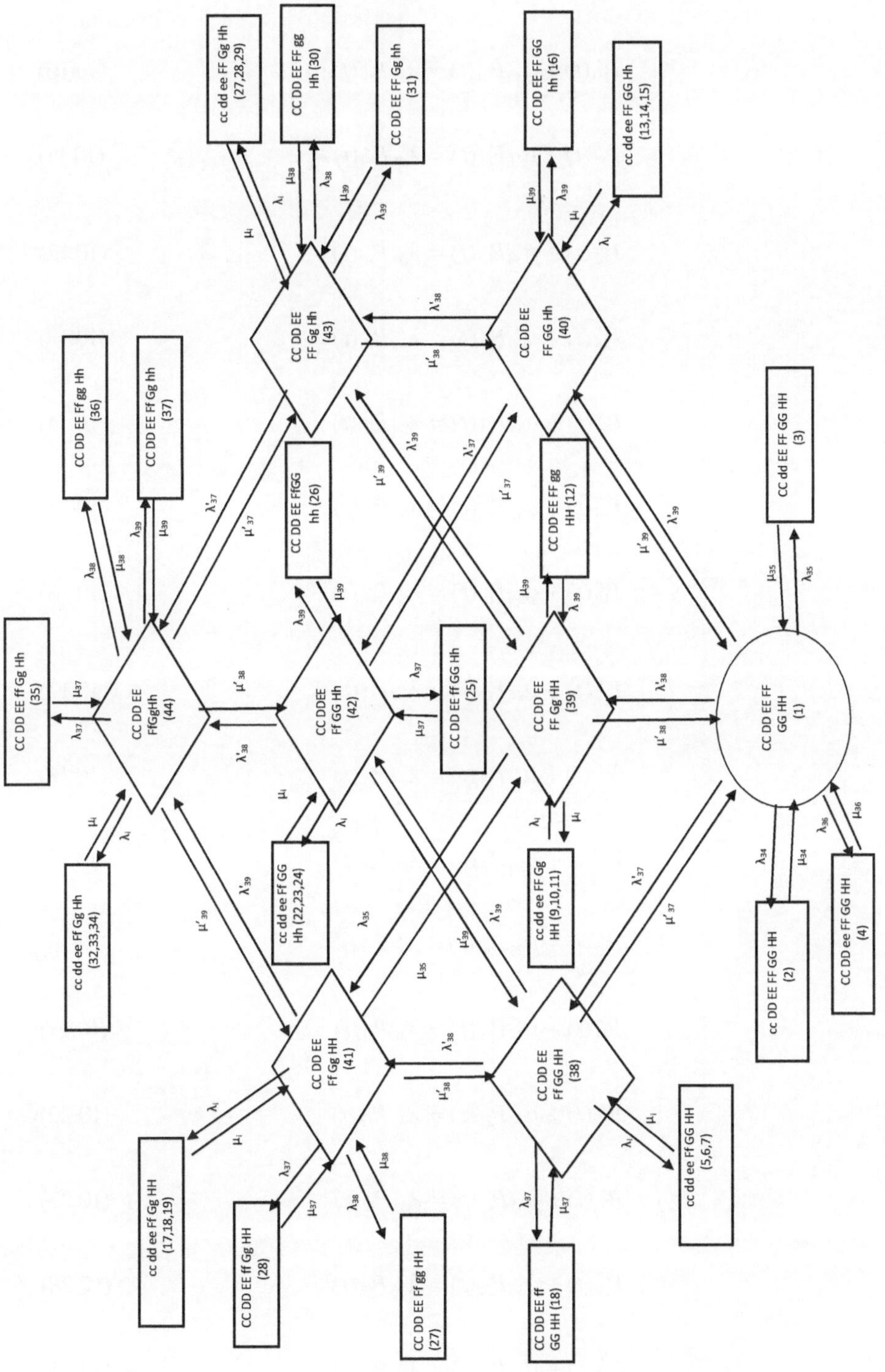

FIGURE 10.2 States transition diagram of RFGS.

$$P_9'(t) + \mu_{34} P_9(t) = \lambda_{34} P_{39}(t) \tag{10.9}$$

$$P_{10}'(t) + \mu_{35} P_{10}(t) = \lambda_{35} P_{39}(t) \tag{10.10}$$

$$P_{11}'(t) + \mu_{36} P_{11}(t) = \lambda_{36} P_{39}(t) \tag{10.11}$$

$$P_{12}'(t) + \mu_{38}' P_{12}(t) = \lambda_{38}' P_{39}(t) \tag{10.12}$$

$$P_{13}'(t) + \mu_{39}' P_{13}(t) = \lambda_{39}' P_{40}(t) \tag{10.13}$$

$$P_{14}'(t) + \mu_{34} P_{14}(t) = \lambda_{34} P_{40}(t) \tag{10.14}$$

$$P_{15}'(t) + \mu_{35} P_{15}(t) = \lambda_{35} P_{40}(t) \tag{10.15}$$

$$P_{16}'(t) + \mu_{36} P_{16}(t) = \lambda_{36} P_{40}(t) \tag{10.16}$$

$$P_{17}'(t) + \mu_{34} P_{17}(t) = \lambda_{34} P_{41}(t) \tag{10.17}$$

$$P_{18}'(t) + \mu_{35} P_{18}(t) = \lambda_{35} P_{41}(t) \tag{10.18}$$

$$P_{19}'(t) + \mu_{36} P_{19}(t) = \lambda_{36} P_{41}(t) \tag{10.19}$$

$$P_{20}'(t) + \mu_{37}' P_{20}(t) = \lambda_{37}' P_{41}(t) \tag{10.20}$$

$$P_{21}'(t) + \mu_{38}' P_{21}(t) = \lambda_{38}' P_{41}(t) \tag{10.21}$$

$$P_{22}'(t) + \mu_{34} P_{22}(t) = \lambda_{34} P_{42}(t) \tag{10.22}$$

$$P_{23}'(t) + \mu_{35} P_{23}(t) = \lambda_{35} P_{42}(t) \tag{10.23}$$

$$P_{24}'(t) + \mu_{36} P_{24}(t) = \lambda_{36} P_{42}(t) \tag{10.24}$$

$$P_{25}'(t) + \mu_{39}' P_{25}(t) = \lambda_{39}' P_{42}(t) \tag{10.25}$$

$$P'_{26}(t) + \mu'_{37} P_{26}(t) = \lambda'_{37} P_{42}(t) \quad (10.26)$$

$$P'_{27}(t) + \mu_{34} P_{27}(t) = \lambda_{34} P_{43}(t) \quad (10.27)$$

$$P'_{28}(t) + \mu_{35} P_{28}(t) = \lambda_{35} P_{43}(t) \quad (10.28)$$

$$P'_{29}(t) + \mu_{36} P_{29}(t) = \lambda_{36} P_{43}(t) \quad (10.29)$$

$$P'_{30}(t) + \mu'_{38} P_{30}(t) = \lambda'_{38} P_{43}(t) \quad (10.30)$$

$$P'_{31}(t) + \mu'_{39} P_{31}(t) = \lambda'_{39} P_{43}(t) \quad (10.31)$$

$$P'_{32}(t) + \mu_{34} P_{32}(t) = \lambda_{34} P_{44}(t) \quad (10.32)$$

$$P'_{33}(t) + \mu_{35} P_{33}(t) = \lambda_{35} P_{44}(t) \quad (10.33)$$

$$P'_{34}(t) + \mu_{36} P_{34}(t) = \lambda_{36} P_{44}(t) \quad (10.34)$$

$$P'_{35}(t) + \mu'_{37} P_{35}(t) = \lambda'_{37} P_{44}(t) \quad (10.35)$$

$$P'_{36}(t) + \mu'_{38} P_{36}(t) = \lambda'_{38} P_{44}(t) \quad (10.36)$$

$$P'_{37}(t) + \mu'_{39} P_{37}(t) = \lambda'_{39} P_{44}(t) \quad (10.37)$$

$$P_{38}{}'(t) + (\textstyle\sum_{i=34}^{36} \lambda_i)_{38}(t) + \lambda_{38} P_{38}(t) + \lambda_{39} P_{38}(t) + \lambda'_{37} P_{38}(t) + \mu_{37} P_{38}(t) = \mu_{38} P_{41}(t) +$$
$$\mu_{39} P_{42}(t) + \mu_{37} P_1(t) + \mu_{36} P_7(t) + \mu'_{37} P_8(t) + \mu_{34} P_5(t) + \mu_{35} P_6(t) \quad (10.38)$$

$$P_{39}{}'(t) + (\textstyle\sum_{i=35}^{38} \lambda_i)_{39}(t) + \lambda'_{38} P_{39}(t) + \mu_{38} P_{39}(t) = \lambda_{38} P_1(t) + \lambda'_{38} P_{12}(t) + \mu_{39} P_{43}(t) +$$
$$\mu_{37} P_{41}(t) + \mu_{34} P_9(t) + \mu_{35} P_{13}(t) + \mu_{36} P_{11}(t) \quad (10.39)$$

$$P_{40}{}'(t) + (\textstyle\sum_{i=34}^{38} \lambda_i)_{40}(t) + \lambda'_{39} P_{40}(t) + \mu_{39} P_{40}(t) = \lambda_{39} P_1(t) + \mu'_{39} P_{13}(t) + \mu_{37} P_{42}(t) +$$
$$\mu_{38} P_{43}(t) + \mu_{34} P_{14}(t) + \mu_{35} P_{15}(t) + \mu_{36} P_{16}(t) \quad (10.40)$$

$$P_{41}'(t) + (\textstyle\sum_{i=34}^{36} \lambda_i)_{41}(t) + \lambda_{38}' P_{41}(t) + \lambda_{37}' P_{41}(t) = \lambda_{39} P_{41}(t) + \mu_{37} P_{41}(t) + \mu_{38} P_{41}(t) = \lambda_{37} P_{39}(t) + \lambda_{38} P_{38}(t) + \mu_{38}' P_{21}(t) + \mu_{39} P_{44}(t) + \mu_{37}' P_{20}(t) + \mu_{34} P_{17}(t) + \mu_{35} P_{18}(t) + \mu_{36} P_{19}(t) \quad (10.41)$$

$$P_{42}'(t) + (\textstyle\sum_{i=34}^{36} \lambda_i)_{42}(t) + \lambda_{39}' P_{42}(t) + \lambda_{37}' P_{42}(t) = \lambda_{38} P_{42}(t) + \mu_{37} P_{42}(t) + \mu_{39} P_{42}(t) = \lambda_{37} P_{42}(t) + \lambda_{39} P_{38}(t) + \mu_{38} P_{44}(t) + \mu_{39}' P_{25}(t) + \mu_{37}' P_{26}(t) + \mu_{34} P_{22}(t) + \mu_{35} P_{23}(t) + \mu_{36} P_{44}(t) \quad (10.42)$$

$$P_{43}'(t) + (\textstyle\sum_{i=34}^{37} \lambda_i)_{41}(t) + \lambda_{39}' P_{43}(t) + \lambda_{38}' P_{43}(t) = \mu_{38} P_{43}(t) + \mu_{39} P_{43}(t) = \lambda_{38} P_{40}(t) + \mu_{34} P_{27}(t) + \mu_{39}' P_{31}(t) + \mu_{38}' P_{30}(t) + \mu_{35} P_{28}(t) + \mu_{36} P_{29}(t) + \mu_{37} P_{44}(t) + \lambda_{39} P_{39}(t) \quad (10.43)$$

$$P_{44}'(t) + (\textstyle\sum_{i=34}^{36} \lambda_i)_{44}(t) + \lambda_{39}' P_{44}(t) + \lambda_{38}' P_{44}(t) = \lambda_{37}' P_{44}(t) + \mu_{37} P_{44}(t) + \mu_{38} P_{44}(t) + \mu_{39} P_{44}(t) = \lambda_{37} P_{43}(t) + \lambda_{38} P_{42}(t) + \lambda_{39} P_{41}(t) + \mu_{34} P_{32}(t) + \mu_{35} P_{33}(t) + \mu_{36} P_{34}(t) + \mu_{37}' P_{35}(t) + \mu_{37}' P_{35}(t) + \mu_{38}' P_{36}(t) + \mu_{39}' P_{37}(t) \quad (10.44)$$

Appling boundary conditions, i.e. when time, $t = 0$

$$P_i(t) = 1, \text{ for } i = 1 \text{ and } P_i(t) = 0 \text{ for } i \neq 1 \quad (10.45)$$

10.5 STEADY STATE AVAILABILITY (SSA) OF RFGS

SSA behavior of the RFGS is examined by putting derivative $d/dt = 0$ and time $(t) = \infty$; the equations from (10.1) to (10.44) can be rewritten as:

$$\left(\sum_{i=34}^{36} \lambda_i\right) P_{38} + \lambda_{38} P_{38} + \lambda_{39} P_{38} + \lambda_{37}' P_{38} + \mu_{37} P_{38} = \mu_{38} P_{41} + \mu_{39} P_{42} + \lambda_{37} P_1 + \mu_{36} P_7 + \mu_{74}' P_8 + \mu_{34} P_5 + \mu_{35} P_6$$

$$A_1 P_{38} = \mu_{38} P_{41} + \mu_{39} P_{42} + \lambda_{37} P_1 \qquad A_2 P_{39} = \mu_{39} P_{43} + \mu_{37} P_{41} + \lambda_{38} P_1$$

$$A_3 P_{40} = \mu_{37} P_{42} + \mu_{38} P_{43} + \lambda_{39} P_1 \qquad A_4 P_{41} = \lambda_{37} P_{39} + \mu_{39} P_{44} + \lambda_{38} P_{38}$$

$$A_5 P_{42} = \lambda_{37} P_{40} + \mu_{38} P_{44} + \lambda_{39} P_{38} \qquad A_6 P_{43} = \lambda_{38} P_{40} + \mu_{37} P_{44} + \lambda_{39} P_{39}$$

$$A_7 P_{44} = \lambda_{37} P_{43} + \lambda_{38} P_{42} + \lambda_{39} P_{41} \quad (10.46)$$

Where

$$A_1 = \mu_{37} + \lambda_{39} + \lambda_{38} \qquad A_2 = \lambda_{37} + \lambda_{38} + \mu_{39} \qquad A_3 = \mu_{39} + \lambda_{37} + \lambda_{38}$$

$$A_4 = \mu_{37} + \mu_{38} + \lambda_{39} \qquad A_5 = \mu_{37} + \mu_{39} + \lambda_{38} \qquad A_6 = \mu_{38} + \mu_{39} + \lambda_{37}$$

$$A_7 = \mu_{37} + \mu_{38} + \mu_{39}$$

Now

On solving Equations 10.1–10.44 recursively, we get,

$$P_2 = Z_1 P_1 \qquad P_3 = Z_2 P_1 \qquad P_4 = Z_3 P_1$$

$$P_5 = Z_1 P_{38} = Z_1 K_9 P_1 \qquad P_6 = Z_2 P_{38} = Z_2 K_9 P_1 \qquad P_7 = Z_3 P_{38} = Z_3 K_9 P_1$$

$$P_8 = Z_4' P_{38} = Z_4' K_9 P_1 \qquad P_9 = Z_1 P_{39} = Z_1 K_{10} P_1 \qquad P_{10} = Z_2 P_{39} = Z_2 K_{10} P_1$$

$$P_{11} = Z_2 P_{39} = Z_3 K_{10} P_1 \qquad P_{12} = Z_5' P_{39} = Z_5' K_{10} P_1 \qquad P_{13} = Z_6' P_{40} = Z_6' K_{11} P_1$$

$$P_{14} = Z_1 P_{40} = Z_1 K_{11} P_1 \qquad P_{15} = Z_2 P_{40} = Z_2 K_{11} P_1 \qquad P_{16} = Z_3 P_{40} = Z_3 K_{11} P_1$$

$$P_{17} = Z_1 P_{41} = Z_1 K_{12} P_1 \qquad P_{18} = Z_2 P_{41} = Z_2 K_{12} P_1 \qquad P_{19} = Z_3 P_{41} = Z_3 K_{12} P_1$$

$$P_{20} = Z_4' P_{41} = Z_4' K_{12} P_1 \qquad P_{21} = Z_5' P_{41} = Z_5' K_{12} P_1 \qquad P_{22} = Z_1 P_{42} = Z_1 K_{13} P_1$$

$$P_{23} = Z_2 P_{42} = Z_2 K_{13} P_1 \qquad P_{24} = Z_3 P_{42} = Z_3 K_{13} P_1 \qquad P_{25} = Z_6' P_{42} = Z_6' K_{13} P_1$$

$$P_{26} = Z_4' P_{42} = Z_4' K_{13} P_1 \qquad P_{27} = Z_1 P_{43} = Z_1 K_{14} P_1 \qquad P_{28} = Z_2 P_{43} = Z_2 K_{14} P_1$$

$$P_{29} = Z_3 P_{43} = Z_3 K_{14} P_1 \qquad P_{30} = Z_5' P_{43} = Z_5' K_{14} P_1 \qquad P_{31} = Z_6' P_{43} = Z_6' K_{14} P_1$$

$$P_{32} = Z_1 P_{44} = Z_1 K_{15} P_1 \qquad P_{33} = Z_2 P_{44} = Z_2 K_{15} P_1 \qquad P_{34} = Z_3 P_{44} = Z_3 K_{15} P_1$$

$$P_{35} = Z_4' P_{44} = Z_4' K_{15} P_1 \qquad P_{36} = Z_5' P_{44} = Z_5' K_{15} P_1 \qquad P_{37} = Z_6' P_{44} = Z_6' K_{15} P_1$$

$$P_{38} = K_9 P_1 \qquad P_{39} = K_{10} P_1 \qquad P_{40} = K_{11} P_1 \qquad P_{41} = K_{12} P_1$$

$$P_{42} = K_{13} P_1 \qquad P_{43} = K_{14} P_1 \qquad P_{44} = K_{15} P_1$$

Where

$Z = \mu_i$; i = 34, 35, 36, 37, 38, and 39

P_1 is the probability of working state that comes after applying normalizing conditions, i.e.

$\sum_{i=1}^{44} P_i = 1$ gives:

$$P_1 = 1 + Z_1K_9 + Z_2K_9 + Z_3K_9 + Z_4'K_9 + Z_1K_{10} + Z_2K_{10} + Z_3K_{10} + Z_5'K_{10} + Z_6'K_{11} + Z_1K_{11} + Z_2K_{11} + Z_3K_{11} + Z_1K_{12} + Z_2K_{12} + Z_3K_{12} + Z_4'K_{12} + Z_5'K_{12} + Z_1K_{13} + Z_2K_{13} + Z_3K_{13} + Z_6'K_{13} + Z_4'K_{13} + Z_1K_{14} + Z_2K_{14} + Z_3K_{14} + Z_5'K_{14} + Z_6'K_{14} + Z_1K_{15} + Z_2K_{15} + Z_3K_{15} + Z_4'K_{15} + Z_5'K_{15} + Z_6'K_{15} + K_9 + K_{10} + K_{11} + K_{12} + K_{13} + K_{14} + K_{15} \quad (10.47)$$

$$A_{SSA} = 1 + P_{38} + P_{39} + P_{40} + P_{41} + P_{42} + P_{43} + P_{44}$$
$$= [1 + K_9 + K_{10} + K_{11} + K_{12} + K_{13} + K_{14} + K_{15}]P_1$$

The SSA expression for RFGS may be found by the sum of probability of states 1, 38, 39, 40, 41, 42, 43, and 44 as shown in TD and represented by 48.

10.6 AVAILABILITY OPTIMIZATION

The SSA of RFGS is highly affected by the FRR of every subsystem. In the present work, GA and PSO are proposed to find out optimal FRR of all subsystems for achieving optimal availability levels. First, GA is applied to optimize the ASSA of the systems chosen for study, which is carried out in two steps.

a. First is by varying to a maximum the number of generations and taking mutation, crossover, and population size constant.
b. Second is by maximizing the number of population and taking mutation, crossover, and generation size constant.

   After performing simulation, the optimal value of SSA has been selected along with the best feasible combination of FRR from the pool of search results and the ranges of FRR.

   Similarly, optimal SSA is obtained using PSO in two steps:

c. Firstly, the maximum number of generations varied, taking the number of particles constant, and other operating parameters are generated randomly.
d. Secondly, the number of particles was varied, taking the size of the generation constant, and other operating parameters are generated randomly.

After performing the simulation run, the optimum value of SSA has been selected along with the best feasible combination of FRR from the pool of search results and the ranges of FFR.

10.7 RESULTS AND DISCUSSION

The optimization of SSA of RFGS that has been performed, to find out the optimum number of generations, varies from 20 to 100 in an increment of 10. The size of population for this simulation is kept constant and is 100. It is clearly seen from Table 10.1 that when the number of the generation is 70, the ASSA of rice finishing and grading system is 87.032%, which is the optimized availability of the subsystem. The corresponding values of FRR are at λ_{34} =0.000344, μ_{34} = 0.0309, $\lambda_{35} = 0.001013$, $\mu_{35} = 0.033$, $\lambda_{36} = 0.00155$, $\mu_{36} = 0.0476$, $\lambda_{37} = 0.00357$, $\mu_{37} = 0.0379$, $\lambda_{38} = 0.0275$, $\mu_{38} = 0.0189$, $\lambda_{39} = 0.00594$, $\mu_{39} = 0.394$, λ'_{37}=0.005, μ'_{37}= 0.0209, $\lambda'_{38} = 0.003$, $\mu'_{38} = 0.058$, $\lambda'_{39} = 0.00226$, and $\mu'_{39} = 0.063$ which is the optimized value of these parameters for the maximum availability. The optimization is performed again by changing the size of population from 20 to 110 in an increment of 10 and taking the size of generation constant at 100. The optimum values of system ASSA is 86.83% at the population size 90 and for which the value of FRR are λ_{34} = 0.00036, $\mu_{34} = 0.031$, $\lambda_{35} = 0.001$, $\mu_{35} = 0.035$, $\lambda_{36} = 0.00175$, $\mu_{36} = 0.0499$, $\lambda_{37} = 0.00316$, $\mu_{37} = 0.037$, $\lambda_{38} = 0.025$, $\mu_{38} = 0.018$, $\lambda_{39} = 0.00501$, $\mu_{39} = 0.37$, $\lambda'_{37} = 0.00509$, $\mu'_{37} = 0.0205$, $\lambda'_{38} = 0.00383$, $\mu'_{38} = 0.059$, $\lambda'_{39} = 0.00241$, and $\mu'_{39} = 0.06342$ as shown in Table 10.2.

Initially, the simulation test has been applied to obtain the optimum number of generations, changed from 5 to 50 with an increment of 5, keeping the number of birds (particles) as 5. The SSA has been affected by variation of number of generations which has been presented in Table 10.3. The SSA comes out to be 87.55% and the best feasible values of FRR are: λ_{34} = 0.00044, $\lambda_{35} = 0.00194$, $\lambda_{36} = 0.00367$, $\lambda_{37} = 0.00603$, $\lambda_{38} = 0.0446$, $\lambda_{39} = 0.01903$, λ'_{37}=0.00747 $\lambda'_{38} = 0.00729$, $\lambda'_{39} = 0.005525$, $\mu_{34} = 0.08593$, $\mu_{35} = 0.04993$, $\mu_{36} = 0.08940$, $\mu_{37} = 0.08081$, $\mu_{38} = 0.07321$, $\mu_{39} = 0.57385$, $\mu'_{37} = 0.02939$, $\mu'_{38} = 0.155191$, and $\mu'_{39} = 0.47560$.

In the same way, the results of PSO has been revealed in Table 10.4. The simulation has been performed to search out optimum no. of birds in the group, varied from 5 to 25 in an increment of 5 keeping size of generation (30).The SSA comes out to be 89.66% and the best feasible values of FRR are: $\lambda_{34} = 0.00078$, $\lambda_{35} = 0.00188$, $\lambda_{36} = 0.00208$, $\lambda_{37} = 0.00361$, $\lambda_{38} = 0.02588$, $\lambda_{39} = 0.01858$, $\lambda'_{37} = 0.00506$, $\lambda'_{38} = 0.00520$, $\lambda'_{39} = 0.00624$, $\mu_{34} = 0.09851$, $\mu_{35} = 0.04526$, $\mu_{36} = 0.08117$, $\mu_{37} = 0.07528$, $\mu_{38} = 0.069887$, $\mu_{39} = 0.181912$, $\mu'_{37} = 0.027696$, $\mu'_{38} = 0.13503$, and $\mu'_{39} = 0.5119$. The same is also shown in Table 10.4.

Steady State Availability Optimization of Rice Finishing and Grading System Using GA

Population size: 100	Number of generations: 50–500	Selection: Stochastic Uniform
Number of coded variables: 18	Probability of mutation: 1%	Probability of crossover: 80%

Parameters	λ_{34}	μ_{34}	λ_{35}	μ_{35}	λ_{36}	μ_{36}	λ_{37}	μ_{37}	λ_{38}	μ_{38}	λ_{39}	μ_{39}	λ'_{37}	μ'_{37}	λ'_{38}	μ'_{38}	λ'_{39}	μ'_{39}
Minimum	0.0003	0.025	0.001	0.020	0.0015	0.035	0.003	0.026	0.025	0.013	0.005	0.10	0.005	0.015	0.0038	0.045	0.002	0.05
Maximum	0.0012	0.105	0.004	0.05	0.0090	0.092	0.0083	0.10	0.050	0.075	0.020	0.98	0.0075	0.030	0.009	0.165	0.008	0.52

TABLE 10.1

Number of Generations vs Availability Rice Finishing and Grading System Using G.A

Parameters No. of Generations	λ_{34}	λ_{35}	λ_{36}	λ_{37}	λ_{38}	λ_{39}	λ'_{37}	λ'_{38}	λ'_{39}	μ_{34}	μ_{35}	μ_{36}	μ_{37}	μ_{38}	μ_{39}	μ'_{37}	μ'_{38}	μ'_{39}	Ass
50	0.00032	0.00103	0.00186	0.00368	0.03068	0.00576	0.00502	0.00417	0.00394	0.029	0.032	0.0493	0.0377	0.0188	0.326	0.0208	0.0598	0.059	0.8613
100	0.00046	0.00100	0.00151	0.00350	0.02650	0.00654	0.00524	0.00384	0.00212	0.030	0.034	0.0483	0.0346	0.0190	0.395	0.0209	0.0599	0.064	0.8660
150	0.00034	0.00114	0.00152	0.00319	0.02582	0.00504	0.00504	0.00399	0.00251	0.031	0.033	0.0486	0.0379	0.0189	0.379	0.0209	0.0600	0.0609	0.8701
200	0.00030	0.00100	0.00150	0.00300	0.02500	0.00586	0.00510	0.00532	0.00200	0.031	0.035	0.0475	0.0380	0.0190	0.400	0.0210	0.0600	0.0650	0.8720
250	**0.00030**	**0.00106**	**0.00150**	**0.00334**	**0.02616**	**0.00517**	**0.00500**	**0.00396**	**0.00207**	**0.031**	**0.035**	**0.0489**	**0.0380**	**0.0190**	**0.397**	**0.0210**	**0.0600**	**0.0650**	**0.8747**
300	0.00034	0.00104	0.00164	0.00320	0.02555	0.00539	0.00501	0.00398	0.00279	0.031	0.035	0.0500	0.0380	0.0190	0.399	0.0210	0.0600	0.0650	0.8733
350	0.00041	0.00108	0.00153	0.00315	0.02547	0.00537	0.00502	0.00390	0.00227	0.031	0.035	0.0500	0.0380	0.0190	0.400	0.0210	0.0600	0.0649	0.8725
400	0.00034	0.00122	0.00157	0.00314	0.02523	0.00521	0.00503	0.00391	0.00208	0.031	0.035	0.0500	0.0380	0.0190	0.400	0.0210	0.0600	0.0650	0.8709
450	0.00030	0.00106	0.00172	0.00382	0.02512	0.00509	0.00510	0.00383	0.00208	0.031	0.035	0.0500	0.0380	0.0190	0.400	0.0210	0.0600	0.0650	0.8698

Steady State Availability Optimization of Rice Finishing and Grading System Using GA

Population size: 10-100	Number of generations: 100	Selection: Stochastic
Uniform number of coded variables: 18	Probability of mutation: 1%	Probability of crossover: 80%

TABLE 10.2
Population Size vs Availability of Rice Finishing and Grading System Using G.A

Parameters No. of Particles	λ_{34}	λ_{35}	λ_{36}	λ_{37}	λ_{38}	λ_{39}	λ'_{37}	λ'_{38}	λ'_{39}	μ_{34}	μ_{35}	μ_{36}	μ_{37}	μ_{38}	μ_{39}	μ'_{37}	μ'_{38}	μ'_{39}	Ass
10	0.00037	0.00179	0.00263	0.00317	0.03000	0.00780	0.00538	0.00543	0.00357	0.029	0.029	0.044	0.037	0.0187	0.35	0.020	0.054	0.053	0.8158
20	0.00035	0.00150	0.00346	0.00357	0.03204	0.00740	0.00526	0.00404	0.00248	0.031	0.035	0.046	0.035	0.0189	0.29	0.019	0.057	0.064	0.8229
30	0.00054	0.00104	0.00325	0.00392	0.02958	0.00547	0.00503	0.00409	0.00230	0.029	0.032	0.049	0.031	0.0189	0.37	0.021	0.059	0.054	0.8378
40	0.0003	0.00117	0.00150	0.00330	0.02500	0.00500	0.00515	0.00380	0.00253	0.031	0.035	0.050	0.038	0.0189	0.20	0.021	0.060	0.064	0.8708
50	0.0003	0.001	0.00150	0.00312	0.02500	0.00564	0.00500	0.00423	0.00222	0.029	0.035	0.050	0.037	0.0161	0.18	0.021	0.055	0.060	0.8718
60	**0.00036**	**0.001**	**0.00150**	**0.00300**	**0.02606**	**0.00635**	**0.00501**	**0.00395**	**0.00200**	**0.031**	**0.034**	**0.049**	**0.038**	**0.0190**	**0.36**	**0.021**	**0.060**	**0.064**	**0.8756**
70	0.00033	0.00107	0.00160	0.00322	0.02575	0.00518	0.00517	0.00423	0.00216	0.029	0.034	0.047	0.037	0.0189	0.31	0.019	0.059	0.063	0.8638
80	0.00037	0.00114	0.00224	0.00330	0.02679	0.00561	0.00501	0.00394	0.00234	0.030	0.033	0.050	0.037	0.0189	0.39	0.020	0.059	0.062	0.8607
90	0.00037	0.00107	0.00194	0.00329	0.02593	0.00569	0.00520	0.00405	0.00337	0.030	0.032	0.048	0.038	0.0177	0.38	0.021	0.059	0.064	0.8603

Steady State Availability Optimization of Rice Finishing and Grading System using PSO

Number of variables: 18	Number of birds: 5	Social component: 1–2
Number of generations: 5–50	Inertia weight: 0–1	Cognitive component: 1–2

Parameters	λ_{34}	μ_{34}	λ_{35}	μ_{35}	λ_{36}	μ_{36}	λ_{37}	μ_{37}	λ_{38}	μ_{38}	λ_{39}	μ_{39}	λ'_{37}	μ'_{37}	λ'_{38}	μ'_{38}	λ'_{39}	μ'_{39}
Minimum	0.0003	0.025	0.001	0.020	0.0015	0.035	0.003	0.026	0.025	0.013	0.005	0.10	0.005	0.015	0.0038	0.045	0.002	0.05
Maximum	0.0012	0.105	0.004	0.05	0.0090	0.092	0.0083	0.10	0.050	0.075	0.020	0.98	0.0075	0.030	0.009	0.165	0.008	0.52

TABLE 10.3

Number of Generations vs Availability of Rice Finishing and Grading System Using P.S.O

Parameters No. of Generations	λ_{34}	λ_{35}	λ_{36}	λ_{37}	λ_{38}	λ_{39}	λ'_{37}	λ'_{38}	λ'_{39}	μ_{34}	μ_{35}	μ_{36}	μ_{37}	μ_{38}	μ_{39}	μ'_{37}	μ'_{38}	μ'_{39}	**Ass**
5	0.0012	0.0036	0.0036	0.0074	0.0376	0.0168	0.0063	0.0078	0.0068	0.078	0.043	0.090	0.092	0.063	0.95	0.023	0.125	0.39	0.8741
10	0.0007	0.0037	0.0061	0.0049	0.0489	0.0078	0.0075	0.0088	0.0060	0.036	0.049	0.076	0.095	0.060	0.84	0.029	0.077	0.30	0.8222
15	0.0009	0.0038	0.0046	0.0060	0.0296	0.0198	0.0061	0.0053	0.0077	0.098	0.048	0.081	0.096	0.075	0.97	0.028	0.097	0.47	0.8437
20	0.0007	0.0037	0.0061	0.0049	0.0489	0.0078	0.0075	0.0088	0.0060	0.036	0.049	0.076	0.095	0.060	0.84	0.029	0.077	0.30	0.8222
25	0.0006	0.0013	0.0090	0.0050	0.0498	0.0162	0.0065	0.0083	0.0071	0.089	0.050	0.091	0.095	0.074	0.97	0.029	0.131	0.23	0.8530
30	**0.0004**	**0.0019**	**0.0037**	**0.0060**	**0.0446**	**0.0190**	**0.0075**	**0.0073**	**0.0055**	**0.085**	**0.049**	**0.089**	**0.080**	**0.073**	**0.57**	**0.029**	**0.155**	**0.47**	**0.8755**
35	0.0010	0.0034	0.0046	0.0053	0.0445	0.0176	0.0066	0.0078	0.0033	0.068	0.031	0.091	0.086	0.047	0.95	0.029	0.164	0.52	0.8438
40	0.0011	0.0011	0.0057	0.0051	0.0497	0.0123	0.0071	0.0089	0.0063	0.081	0.024	0.083	0.095	0.067	0.96	0.023	0.156	0.51	0.8360
45	0.0012	0.0036	0.0036	0.0074	0.0376	0.0168	0.0063	0.0078	0.0068	0.078	0.041	0.090	0.092	0.063	0.95	0.023	0.125	0.39	0.8741
50	0.0007	0.0037	0.0061	0.0049	0.0489	0.0078	0.0075	0.0088	0.0060	0.036	0.049	0.076	0.095	0.060	0.84	0.029	0.077	0.30	0.8222

Availability Optimization of Rice Finishing and Grading System Using PSO

Number of variables: 18	Number of birds: 5–25	Social component: 1–2
Number of generations: 30	Inertia weight: 0–1	Cognitive component: 1–2

TABLE 10.4
Number of Particles vs Availability of Rice Finishing and Grading System Using P.S.O

Parameters No. of Particles	λ_{34}	λ_{35}	λ_{36}	λ_{37}	λ_{38}	λ_{39}	λ'_{37}	λ'_{38}	λ'_{39}	μ_{34}	μ_{35}	μ_{36}	μ_{37}	μ_{38}	μ_{39}	μ'_{37}	μ'_{38}	μ'_{39}	Ass
5	0.0004	0.0019	0.0037	0.0060	0.0446	0.0190	0.0075	0.0073	0.0055	0.085	0.049	0.089	0.080	0.073	0.57	0.029	0.155	0.47	**0.8755**
10	0.0010	0.0023	0.0062	0.0083	0.0421	0.0164	0.0073	0.0074	0.0059	0.099	0.050	0.066	0.094	0.073	0.59	0.029	0.125	0.45	**0.8367**
15	0.0012	0.0010	0.0074	0.0070	0.0479	0.0113	0.0060	0.0050	0.0071	0.101	0.047	0.091	0.097	0.043	0.96	0.028	0.084	0.274	**0.8615**
20	0.0009	0.0019	0.0048	0.0082	0.0411	0.0148	0.0063	0.0051	0.0068	0.060	0.049	0.085	0.082	0.060	0.96	0.029	0.143	0.45	**0.8778**
25	0.0008	0.0019	0.0021	0.0036	0.0259	0.0186	0.0051	0.0052	0.0062	0.098	0.045	0.081	0.075	0.069	0.81	0.027	0.135	0.51	**0.8967**
30	0.0009	0.0019	0.0024	0.0080	0.0485	0.0104	0.0067	0.0053	0.0076	0.072	0.048	0.091	0.084	0.062	0.60	0.028	0.131	0.46	**0.8756**
35	0.0006	0.0020	0.0022	0.0064	0.0430	0.0198	0.0060	0.0083	0.0078	0.091	0.031	0.089	0.070	0.073	0.57	0.028	0.138	0.215	**0.8741**

10.8 CONCLUSION

The present research work highlights the applications of two evolutionary approaches, i.e. PSO and GA to find the optimal SSA of an RFGS of a rice milling plant. First, the SSA governing equation of the system concerned has been developed using MA. Thereafter, SSA has been optimized using two evolutionary techniques, i.e.GA and PSO. The impact of the number. of generations versus number of birds (particles)/ population size on the SSA of the system has been studied. The latter technique shows a better result as compared to the first approach in the present case study. PSO technique gives 2.5% more availability at 30th generation and 25 particle size as compared to GA at 100th generation and 70 population size. The outcome of the present research work was discussed with management and found beneficial for the improvement of SSA.

10.9 SCOPE FOR FUTURE WORK

The research work and methods of the analysis reported in the present case study may be further extended in the following ways:

- A similar SSA analysis and optimization may be addressed to other capital-intensive industries, such as food processing, paper, power plant, and manufacturing industries.
- GA and PSO techniques have been used as an optimization tool, other techniques such as Evolutionary Computation, Neural Network, Ant Colony optimization can also be used and the results may be compared with the present work for further improvements.
- Problem targeted in the present work used single objective multi constraint function. It may further be extended to multi-objective, multi constraint problems by considering other objectives such as reliability, availability and maintainability costs, spare requirement estimation, number of maintenance personnel, etc.

ACKNOWLEDGMENT

The authors express heartfelt gratitude to the management of plant and maintenance personnel for their full cooperation without which the present task cannot be completed. The rice plant is situated in the northern part of India.

REFERENCES

Aggarwal, A.K., Kumar, S., Singh, V. & Gupta, T. (2015). Markov modeling and reliability analysis of urea synthesis system of a fertilizer plant. *Journal of Industrial Engineering International*, *11*(1), 1–14.

Aggarwal, A.K., Singh, V., & Kumar, S. (2014). Availability analysis and performance optimization of a butter oil production system: a case study. *International Journal of System Assurance Engineering and Management*, *8*, 1–17.

Ahmadi, S., Moosazadeh, S., Hajihassani, M., Moomivand, H. & Rajaei, M. M. (2019). Reliability, availability and maintainability analysis of the conveyor system in mechanized tunneling. *Measurement, 145*, 756–764. Doi: 10.1016/j.measurement.2019.06.009.

Arabi, A.A.Y. & Jahromi, A.E. (2012). Developing a new model for availability of a series repairable system with multiple cold-standby subsystems and optimization using simulated annealing considering redundancy and repair facility allocation. *International Journal of System Assurance Engineering and Management, 3*(4), 310–322.

Azaron, A., Perkgoz, C., Katagiri, H., Kato, K., & Sakawa, M. (2009). Multi-objective reliability optimization for dissimilar-unit cold-standby systems using a genetic algorithm. *Computers & Operations Research, 36*(5), 1562–1571.

Bahl, A., Sachdeva, A. and Garg, R.K. (2018). Availability analysis of distillery plant using petri net. *International Journal of Quality & Reliability Management, 35*(10), 2373–2387. Doi: 10.1108/IJQRM-06-2017-0108.

Bose, D., Chattopadhyay, S., Bose, G., Adhikary, D., & Mitra, S. (2012). RAM investigation of coal fired thermal power plants: a case study. *International Journal of Industrial Engineering Computations, 3*(3), 423–434.

Dhillon, B. S., & Singh, C. (1981). *Engineering Reliability: New Techniques and Applications* (4th ed.). New York: Wiley.

Diyaley, S. & Chakraborty, S. (2019). Optimization of multi-pass face milling parameters using metaheuristic algorithms. *FactaUniversitatis– Series: Mechanical Engineering, 17*(3), 365–383.

Garg, H. (2013). Performance analysis of complex repairable industrial systems using PSO and fuzzy confidence interval based methodology, *ISATransactions, 52*(2), 171–183.

Juang, Y. S., Lin, S. S., & Kao, H. P. (2008). A knowledge management system for series-parallel availability optimization and design. *Expert Systems with Applications, 34*(1), 181–193.

Kachitvichyanukul, V. (2012). Comparison of three evolutionary algorithms: GA, PSO, and DE. *Industrial Engineering and Management Systems, 11*(3), 215–223.

Khanduja, R., Tewari, P. C. & Chauhan, R. S. (2012). Performance modeling and optimization for the stock preparation unit of a paper plant using genetic algorithm. *International Journal of Quality Reliability and Management, 28*(6), 688–703.

Kumar, N., Tewari, P.C. & Sachdeva, A. (2019). Performance modeling and analysis of refrigeration system of a milk processing plant using petri nets. *International Journal of Performability Engineering, 15*(7), 1751–1759.

Kumar, P. & Tewari, P.C. (2017). Performance analysis and optimization for CSDGB filling system of a beverage plant using particle swarm optimization. *International Journal of Industrial Engineering Computation, 8*(3), 303–314.

Kumar, R., Sharma, A., & Tewari, P. (2012). Markov approach to evaluate the availability simulation model for power generation system in a thermal power plant. *International Journal of Industrial Engineering Computations, 3*(5), 743–750.

Kumar, S. & Tewari, P.C. (2011). Mathematical modeling and performance optimization of CO2 cooling system of a fertilizer plant. *International Journal of Industrial Engineering Computation, 2*(3), 689–698.

Kumar, V. & Reji, M. (2018). Performance analysis and optimization of cooling tower. *International Journal of Scientific and Research Publications, 8*(4), 225–237. ISSN 2250-315.3.

Li, Y.F. & Peng, R. (2014). Availability modeling and optimization of dynamic multi-state series–parallel systems with random reconfiguration. *Reliability Engineering & System Safety, 127*, 47–57.

Malik, S. & Tewari, P.C. (2018). Performance modeling and maintenance priorities decision for the water flow system for a coal based thermal power plant. *International Journal of Quality & Reliability Management, 35*(4), 996–1010. Doi: 10.1108/IJQRM-03-2017-0037.

Malik, S. & Tewari, P.C. (2020). Optimization of coal handling system performability for a thermal power plant using PSO algorithm. *International Journal of Quality & Reliability Management*, *10*(3), 359–376. Doi: 10.1108/GS-01-2020-0002.

Modgil, V., Sharma, S.K., & Singh, J. (2013). Performance modeling and availability analysis of shoe upper manufacturing unit. *International Journal of Quality & Reliability Management*, *30*(8), 816–831.

Rabbani, M., Mohammadi, S. & Mobini, M. (2018). Optimum design of a CCHP system based on economical, energy and environmental considerations using GA and PSO. *International Journal of Industrial Engineering Computations*, *9*(1), 99–122.

Usubamatov, R., Ismail, K.A., & Sah, J.M. (2013). Mathematical models for productivity and availability of automated lines. *The International Journal of Advanced Manufacturing Technology*, *66*(1–4), 59–69.

Waintraub, M., Schirru, R., & Pereira, C.M. (2009). Multiprocessor modeling of parallel particle swarm optimization applied to nuclear engineering problems. *Progress in Nuclear Energy*, *51*(6), 680–688.

11 Analysis of Sign Language Recognition System for Society 5.0 for Sensory-Impaired People

Jatinder Kaur, Nitin Mittal, and Sarabpreet Kaur

CONTENTS

11.1 INTRODUCTION

Gesture is a sign of emotional speech or physical behavior. This includes motions of the body and hand gestures. Two classifications fall into it: static gesture and dynamic gesture. The body posture or the hand gesture depicts a sign for the former (Al-Hammadi et al., 2020). Several researchers have been working in recent decades to improve the technology for hand gesture recognition (Nikam & Ambekar, 2016). In many applications, hand gesture recognition is an intelligent system that has great importance, like the recognition of sign language, augmented or virtual reality, sign language interpreters, and robot control. The importance of gesture recognition has increased at a very fast pace owing to the new generation of gesture interface technology. Sign language is the main mode of communication for the deaf community who cannot use spoken languages to communicate with others (Chong & Lee, 2018). The use of hand signs to interact becomes inconvenient to express one's feelings readily if a common hand sign language is not followed (Winkelmann et al., 2005). A standard sign language includes a predefined array of signs and their meanings, making them easy to understand (Rahagiyanto et al., 2017). The sign language too uses different gestures for communication, mostly hand gestures. Sign languages

DOI: 10.1201/9781003158165-11

are different from spoken languages. The use of hand gestures for interaction either with humans or with machines is comparatively higher than other body gestures like head and eyes since hands send more clearer signals, and the gestures can be made spontaneously. Nowadays, deaf people are becoming more and more outgoing, and unlike olden days, they do not depend on anybody for communication (Kılıboz & Güdükbay, 2015). So, for such people, it is important that the public around them must also be able to understand what they want to tell them using sign language. We build the hand gesture recognition program so that even if someone does not know the meanings of the signs, they can translate the sign language using this program. Hence, deaf people will have no barrier in their communication (Misra et al., 2018).

The sign language recognition system (SLRS) helps to improve the communication environment between hearing people and sensory-impaired people of our society. The contributions of the SLRS for society are given below:

- Review of the state of the art of hand gesture recognition.
- Development of optimal feature vector by minimizing the redundant information.
- Analysis of an intelligent system that helps the sensory-impaired population.

11.2 LITERATURE SURVEY

Nowadays, every electronic device around us can be controlled without any need to touch it, due to the surge of voice and gesture controls. Voice controls recognize the words spoken by a person, while in gesture controls, the device is controlled by the movement of our hand in a specific pattern for the device to perform a particular task (Murakami & Taguchi, 1991). Researchers have paid attention for designing a fast and efficient hand gesture recognition system. There are many experiments for recognizing hand gestures, and such programs are also used as parts of other programs such as controlling a device using hand gestures. There are many ways to recognize a hand gesture from an image (Ohn-Bar & Trivedi, 2014). The deaf community use hand sign language to communicate, which is hard to understand for those who have not learned it. Therefore, a program module that can translate gestures into text and speech needs to be created. This would be a big step in making it possible for deaf people to communicate with the hearing public (Sood & Mishra, 2016). Recognition of sign language involves hand gesture classification. For American Sign Language (ASL) finger spelling alphabets and digits, a method for static hand gesture classification has been presented. The model utilizes segmentation based on skin color which includes little post-processing (Rahim et al., 2019). As the features characterizing the hand movements, the averages of central moments of order 2–9 have been extracted. For recognition, the neural network classifier was used which provided respectable classification results of 73.68% with a limited feature vector size containing 8 characteristics (Sahoo et al., 2018). The workflow for hand symbol recognition is defined first; from the original pictures of the input devices, a hand region is detected. Various features are then extracted to clarify the gestures of hands (Gamage et al., 2011). Finally, hand movements are defined by calculating the relationship of the feature information with the hand gestures (Sood & Mishra, 2016). Gesture is the most primitive technique for conversation among human beings. Presently, innovation gesture recognition affects the world differently, from the physically challenged individuals. Hand gestures are significant body movements that are actions of the hands, arms, or fingers (Al-Hammadi et al., 2020). Hand gestures identify the level of evidence from the static gestures and dynamic movements to reflect the feelings of the person (Avola et al., 2019).

The hand movements are used as inputs to the machine, and there is no need for the exchange of verbal explanation as an intermediary medium for gesture recognition (Chang & Wu 1998). Zhi et al. (2018) investigated the essential components of communication and distinguished the strategies that could help for gesture-based communication (Zhang et al., 2019). The main purpose was to report the relevant challenges and likely arrangements in the realistic implementation of the translation of sign languages. Itkarkar et al. (2017) examined the hand gestures for sign language (numerically from 0 to 9). In this work, finger patterns were employed for the hand gesture recognition, and Leap Motion Controller is used for finger detection. Hand gesture recognition is a subject matter that is currently examined by numerous experts for many useful aspects (Stergiopoulou & Papamarkos, 2009). Two different methods, viz. threshold and Artificial Neural Network (ANN), were performed for modeling 0 of finger direction pattern. The output depicted that the ANN model yields much better results than the threshold model. The 98% outcome in terms of accuracy was achieved through the ANN method. Khan and Ibraheem (2012) tried to propose a highly effectual deep convolutional neural network (CNN) technique for recognition of hand gestures. The proposed technique was applied on three hand gesture datasets having three classes, containing 40, 23, and 10, respectively. Just in the case of the signer-independent mode, the result showed that the recognition rates in class 1 and class 3 are 84.38% and 70%, respectively. Class 2 has the lowest recognition rate of 34.9%. The proposed approach is also used on the signer-dependent mode to indicate the usefulness of work. It is strongly encouraged to make use of both spatial and temporal features for the hand gesture recognition process. Chen et al. (2020) tried to improve the ease of communication in deaf people. A total of 24 hand signals of ASL were included in the study, and CNN was applied to enhance the communication pattern among deaf people. The database was taken directly from kaggle.com, and OpenCV was applied as an image extraction technique. The results showed that the CNN model provides an accuracy of 99.7% for the above-mentioned dataset. Delaye and Anquetil (2013) proposed a method to train recurrent neural network (RNN) on the features extracting from the bone angle of hand finger (Table 11.1).

Rahim et al. (2019) proposed a hybrid segmentation method to identify hand gestures and retrieve the features using a feature fusion from the convoluted neural network, like YCbCr and skin mask segmentation. Finally, to identify the hand movements of a sign phrase, a multiclass support vector machine (SVM) classifier was used. Ahuja et al. (2019) tried to apply CNN on 24 ASL hand signals to increase the level of communication and upgraded the recognition accuracy using the database available at kaggle.com. Chong and Lee (2018) showed that the hand gesture recognition results for the 26 letters using an SVM and a Deep Neural Network (DNN) are 80.30% and 93.81%, respectively. The suggested prototype acts as a translator in the retail industry, such as at the bank or post office, for the deaf in everyday

TABLE 11.1
Parameterization-Based Literature Review

Author	Dataset/Language	Performance Metric	Results (%)
Ohn-Bar and Trivedi (2014)	ASL	Recognition accuracy	92.8–98.4
Kılıboz and Güdükbay (2015)	ASL	Accuracy	73
Stergiopoulou and Papamarkos (2009)	180 test hand images 1800 times	Recognition rate	90.45
Murakami and Taguchi (1991)	Japanese sign language	Recognition rate	96

life. Misra et al. (2018) performed an analysis to identify the difference between the five classifiers, namely K-Nearest Neighbor (KNN), SVM, ANN, naïve Bayes, and Extreme Learning Machine (ELM). An optimum accuracy of 96.95% was reached using the combination of current and planned features. Sahoo et al. (2018) used a wavelet packet transition and Fisher proportion in an unregulated setting to distinguish hand movements.

11.3 SIGN LANGUAGE RECOGNITION SYSTEM

From a video series, we are going to recognize hand movements and another way is to select the dataset related with our problem. To recognize these movements from a live video series, we first need to delete all the unwanted portions in the video series by subtracting the hand region separately. The next step includes the segmentation steps related to the hand region. Thus, by using various simple steps, the entire problem could be solved. Finally, we will convert the obtained data into text and display the text. Many in the deaf community cannot speak; hence, they often use hand signals to communicate among themselves. But when it comes to communicating with hearing people, it proves to be difficult, as most hearing people do not understand sign language. So, we are developing this system to make it easier for the deaf to communicate with people who do not understand their sign language. Our program will detect and translate the hand signs made by the user into text and display it on the screen for the hearing people to read. Thus, this work will make it easier for deaf people to interact with other people (Koppu & Rajgopal, 2017).

In this project, the main stage is the conversion of hand gestures of deaf people into text. The main theme of the proposed work is to detect and recognize the hand gestures, and show the output in the form of accurate text through various classifiers. The end user needs to perform various hand signs in front of the camera. As the user performs gestures, our program will detect and translate the gestures into text in parallel. There will be two displays on the screen: one will show a hand gesture, and the other will show the text representing the gesture. Our project works as a translator for sign language. It reduces many problems such as the need of a human translator. Deaf people will be able to express themselves through our program. Overall, the first need is to collect hand gesture samples from video sequence related to some sign language. For the recognition of hand signs, after the collection of dataset, the next level consists of subtraction of unwanted regions from the hand portion. For the same, in some methods, we need to define the threshold. After the segmentation level, next is to check the accuracy of recognition. In the segmentation level, we will match the segmented part to the available dataset. Then, we will select the most accurate data from the dataset and assign a weight for the next comparison, for faster selection of data. Finally, we will convert the obtained data into text and display the text.

Although the below-mentioned procedure is sufficient to attain a considerable level of precision, they still carry a wide margin in terms of the following:

- **Authentication of sample inputs of the dataset.**
- **Availability of Samples of Every Class**: To check the flexibility and robustness of the SLRS.
- **Conditioning of Image of Hand Gestures**: To make an input sample image appropriate.
- **Elimination of Redundant Information**: To eliminate the redundant information by considering the most relevant features.
- **Optimization of Feature Vector**: To prepare an adequate feature vector that helps to improve the accuracy of the classification system.
- **Systematic Classification**: To analyze various classifiers on the optimal feature vector.

Figure 11.1 depicts the process that is used for sign image detection, and the classification begins with acquisition of image by using the image acquisition device or using the images from a benchmark dataset. This image needs to be passed through certain preprocessing filters to improve its quality by making noise free, high in contrast, and ghost free. An appropriate segmentation technique needs to be employed to

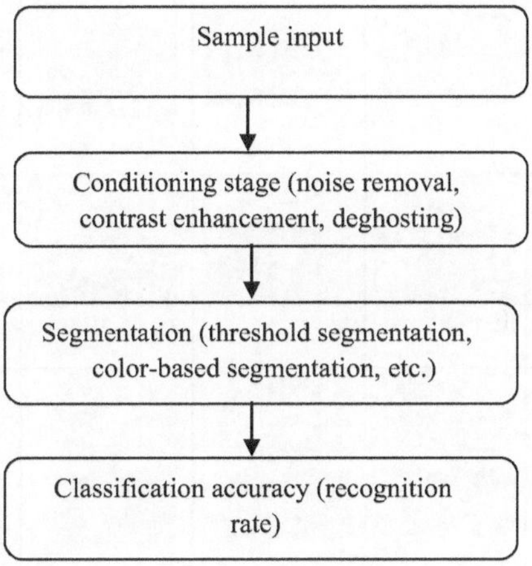

FIGURE 11.1 Block structure of sign language recognition system.

extract region of interest from the improved image. The various features are extracted from enhanced as well as segmented image to make a feature vector and thus a codebook. This feature vector trains the classifier through transfer learning, and then, the whole system is quantified by evaluating performance on the test images.

11.3.1 Validated Dataset

For the analysis of the SLRS, the main thing to focus is finding the validated dataset (Figure 11.2).

The beginning point of this work was to create a database or use an existing dataset with all the samples that would be required for classification (training and testing). The segmentation has been used on hand symbol sample images taken from the publicly available database of Senz3D and the Kaggle dataset. The image database includes various gestures (static) taken by using the Creative Senz3D camera as mentioned in literature and has been referenced (Memo & Zanuttigh, 2018). The dataset includes several static gestures acquired with the Creative Senz3D camera (Buonocore & Gao, 1997).

The dataset consists of the gestures performed by four individuals that are different, each one given the performance for eleven distinctive gestures that are repeated thirty times each, for a maximum of 1320 input samples. For each sample input,

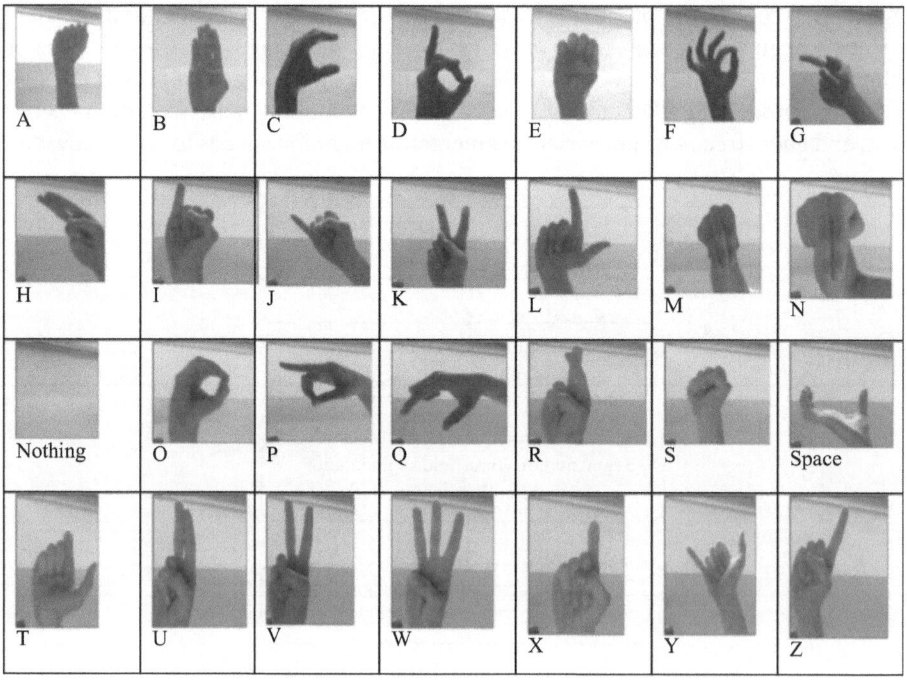

FIGURE 11.2 Exemplary dataset.

TABLE 11.2
Comparative Analysis in Terms of PSNR with State-of-the-Art Systems

Reference	Method	Reported Value	PSNR (By Median Filter)
Stergiopoulou and Papamarkos (2009)	Color-based segmentation	29.7757	81.17%

color, confidence, and depth frames can be found. Intrinsic parameters because of the Creative Senz3D can also be provided. Another training dataset utilized within this effort is the Kaggle dataset, which includes 87,000 pictures that are 200×200 pixels. You will find 29 classes, of which 26 are for all the letters A–Z and three classes for SPACE, DELETE, and NOTHING. These three classes are extremely good in real-time programs and category (Oyedotun & Khashman, 2017). The test dataset has a mere 29 images, to motivate using actual evaluation images (Table 11.2).

11.3.2 Conditioning Stage

A better contrast image, noise removal, and deghosting are needed to devise a highly precise system for identification and classification purposes. Noise filtering, contrast enhancement, and deghosting are the main operations at the preprocessing stage for the conditioning of input image. Preprocessing is a high-priority area in digital image processing, with an enormous number of image processing applications such as pattern recognition and image analysis relying heavily on the quality upgrading of contaminated images. The goal of preprocessing is to upgrade the quality of an image. Image contrast enhancement, ghost artifact removal, and noise reduction are all done at this stage.

To smooth the contour area of hand region, different filters are applied on the image captured through camera or the sample image taken from the validated dataset. Filtering may help to alleviate the noisy information from the image. Noise can occur through different causes, such as defective sensors, during image capture, and poor lighting conditions, during transmission.

Preprocessing contrast depicts the pixel intensity that will help to distinguish one object from the other object or distinguish between foreground and background of an image. In grayscale images, contrast is calculated by finding the difference between the object brightness and its external environment (surroundings). Contrast enhancement is one of the essential steps in image analysis that is significantly used in image processing, pattern recognition, and many other applications. Contrast enhancement sharpens the edges of objects in an image, making it easier to extract objects and extract more information from the quality improved image. Actually, the main objective of image contrast enhancement is to make an image suitable for a particular application by improving the quality of it.

Deghosting is a topic that has gotten a lot of attention in the last decade. Tursun et al. (2015) made an attempt to segment HDR image deghosting approaches into four categories based on selection and registration of dynamic object. Ward suggested a multi-resolution analysis method for correcting translational misalignment between captured images using pixel median values, based on the global exposure registration methodology (Ward, 2003). Cerman et al. (2019) proposed employing frequency domain correlation to reduce both translation- and rotation-based misalignments. Im et al. (2011) proposed minimizing the sum of squared errors to estimate affine transformation between multi-exposure images. Akyuz (2011) proposed a method for testing pixel order relations that was inspired by Ward's method. They observed that pixels with less intensity value than their bottom neighbors and much more intensity as compared to their right neighbors should have the same association across exposures. They discovered a link between these types of relationships and reduced the hamming distance between correlation maps for alignment (Kim et al., 2008). These ghost artifacts appear as shadows of the original structure in DTI acquisitions in various diffusion gradient directions. This can make it harder to see the true anatomy of the spinal cord and make the true position of the cord more ambiguous (Gevrekci & Gunturk, 2007). In the literature, several approaches for correcting echo misalignment have been proposed. Reference scan-based approaches are now employed primarily on clinical MRI scanners. These strategies use modest gradients to align echoes after performing a calibration scan to estimate the on-axis gradient/data collection time delay (Li et al., 2013). Reference scans, on the other hand, are susceptible to dynamic changes such as subject mobility, which complicates MRI pulse sequence design (Durand et al., 2001). Parallel imaging techniques are another option to reduce the ghosting phenomenon.

Despite the good performance of the various spatial filters for upgrading image quality, there is a need of hybrid catering denoising, contrast enhancement, and removal of ghost artifacts with improved metric values (Tomaszewska & Mantiuk, 2007). In the stage dedicated to identifying and reducing impulses, the approach can be significantly enhanced. To do this, we combined the stage of image denoising with the NLM filter, which was specifically designed to provide the best noise suppression outcomes. The experiments will depict that the introduced filter improves various attributes such as peak signal-to-noise ratio (PSNR), the mean squared error (MSE), Structural Similarity Index (SSIM), the NAE method, and other state-of-the-art filters. Then, in this work, a new hybrid method called modified Yanowitz–Bruckstein binarization (MYBB) is introduced to tackle ghosting effects of an image. There are two stages to the suggested hybrid technique. A two-step approach based on non-local means filtering and top-hat filtering is used in the first stage of the complete model to reduce the effect of noise and improve image contrast. Then, a hybrid technique based on the Yanowitz–Bruckstein binarization filter is utilized to enhance the metric values of the preprocessed stage.

The following are the main goals of using preprocessing techniques:

- The image must be converted into a high-contrast equalized image.
- Noise pixels should have a minimal gain amplification, and information pixels should have a maximum gain amplification.
- As part of the segmentation stage, the image must be transformed to the proper color space.

TABLE 11.3
Objectives for the Image Quality Improvement

Name of the Objective	Description
Image filtering	To minimize the noise effect from the input image
Contrast enhancement	To improve the contrast of the filtered image
Deghosting	To reduce the interference contributed by the capturing system
Peak signal-to-noise ratio	To maximize the peak signal-to-noise ratio through an adequate filter
Mean squared error	To minimize the mean squared error

The contrast enhancement methods are classified into two categories as follows:

- **Contrast Stretching**: The initial pixel intensity value is linearly mapped onto possibilities, with the linear operator chosen to improve the image. There are three approaches for enhancing linear contrast: maxima–minima piecewise linear stretch, percentage linear stretch, and linear constant search (Dah-Chung & Wen-Rong, 1998).
- **Histogram Equalization and Matching**: Histogram equalization is a technique for evenly spreading all pixels throughout the whole intensity range. The image's contrast will be boosted because of the adjusted histogram.

For the minimization of noise, contrast enhancement, removal of ghost artifacts, and for the improvement of various metric values, different algorithms are utilized to achieve adequate results, and at the final point, a hybrid algorithm is used to yield adequate results. Table 11.3 lists the details of the above-said preprocessing objectives.

11.3.3 Segmentation Stage

The main purpose of the segmentation stage is to extract region of interest. Samples within a dataset that are marked for a specific reason are a region of interest (often abbreviated ROI). Its aim is to perform max pooling to obtain fixed-size feature maps on inputs of nonuniform sizes. The design for object detection is broken down into the following phases.

Convex Hull: A convex object is one that is greater than 180° without interior angles. Non-convex is the name of a form which is not convex. In computational geometry, the convex hull is a ubiquitous structure. While it is a helpful method, it is also useful in creating other structures, such as Voronoi diagrams, and in applications such as unsupervised image analysis.

Contour Extraction: Usually, edge detectors generate small, disjoint segments of the edge. Before they are aggregated into extended edges, these segments are typically of little use. Two major method categories are local techniques (extended edges by looking for the most "compatible" candidate edge in a neighborhood) and global techniques (it is possible to integrate more computationally costly domain information into their cost function).

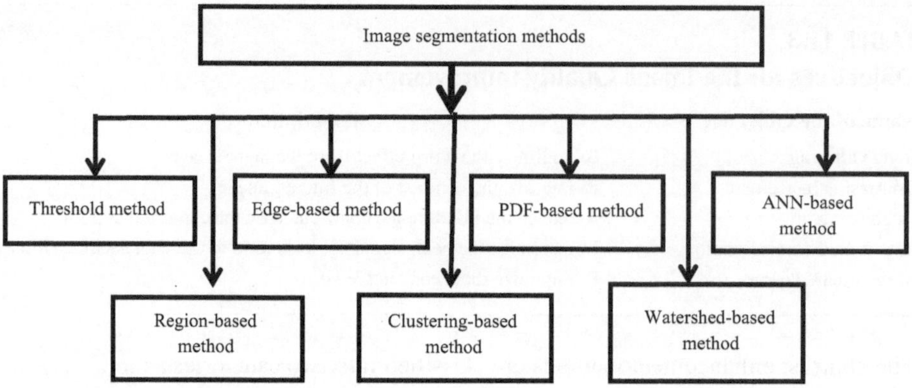

FIGURE 11.3 Various algorithms toward image segmentation.

Various segmentation techniques are shown in Figure 11.3, which will help to extract region of interest. The edge-based methods work upon finding the edges in the images and combining the edges to form structures and objects in an image. The threshold-based segmentation methods depend upon comparing the intensity value to a threshold value and then assigning either 1 or 0 to the pixel. The region-based segmentation methods either combine the pixels to form the regions or split the larger region into subsequent regions.

Thresholding is a method of segmenting images and is commonly used to produce binary images (Jani et al., 2018). Easy thresholding and adaptive thresholding are two ways of thresholding. Simple Thresholding: In a simple thresholding procedure, the standard value is allocated to pixels whose values are greater than the threshold value defined.

Thresholding strategies turn a set of gray picture colors into two colors, i.e., white and black. It works by choosing a threshold estimate. The grouping of pixels into two clusters depends on the pixel's intensity value. All pixels with a gray value less than the gray value of one cluster would be the value of one cluster. Both pixels having a gray will consist of the threshold and the second cluster.

The thresholding segmentation technique for segmentation of images is a simple and effective procedure. By selecting an appropriate intensity value as threshold, a comparison of individual pixel is carried out with the threshold and the pixels are grouped together to form one region. This algorithm concentrates mainly on how to subtract the front portion from the background region. It is divisible into three groups. They are as follows:

1. **Global Thresholding**: One intensity value is serving as threshold, and it is applied to all the pixels of the image by using the following equation:

$$Y(m, n) = \begin{cases} 1 \text{ if } I(m, n) > T \\ 0 \text{ if } I(m, n) < T \end{cases}$$

2. **Variable Thresholding**: In this type of thresholding, the intensity value selected as threshold varies throughout the image; i.e., no fixed intensity value is defined as threshold.
3. **Multiple Thresholding**: The multiple intensity values are selected as threshold, and the pixel intensities are compared with these thresholds to create partitions in an image (Table 11.4).

11.3.4 Feature Extraction and Selection

After the segmentation stage, the next main step is to build an optimal feature vector. One of the most important aspects of the classification technique is feature extraction. It has become particularly relevant in the domain of pattern recognition. The consistency of the segmented image decides which features are selected. The extracted features are compiled into a codebook that a classifier may use as a reference. The extracted features must be chosen, so there is less redundancy in the applied parameter.

The most important aspects of image features are as follows:

1. **Features Must Include All Types of Information That Will Help to Distinguish Different Classes**: For designing optimal feature vector, the feature set must involve all types of features that exhibit complete information about an image.
2. **Adequate Number of Features for Yielding the Best Results**: It involves the quantitative aspect regarding features. It means to consider enough features for designing adequate feature vector.
3. **Select Appropriate Features for Avoiding Redundant Information**: For avoiding redundant information, choose an appropriate method that helps to consider maximum relevance features.

The details of image feature extraction are shown in Figure 11.4.

11.3.5 Classification Stage

The classification stage includes the training and testing phase that needs a wide range of decision theoretical methods to object recognition. It is the process of arranging pixels into groups and assigning labels to each group. The various classifiers used for classification purpose include SVM and naïve Bayes classifier.

SVM is an algorithm for supervised learning which can be used to classify or to regress. However, much of it is used for problems of classification. The value of each function is a given coordinate's value. Then, we define the hyperplane that makes a strong distinction between the two groups. Table 11.5 depicts that SVM classifier has been used in various application fields including image analysis, classification, prediction/forecasting, and optimization problems (Figure 11.5).

One of the probability-based methods is used as classification algorithm. The naïve Bayes technique comes under the category of learning technique in which the target value is known, i.e., supervised learning (Table 11.6).

TABLE 11.4
Analysis of Different Segmentation Methods

Sr. No.	Underlying Method	Depiction	Categories	Benefits	Drawbacks	Applications
1	Region-based Senthilkumaran and Rajesh (2007)	Shape of sub-regions in a cluster of neighboring pixels based on models of similar intensity	1. Region-based merging or splitting 2. Watershed-based transformation	Very simple algorithm and best for noise removal	Time-consuming process	Pixel grouping, piecewise constant radiance, 3D reconstruction
2	Thresholding segmentation Zhang (2006)	Thresholding algorithm divides an image into two different parts based on intensity values	1. Segmentation based on local thresholding 2. Global thresholding 3. Dynamic thresholding	Lesser computational complexity, easy data management	Thresholding segmentation methods are sensitive to noise	Medical image analysis (e.g., to locate tumor)
3	Edge segmentation Al-Amri et al. (2010)	This detects the abrupt changes in the intensity value at the edges	1. Gray histogram algorithm. Gradient-based edge segmentation method	The method can retrieve information easily	Suitable even under noise effect	Image identification and analysis algorithm
4	Theory-based Haralick et al. (1985)	In this technique, the derivatives from various regions are considered	1. Clustering-based 2. Neural network-based	Straightforward for classification. Implementation is quite simple	Higher computation cost. Features also rely on image. The criterion for choosing the feature is unclear	The volume of tissues is accurately measured

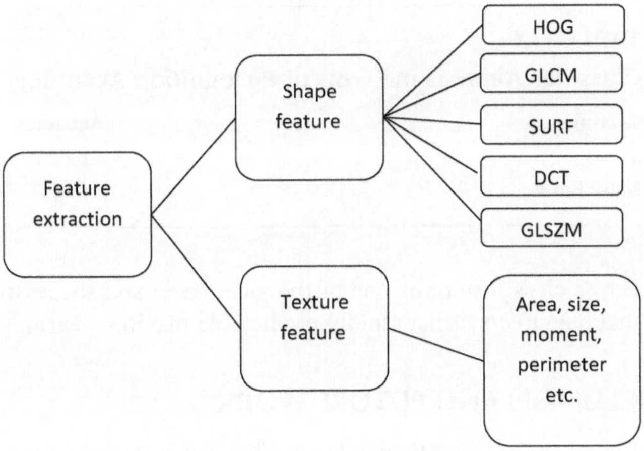

FIGURE 11.4 Different image features.

TABLE 11.5
Different SVM Models Used in Different Areas

Author	Model Name	Basis Function	Application Field
Tsai (2005)	SVM	Polynomial	Image classification
Shao and Deng (2012)	CDMTSVM	Quadratic	Classification
Lin et al. (2006)	SVFNN	Polynomial	Pattern classification
Kim (2004)	SVM	Polynomial	Prediction

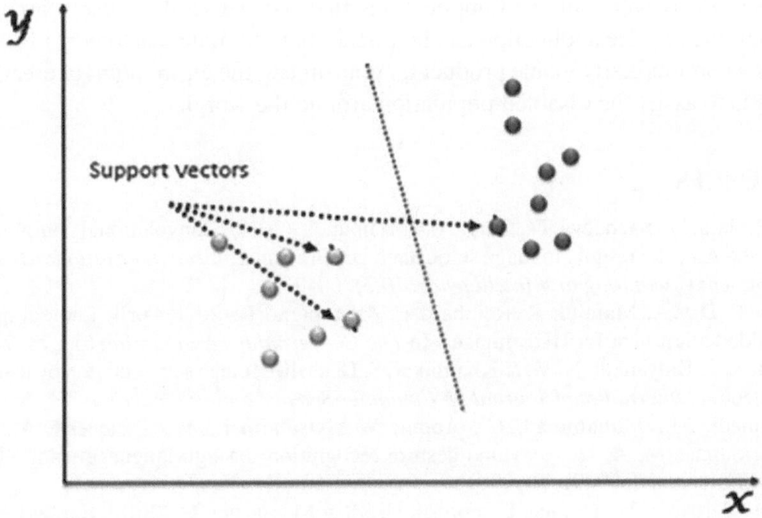

FIGURE 11.5 SVM classifier.

TABLE 11.6
Classifier Analysis in Terms of Recognition Accuracy

Classifier	Accuracy
SVM	98.4%
Naïve Bayes	92.2%

The naïve Bayes classification is one of the most basic and successful classification methods that help to construct rapidly predictable machine learning models.

11.4 CONCLUSION AND FUTURE WORK

The suggested hand recognition device is very helpful because it can be used as a human–computer interface and can assist paralyzed individuals. The movement of the hand gesture can be easily regulated in future software. The sensor techniques were used in previous concepts, but no extra sensor is needed in this framework. The deep CNN algorithm will accomplish this. The strong dataset is solved by deep learning. The first is the extraction of coevolutionary neural network features, and the second is the classification process with fully linked layers. For a wide collection of data, neural network training is done, and the aim is achieved. The current work can be upgraded by applying some optimization technique during selection of feature vector of hand symbols. Even at the end stage, the system can be appended with some Internet-of-Things (IoT) system for upgrading the human–computer interaction. The optimization methods can enhance the accuracy rate of the system. An alternative stress may be placed on the usage of the application in the fields of drugs, military, governance, etc. A genuine blend of various technologies in the listed fields could make way for power tools and applications that will serve the society around the world. Ultimately, the application can be further built to make customers more available. As a commercially viable product for consumers, the entire point of creating the solution is to assist the disabled population around the world.

REFERENCES

Ahuja, R., Jain, D., Sachdeva, D., Garg, A., & Rajput, C. (2019). Convolutional neural network based American sign language static hand gesture recognition. *International Journal of Ambient Computing and Intelligence*, *10*(3), 60–73.

Akyüz, A O, Day, A., Mantiuk, R., Reinhard, E., & Scopigno, R. (2011, April). Photographically guided alignment for HDR images. In *The Eurographics Association* (pp. 73–74).

Al-Amri, S. S., Kalyankar, N. V., & Khamitkar, S. D. (2010). Image segmentation by using edge detection. *International Journal on Computer Science and Engineering*, *2*(3), 804–807.

Al-Hammadi, M., Muhammad, G., Abdul, W., Alsulaiman, M., Bencherif, M. A., & Mekhtiche, M. A. (2020). Hand gesture recognition for sign language using 3DCNN. *IEEE Access*, *8*, 79491–79509.

Avola, D., Bernardi, M., Cinque, L., Foresti, G. L., & Massaroni, C. (2018). Exploiting recurrent neural networks and leap motion controller for the recognition of sign language and semaphoric hand gestures. *IEEE Transactions on Multimedia*, *21*(1), 234–245.

Buonocore, M. H., & Gao, L. (1997). Ghost artifact reduction for echo planar imaging using image phase correction. *Magnetic Resonance in Medicine*, *38*(1), 89–100.

Cerman, L., & Hlavac, V. (2006, February). Exposure time estimation for high dynamic range imaging with hand held camera. In *Proceedings of Computer Vision Winter Workshop*, Czech Republic, Ondrej Chum, Vojtêch Franc (eds.).

Chang, D. C., & Wu, W. R. (1998). Image contrast enhancement based on a histogram transformation of local standard deviation. *IEEE Transactions on Medical Imaging*, *17*(4), 518–531.

Chen, H., Zhang, Y., Li, G., Fang, Y., & Liu, H. (2020). Surface electromyography feature extraction via convolutional neural network. *International Journal of Machine Learning and Cybernetics*, *11*(1), 185–196.

Chen, Z. H., Kim, J. T., Liang, J., Zhang, J., & Yuan, Y. B. (2014). Real-time hand gesture recognition using finger segmentation. *The Scientific World Journal*, *2014*, 1–9.

Chong, T. W., & Lee, B. G. (2018). American sign language recognition using leap motion controller with machine learning approach. *Sensors*, *18*(10), 3554.

Delaye, A., & Anquetil, E. (2013). HBF49 feature set: a first unified baseline for online symbol recognition. *Pattern Recognition*, *46*(1), 117–130.

Durand, E., van de Moortele, P. F., Pachot-Clouard, M., & Le Bihan, D. (2001). Artifact due to B0 fluctuations in fMRI: correction using the k-space central line. *Magnetic Resonance in Medicine: An Official Journal of the International Society for Magnetic Resonance in Medicine*, *46*(1), 198–201.

Gamage, N., Kuang, Y. C., Akmeliawati, R., & Demidenko, S. (2011). Gaussian process dynamical models for hand gesture interpretation in sign language. *Pattern Recognition Letters*, *32*(15), 2009–2014.

Gevrekci, M., & Gunturk, B. K. (2007, April). On geometric and photometric registration of images. In *2007 IEEE International Conference on Acoustics, Speech and Signal Processing-ICASSP'07* I-1261-I-1264, Honolulu, HI, USA.

Haralick, R. M., & Shapiro, L. G. (1985). Image segmentation techniques. *Computer Vision, Graphics, and Image Processing*, *29*(1), 100–132.

Im, J., Jang, S., Lee, S., & Paik, J. (2011, September). Geometrical transformation-based ghost artifacts removing for high dynamic range image. In *2011 18th IEEE International Conference on Image Processing*, Brussels, Belgium, 357–360.

Itkarkar, R. R., Nandi, A., & Mane, B. (2017). Contour-based real-time hand gesture recognition for Indian sign language. In Behera H., Mohapatra D. (eds) *Computational Intelligence in Data Mining. Advances in Intelligent Systems and Computing*, vol. 556. Springer, Singapore (pp. 683–691).

Jani, A. B., Kotak, N. A., & Roy, A. K. (2018, October). Sensor based hand gesture recognition system for English alphabets used in sign language of deaf-mute people. *2018 IEEE Sensors*, 1–4.

Khan, R. Z., & Ibraheem, N. A. (2012). Survey on gesture recognition for hand image postures. *Computer and Information Science*, *5*(3), 110.

Kılıboz, N. Ç., & Güdükbay, U. (2015). A hand gesture recognition technique for human–computer interaction. *Journal of Visual Communication and Image Representation*, *28*, 97–104.

Kim, D. (2004). Prediction performance of support vector machines on input vector normalization methods. *International Journal of Computer Mathematics*, *81*(5), 547–554.

Kim, Y. C., Nielsen, J. F., & Nayak, K. S. (2008). Automatic correction of echo-planar imaging (EPI) ghosting artifacts in real-time interactive cardiac MRI using sensitivity encoding. *Journal of Magnetic Resonance Imaging: An Official Journal of the International Society for Magnetic Resonance in Medicine*, *27*(1), 239–245.

Koppu, S., & Rajgopal, M.K. (2017). Study of hand gesture recognition and classification. *Asian Journal of Pharmaceutical and Clinical Research*, *10*(13), 25–30.

Li, H., Fox-Neff, K., Vaughan, B., French, D., Szaflarski, J. P., & Li, Y. (2013). Parallel EPI artifact correction (PEAC) for N/2 ghost suppression in neuroimaging applications. *Magnetic Resonance Imaging*, *31*(6), 1022–1028.

Lin, C. T., Yeh, C. M., Liang, S. F., Chung, J. F., & Kumar, N. (2006). Support-vector-based fuzzy neural network for pattern classification. *IEEE Transactions on Fuzzy Systems*, *14*(1), 31–41.

Memo, A., & Zanuttigh, P. (2018). Head-mounted gesture controlled interface for human-computer interaction. *Multimedia Tools and Applications*, *77*(1), 27–53.

Misra, S., Singha, J., & Laskar, R. H. (2018). Vision-based hand gesture recognition of alphabets, numbers, arithmetic operators and ASCII characters in order to develop a virtual text-entry interface system. *Neural Computing and Applications*, *29*(8), 117–135.

Murakami, K., & Taguchi, H. (1991, March). Gesture recognition using recurrent neural networks. In *Proceedings of the SIGCHI Conference on Human Factors in Computing Systems*, Montréal Québec Canada (pp. 237–242).

Nikam, A. S., & Ambekar, A. G. (2016, November). Sign language recognition using image based hand gesture recognition techniques. *2016 Online International Conference on Green Engineering and Technologies-IC-GET)*, Coimbatore, India (pp. 1–5).

Ohn-Bar, E., & Trivedi, M. M. (2014). Hand gesture recognition in real time for automotive interfaces: a multimodal vision-based approach and evaluations. *IEEE Transactions on Intelligent Transportation Systems*, *15*(6), 2368–2377.

Oyedotun, O. K., & Khashman, A. (2017). Deep learning in vision-based static hand gesture recognition. *Neural Computing and Applications*, *28*(12), 3941–3951.

Pal, N. R., & Pal, S. K. (1993). A review on image segmentation techniques. *Pattern Recognition*, *26*(9), 1277–1294.

Rahagiyanto, A., Basuki, A., & Sigit, R. (2017). Moment invariant features extraction for hand gesture recognition of sign language based on SIBI. *EMITTER International Journal of Engineering Technology*, *5*(1), 119–138.

Rahim, M. A., Islam, M. R., & Shin, J. (2019). Non-touch sign word recognition based on dynamic hand gesture using hybrid segmentation and CNN feature fusion. *Applied Sciences*, *9*(18), 3790.

Sahoo, J. P., Ari, S., & Ghosh, D. K. (2018). Hand gesture recognition using DWT and F-ratio based feature descriptor. *IET Image Processing*, *12*(10), 1780–1787.

Senthilkumaran, N. and Rajesh, R. (2009). Edge detection techniques for image segmentation- a survey of soft computing approaches. *International Journal of Recent Trends in Engineering. 1(*2), 250–254.

Shao, Y. H., & Deng, N. Y. (2012). A coordinate descent margin based-twin support vector machine for classification. *Neural Networks*, *25*, 114–121.

Sood, A., & Mishra, A. (2016, September). AAWAAZ: a communication system for deaf and dumb. In *2016 5th International Conference on Reliability, Infocom Technologies and Optimization (Trends and Future Directions)-ICRITO*, Uttar Pradesh, Noida, India, 620–624.

Stergiopoulou, E., & Papamarkos, N. (2009). Hand gesture recognition using a neural network shape fitting technique. *Engineering Applications of Artificial Intelligence*, *22*(8), 1141–1158.

Tomaszewska, A., & Mantiuk, R. (2007). Image registration for multi-exposure high dynamic range image acquisition. In *Proceedings of the 15th International Conference in Central Europe on Computer Graphics, Visualization and Computer Vision 2007*, University of West Bohemia, Pilsen, Czech Republic.

Tsai, C. F. (2005). Training support vector machines based on stacked generalization for image classification. *Neurocomputing*, *64*, 497–503.

Tursun, O. T., Akyüz, A. O., Erdem, A., & Erdem, E. (2015, May). The state of the art in HDR deghosting: a survey and evaluation. *Computer Graphics Forum*, *34* 9(2), 683–707.

Ward, G. (2003). Fast, robust image registration for compositing high dynamic range photographs from hand-held exposures. *Journal of Graphics Tools*, *8*(2), 17–30.

Winkelmann, R., Börnert, P., & Dössel, O. (2005). Ghost artifact removal using a parallel imaging approach. *Magnetic Resonance in Medicine: An Official Journal of the International Society for Magnetic Resonance in Medicine*, *54*(4), 1002–1009.

Zhang, Y.-J. (2006). An overview of image and video segmentation in the last 40 years. In *Advances in Image and Video Segmentation*, IGI Global, USA (pp. 1–16).

Zhang, Z., Yang, K., Qian, J. & Zhang, L. (2019). Real-time surface EMG pattern recognition for hand gestures based on an artificial neural network. *Sensors*, *19*(14), 3170.

Zhi, D., de Oliveira, T. E. A., da Fonseca, V. P., & Petriu, E. M. (2018, June). Teaching a robot sign language using vision-based hand gesture recognition. *2018 IEEE International Conference on Computational Intelligence and Virtual Environments for Measurement Systems and Applications-CIVEMSA*, Ontario, Canada (pp. 1–6).

12 Study and Control of Shrinkage in Gearbox Sand Casting Using Simulation and Experimental Validation

Sarabjit Singh, Rajesh Khanna, and Neeraj Sharma

CONTENTS

12.1 INTRODUCTION

The majority of industrial sectors use castings as the major inputs due to a variety of reasons. Among different types of casting processes, green sand casting is widely used in the industry due to good geometric flexibility. Green sand castings offer numerous advantages such as good dampening properties, compressive strength, ease of machining, ease of casting, and good structural rigidity.

DOI: 10.1201/9781003158165-12

Today, the casting process has become an integral part of many sectors such as automotive, agriculture, heavy engineering, construction, and machinery. Various casting manufacturing processes are being used worldwide on the basis of various factors such as mass production or batch production, economic considerations, surface finish, and design intricacy. The main casting processes include sand casting, die casting, sand-shell casting, and investment casting.

Green sand casting, as the name depicts, is the product of molten metal filled in sand cavity with moisture content. After the addition of different binders and chemicals in typical river bed sand, the moisture level of approx. 4% is maintained to get core hardness of 80–90 Hb and green compressive strength of approx. 1200–1400 g/cm^2. To achieve the desired mold qualities, different sand properties such as AFS (American Foundry Society) number, permeability, compatibility, and moisture content are maintained. After maintaining these properties, the sand is filled in the upper half (cope) and lower half (drag) of molding boxes. The sand in molding boxes is compressed under molding machines with compressive forces varying from 450 – 1200 bar, depending upon molding box size. For parts involving hollow sections or cavities, cores are placed in the molding box sands with the help of core setting fixtures. The use of certain blow candles and chaplets ensures proper positioning of the core in the molding box. The core sand should be collapsible in nature, during the fettling process after metal gets solidified, after core has given the desired shape in the mold cavity. The core scratch hardness is kept in the range 75–80 Hb.

The ingredients of molten metal, in the case of gray cast iron, mainly include pig iron, steel scrapping, and constituent elements including carbon, magnesium, and silicon. The melting process is generally done in induction furnaces, with the target pouring temperature range of 1100°C–1250°C. The inoculation process during the pouring process helps in grain structure refinement of metal to be casted. The target carbon equivalent CEV is kept in the range of 3.5–4.2 for gray cast iron. Other important parameters for the pouring process involve molten metal pressure, molten metal velocity, and Reynolds number (indicator of turbulence). There are certain aspects that need to be considered during the casting process. For the mold cavity, due consideration must be given for the shrinkage allowance of the solidifying metal. The mold material must have a refractory character so that it is not affected by the molten metal in the cavity. In order to ensure that the produced casting is dense and free of defects, provision should be made to permit the escape of any gases or air from the mold before pouring the molten metal in the cavity. After successful molten metal pouring process, the metal is allowed to solidify. It is then subjected to the fettling and grinding process, and finally primer application. Further, machining is done on the cast parts. As far as the performance of the casting part is concerned, the main parameters include chemical composition, hardness, ultimate tensile strength, elongation, microstructure, and flake size.

In any automotive or agriculture industry, the gray cast iron casting parts contribute to more than 30% of total parts being assembled on the vehicle or tractor. Castings, however, bring inherently with them unpredictable defects such as crack, shrinkage porosity, and blow holes. In typical foundries with good process controls, the rejections due to various casting defects, such as crack, shrinkage porosity, blow holes, and unclean areas, can go up to 8% of the total tonnage. With manual controls,

TABLE 12.1
Matrix for Contributing Factors for Casting Defects

Characteristic	Contributing Factor	Type of Defect
Tooling	Pattern dimensions	Mismatch
	Profiles and sharp radius	Excess/less machining margins
Manufacturing process	Molding sand	Mismatch
		Scab
		Excess flash
	Metal pouring temperature	Blowholes
		Cold shuts
		Gas Porosity
		Sand inclusion
	Metal solidification	Shrinkage porosity
		Hot tears
		Crack
		Distortion/swelling
Material	Chemical composition	Chills/hard spots
		Segregation
		Low strength/hardness
		Poor machinability

this rejection percentage can increase up to 15%. Due to shrinking margins and competitions, such rejections decrease the profit of foundries as the returned rejected component value is only that of scrap. In spite of the state-of-the-art infrastructure setup of various foundries, the defect percentage and defect location always remain a big question mark. Table 12.1 shows various contributing factors for various casting defects in green sand castings.

12.2 LITERATURE REVIEW

Different functional advantages and economic benefits of the casting process make it a better option, when compared to other metal forming methods. Right from the component design stage to the final casting production stage, the casting process offers good flexibility and versatility. Foundries offer good products rage and are normally categorized by ferrous foundries (production of various alloys of cast iron and cast steel) and non-ferrous foundries (production of aluminum-base, copper-base, zinc-base, magnesium, and other non-ferrous castings) (Mocellin et al., 2003). Different forms of cast iron include gray iron, white iron, ductile iron, malleable iron, and mottled iron. On the basis of appearance of fracture surfaces, gray iron and white iron have driven such names. Mottled iron is a variety between gray iron and white iron. Ductile iron has good malleable ductility. The good geometric freedom capability and cost-effectiveness have made sand casting a preferred manufacturing process.

Casting defects have been categorized on the basis of their geometry (physical appearance), integrity (location, formation principle), and properties (consequences) (Renukananda et al., 2012). Based on their physical appearance, there are seven different categories, including metallic projection, cavity, discontinuity, defective surface, incomplete casting, incorrect dimension/shape, and inclusion/structural anomaly. Another way of classifying most commonly occurring defects in sand casting is based on the location of defects (surface or subsurface). The defects in cast part may arise at any one of the following steps in the casting process: (i) design of casting and pattern, (ii) molding sand and design of mold and core, (iii) metal composition, (iv) melting and pouring, and (v) gating and riser. Shrinkage porosity has been found as the most persistent and common complaint of casting users (Rajkolhe & Khan et al., 2014). In case of forgings, machined parts, and fabrications, mechanical processing and ingot cast feedstock help in avoiding defects such as porosity. A properly designed gating system successfully delivers metal to part cavity, promotes proper solidification, minimizes turbulence, keeps damaged metal and inclusions away from entering the part, and fills the mold quickly. Geometrical modulus is responsible for controlled solidification. The Young modulus is linked with the solidification in the casting process (Girish et al., 2013). The modulus changes during the solidification process. Geometrical modulus combined with casting modulus and feeder modulus affects the shrinkage porosity in the casting (Havlicek et al., 2011). Riser design is also an important factor to avoid undesirable shrinkage cavities in the castings. The risering may also help in moving the shrinkage to locations where they may be acceptable for the intended application of the casting (Plutshack & Foseco, 1992). In case of metals such as carbon steel, huge shrinkage up to 3% may be encountered. During cooling of the metal under solidification, more shrinkage occurs. These contractions result in porosities, if sufficient risering is not available in the casting gating and feeding system (Hardin & Beckermann, 2002).

In order to study the influence of various factors and runner designs on the flow of molten steel in various types of horizontal and vertical gating systems, different experiments have been conducted and the resulting turbulence noted in the mold cavity. Experiments have also been conducted on shrinkage in gray iron castings, using phenolic urethane no-bake binders (Umezurike & Onche, 2010). Different parameters such as binder level, binder ratio, mixing effect, core washes and core post-baking, casting temperature, and section size have been studied for porosity formation in gray iron castings. The flow rate of molten metal is dependent upon filling sequence, shape and size of elements, location, number and distribution of molten metal into the mold cavity (Renukanand et al., 2012). Defects such as misrun and cold shut may occur due to slow filling. Similarly, defects such as shrinkage porosity and blow holes may occur due to rapid filling.

The process of mold filling of liquid metal involves turbulence, pressure variation, viscous flow, heat transfer in the metal–mold interface, and phase change during casting. Generally, mold filling takes only a small fraction of time that is needed for casting to cool and solidify. In a casting process, there is complex involvement of simultaneous fluid-thermal-mechanics phenomena. There are multiple simulation software packages available for foundries from different companies across the world. The main software packages include MAGMASoft, AutoCAST, ProCAST, SOLIDCast, CastCAE, MAVIS, NovaCast, CAP/WRAFTS, JSCast, PAM-CAST, and RAPIDCAST (Paknikar, 2014). Different foundries using gray iron, steel, aluminum, copper, etc., are using these simulation software packages. These simulation techniques involve the use of physical phenomena such as stress/strain behavior of castings, phase transformation, heat transfer, and flow rate. These physical phenomena can be expressed as governing equations. These equations include continuity, momentum conservation (Navier–Stokes), and energy conservation (Eulerian multimodel) equations for flow, heat transfer, and solidification. One of the design aspects that continue to create a challenge for casting designers is the optimum design of casting feeders (risers) (Ahmad, 2015). The governing equations are generally differential in nature and can be solved using various analytical methods (in case of simple shapes) or numerical methods (in case of complex industrial casting) to predict the behavior of molten metal. Various numerical simulation techniques such as finite difference method (FDM), vector element method (VEM), finite element method (FEM), and finite volume method (FVM) are used to solve the relevant governing equations. These numerical methods are generally used to obtain the temporal values of metal velocity and temperature. MAGMASoft has been validated to be able to produce reliable simulation results that actually reflect the real casting phenomena. The results signify the validity of simulation using MAGMASoft to perform mold design and casting process simulation. The results of mold filling and solidification would be of high fidelity that can be relied upon to make decisions on designing the mold, feeder, sprue, runner, and gating system as well as setting the casting process parameters in order to get the required casting quality.

12.3 RESEARCH MATERIAL AND PROPERTIES

The current research study has been carried at M/s DCM (India) having PLC (Programmable Logic Control)-controlled sand plant and press molding lines.

Figure 12.1 shows the defect location in the gearbox housing main bearing bores. The scanning electron microscope (SEM) technique was used to validate the major casting defect as shrinkage porosity defect as per Figure 12.2.

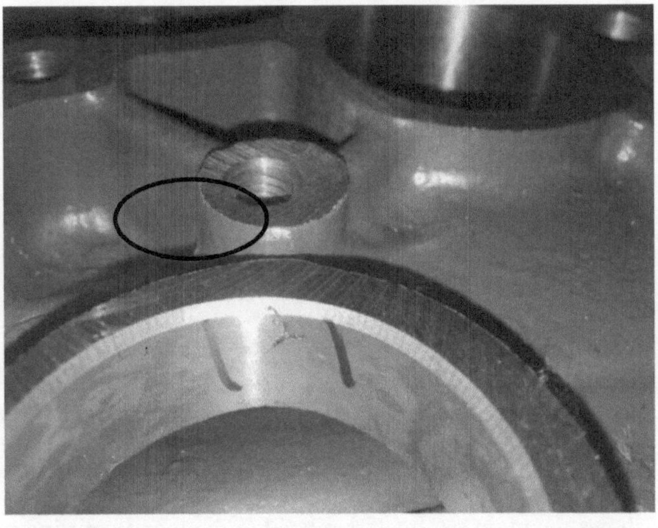

FIGURE 12.1 Defect location in gearbox housing.

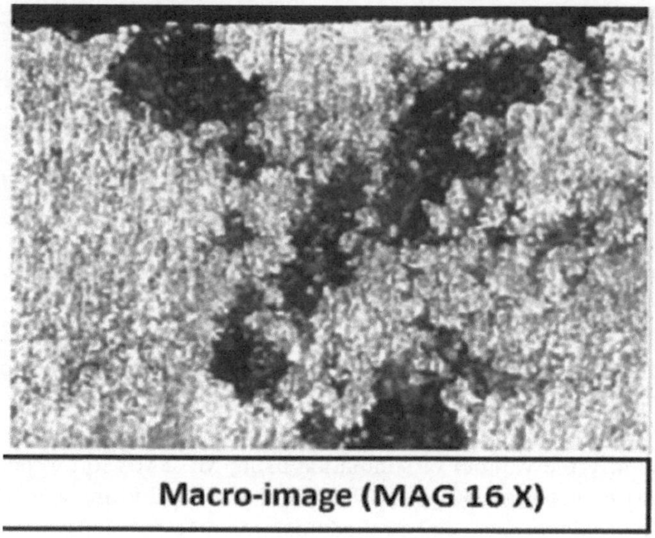

FIGURE 12.2 SEM image of shrinkage defect.

TABLE 12.2
Main Parameters for Gearbox Housing Manufacturing

Sr. No.	Description	Specifications
1	Metal grade	D-25-6
2	Bunch weight	34.3 kg
3	Casting weight	16.1 kg
4	Monthly demand	2500 nos.
5	Sand specification	GCS: 1.5–2.2 kg/cm^2 Compact: 36%–44% Mold hardness: 80~100
6	Line allocation	Line 4
7	Hardness	228–285 BHN
8	Tensile strength (N/mm^2)	497 MPA MIN
9	Matrix	Pearlite – 80% min Ferrite – 20% max Carbide – nil

Table 12.2 shows the main parameters related to the manufacturing process, and Table 12.3a–c shows the component functional properties considered.

TABLE 12.3A
Components' Functional Properties (Chemical Composition and % Age Range)

Chemical Composition	% Age Range
Carbon	3.1–3.3
Silicon	3.9–4.1
Manganese	0.1–0.3
Phosphorus	0.05 max
Sulfur	0.02 max
Chromium	0.05 max
Copper	0.15 max
Magnesium	0.03–0.06

TABLE 12.3B
Components Functional Properties (Stage and Temperature)

Stage	Temperature (°C)
Bath	1550–1590
Transfer	1490–1540
Pouring	1395–1415

TABLE 12.3C
Components' Functional Properties (Stage and Hardness in C Scale)

Stage	Hardness in C Scale
Mold	80~100

12.4 RESEARCH WORK METHODOLOGY TAKEN FOR THE WORK

The research methodology proposed to be used for the research work has been outlined in five stages. In the first stage, physical verification of the rejected machined castings has been done. The type and location of the defect have been noted, and a Pareto diagram (vital-few and trivial-many) has been made to prioritize the defect's type. The second stage involves identification of various parameters affecting the defect and studies the factors influencing the parameters. The output of this stage becomes the input for the third stage, which is numerical simulation for the castings under study. Mass production has been done, and then, machining of these modified castings has been done. Data have been again collected for the rejected castings through concentration diagrams and Pareto diagrams. Finally, in the fifth stage, validation of results has been done for numerical simulation study and experimental study.

12.4.1 Identification of Vital Few Factors through Pareto

The casting machining rejection mainly includes shrinkage, blow holes, and casting unclean. Out of these defects, the shrinkage porosity defect contributes to nearly 60% of the total machining rejection. The first step includes the in-depth study and analysis of current rejection rate in industrial heavy casting housing. This also includes making the concentration diagram for the porosity-related rejection over a period.

12.4.2 Listing of Parameters Responsible for Turbulence

Shrinkage porosity defect increases with an increase in molten metal temperature. However, molten metal should also have sufficient superheat to improve the fluidity of the molten metal. Thinner sections require more superheat as compared to thicker sections. Metal cooling rate is dependent upon many factors such as pouring rate and pouring temperature.

12.4.3 Numerical Simulation for the Existing and Proposed Casting System Parameters

The current research uses MAGMA software for predicting, controlling, and eliminating the shrinkage porosity defect. It is proposed to calculate the Reynolds number through numerical simulation at the problematic and defect potential areas. Various iterations of metal filling time (also percentage) are proposed to be done to obtain the output results. The format used for data collection is depicted in Table 12.4.

Reynolds number is directly related to the molten metal turbulence. Turbulence is the indication of non-smooth cavity filling and causing defects such as shrinkage porosity. This will help in redesigning the casting gating and feeding system. As a thumb rule, porosity value exceeding 10% in numerical simulation may be taken as a basis for carrying the further improvement (on the basis of Indian Foundries data using MAGMAS software. With the optimum design of the casting gating and feeding system, defects such as shrinkage porosity can be reduced by ensuring proper directional solidification in the casting. In the study of MAGMAS key performance indicators (KPIs), the input parameters may be taken as pouring temperature, metal chemical composition, green sand properties, and dimensions of the casting gating and feeding system. Casting filling results will give values of metal pressure, air pressure, air entrapment, sand inclusion, and mold erosion at different levels of filling temperature and molten metal velocity. Casting solidification results will give values of fraction liquid, hotspot, sand burn, sand penetration, and shrinkage porosity. Input parameters in terms of sand properties (sand permeability, moisture content, etc.), chemical composition of molten metal, and casting gating and feeding system dimensions (sprue size, runner size, ingate size, etc.) are used to predict various output parameters in terms of location and intensity of shrinkage porosity, metal temperature, metal velocity, and metal air pressure.

TABLE 12.4
Simulation Format for Carrying Out Trials in MAGMAS Software

No. of Iteration\ MAGMA KPIs	1st	2nd	3rd	4th	5th
Activity done	Only solidification	Without flow off 1. Changed nos. of ingates 2.Changed runner ad D.S. area	1. Changed ingate layout 2. Changed the gating ratio	Used double filter	1. Used double vertical runner 2. Changed ingate area
Filling/solidification	100% only solidification	Filling + solidification 100%	Filling + solidification 100%	Filling + solidification 100% completed	Filling 30% completed
Gating ratio	1364:1250:1137	1364:1250:1137	1364:1091:909	1364:1091:909	1258:1252:1142
Ingate:runner:down sprue	1.2:1.1:1	1.2:1.1:1	1.5:1.2:1	1.5:1.2:1	1.07:1.1:1
Metal temp (°C)	1410	1410	1410	1410	1415
Filling results for shrinkage porosity	Porosity shown at the top area	Due to high velocity, shrinkage porosity can come	Due to high velocity, shrinkage porosity can come	Due to high velocity, shrinkage porosity can come	No shrinkage porosity
Molten metal velocity	1.4 m/sec	1.4 m/sec	1.4 m/sec	1.65 m/sec	1.1 m/s
Reynolds number	1. Nothing required 2. No filling simulation done	Run 1: 4510 Run 2: 5717 Redesign	Run 1: 2899 Run 2: 3677 Redesign	Run 1: 1975 Run 2: 1975	Run 1: 3279 Run 2: 3279 Run 3: 1904 Run 4: 1904
Inferences for activity done	1. Thickness analysis 2. Hotspot area	High velocity at ingate	High velocity at ingate	1. Ingate in the cope side 2. Double filter used 3. Runner made in the drag side	Velocity at ingate is high
Actions decided	Ingate layout changed	Ingate layout changed	Double filter + drag side gating	Double filter + cope side	Horizontal filter cope side gating
Inferences for activity done	Velocity at ingate is slightly high	Velocity target not achieved	Vertical filter should be used	Controlled velocity sand burn is less	-
Actions decided	Horizontal filter ingate layout changed	Horizontal filter	Double filter	Both actual time differences are up to 20 s	--

12.4.4 Tooling Modification and Bulk Production

The concentration diagram is used to highlight the defect's location and quantity over entire surface of part. The concentration diagram will show the location and quantity of the casting defect of a casting under study. On the basis of numerical simulation done and actual defects concentration diagram, it is proposed to undertake the casting pattern and tooling correction and then mass production of different batches of 100 numbers each.

12.4.5 Validation of Experimental and Numerical Study

As per the results obtained from the last step, validation has been carried through numerical simulation study with experimental study. The use of Pareto analysis helped in the identification of main factors contributing to the porosity defect.

12.4.6 Validation for the Various Parameters Responsible for Shrinkage Defect

Figure 12.3 shows the fishbone diagram of various probable causes and effects of shrinkage porosity defect in gearbox housing castings.

Table 12.5 shows the validation matrix for possible causes of shrinkage in gearbox housing, and Table 12.6 shows the actions decided to counter the shrinkage defect.

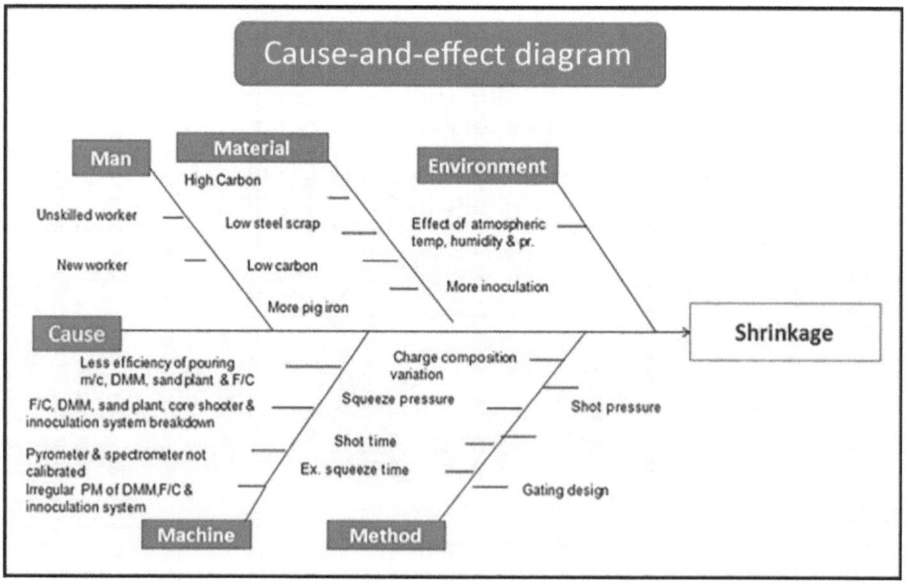

FIGURE 12.3 Cause-and-effect diagram (fishbone diagram) for shrinkage defect in gearbox housing.

TABLE 12.5
Validation Matrix for Possible Causes of Shrinkage in Gearbox Housing

Sr. No.	Possible Causes	Trial Detail	Valid	Invalid
1.	Core assembly not clean	100% cleaning of assembly observed in core setting.	-	Yes
2.	Core fault	100 nos. of core checked and no core found repaired.	-	Yes
3.	Blow hole	Casting machined in-house – 5 nos. No blow hole observed.	-	Yes
4.	Sand drop	100% cleaning of assembly observed in core setting.	-	Yes
5.	Shrinkage	Casting machined in-house – 5 nos. Shrinkage observed after machining in two castings.	Yes	-
6.	High metal temp	Temperature observed – 1410°C–1430°C (within range).	-	Yes
7.	Gating system	Gating system reviewed and ingate location observed at shrinkage-prone area.	Yes	-
9.	Heavy metal section	Casting area reviewed w.r.t. drawing and observed heavy section.	Yes	-
10.	Charge ratio	Steel scrap %age observed up to 60%.	Yes	-
11.	Less wall after drilling	5 nos. rejected casting due to leakage machined in-house and drill wall observed as per specification.	-	Yes

TABLE 12.6
Actions Decided for Various Root Causes of Shrinkage Defect

Sr. No.	Root Cause	Actions Decided
1.	Gating system at shrinkage-prone area	Gating system modified
2.	Heavy metal section area	Profile of the core box changed at thick metal section area
3.	Charge ratio	Charge ratio varied from 60% to 55% after steel scrap addition. Reduction in metal inoculation from 0.25% to 0.20%.

12.5 CASTING PROCESS SIMULATION STUDY

Figure 12.4 shows the molten metal velocity variation with ingate size changes with the present gating system. Figure 12.5 shows the molten metal velocity variation with ingate size changes with the proposed gating system.

Figure 12.6 shows molten metal temperature variation at 10% filling volume. Figure 12.7 shows molten metal temperature variation at 50% filling volume. Figure 12.8 shows molten metal temperature variation at 70% filling volume. Figure 12.9 shows molten metal temperature variation at 100% filling volume.

Figure 12.10 shows molten metal temperature variation at 10% filling volume. Figure 12.11 shows molten metal temperature variation at 50% filling volume. Figure 12.12 shows molten metal temperature variation at 70% filling volume. Figure 12.13 shows molten metal temperature variation at 100% filling volume.

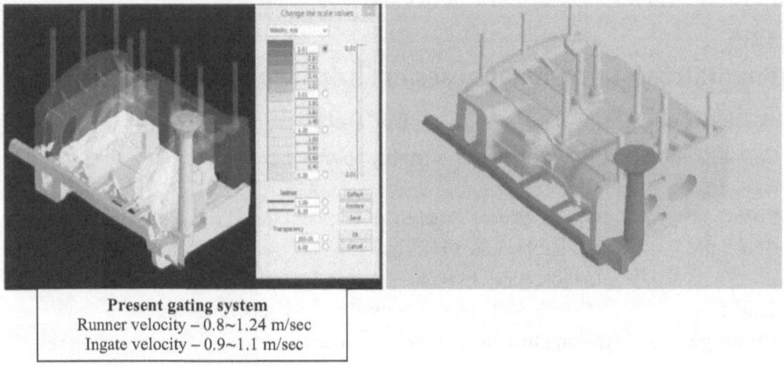

FIGURE 12.4 Molten metal velocity variation with the present gating system of gearbox.

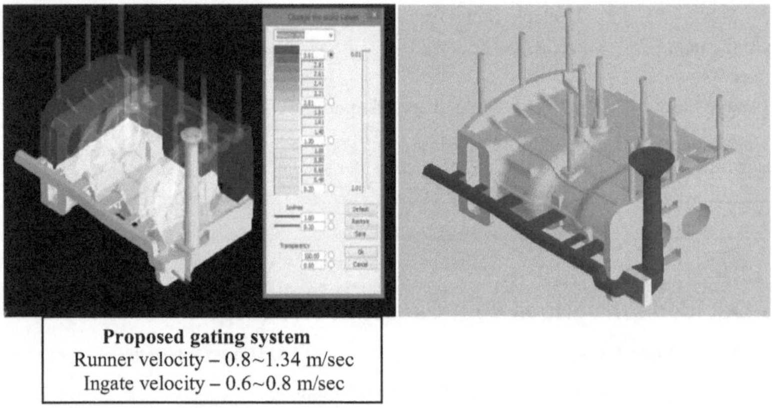

FIGURE 12.5 Molten metal velocity variation with the proposed gating system of gearbox.

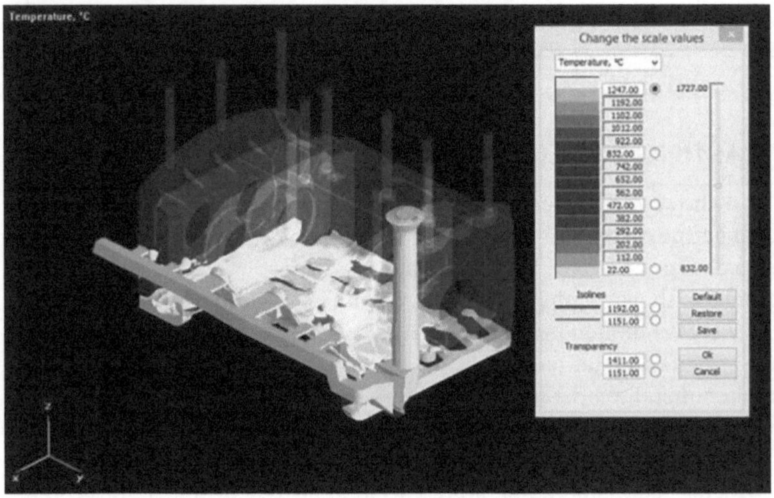

FIGURE 12.6 Filled time – 2.084 s; filled volume – 10%.

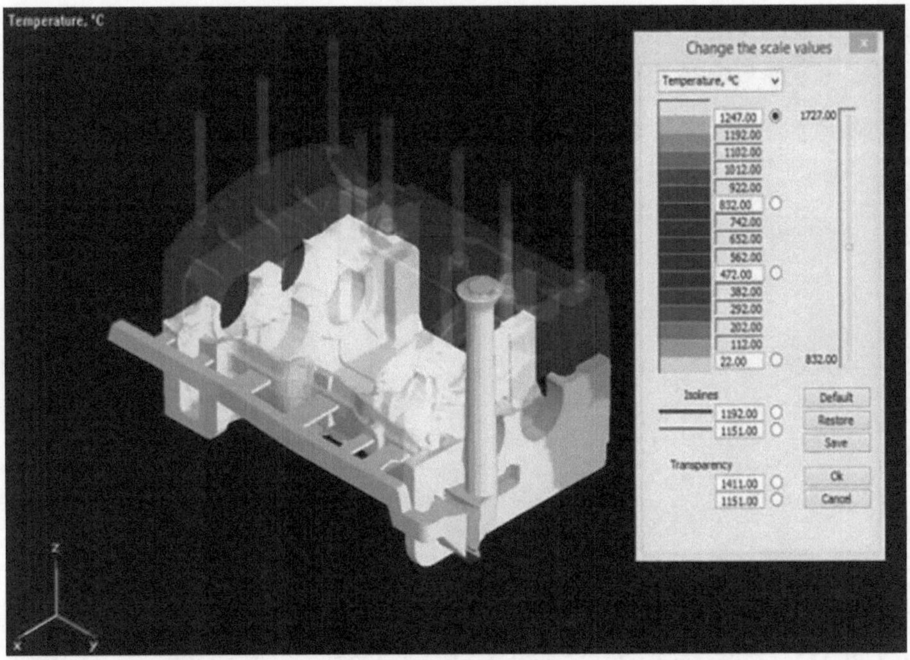

FIGURE 12.7 Filled time – 10.015 s; filled volume – 50%.

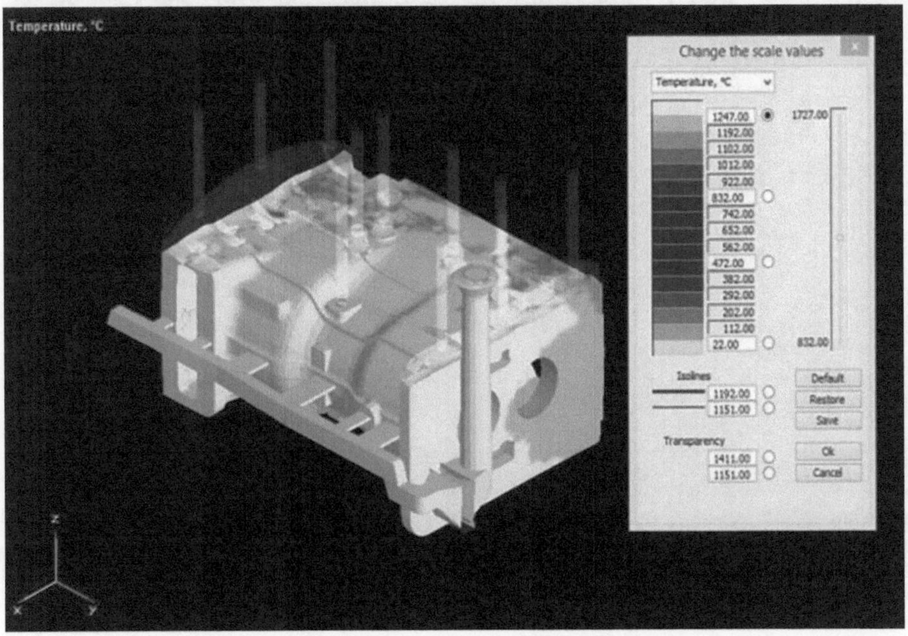

FIGURE 12.8 Filled time – 14.065 s; filled volume – 70%.

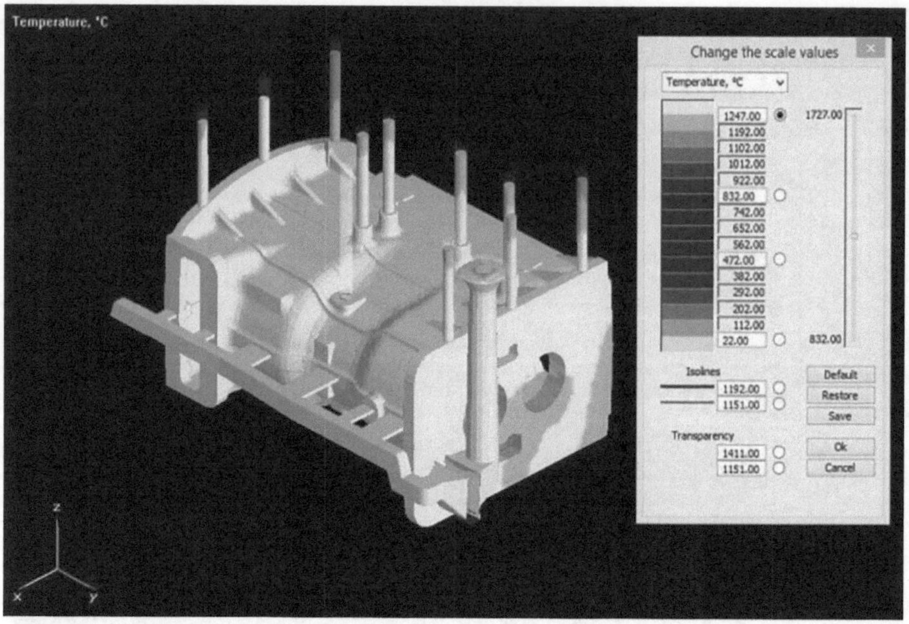

FIGURE 12.9 Filled time – 20.015 s; filled volume – 100%.

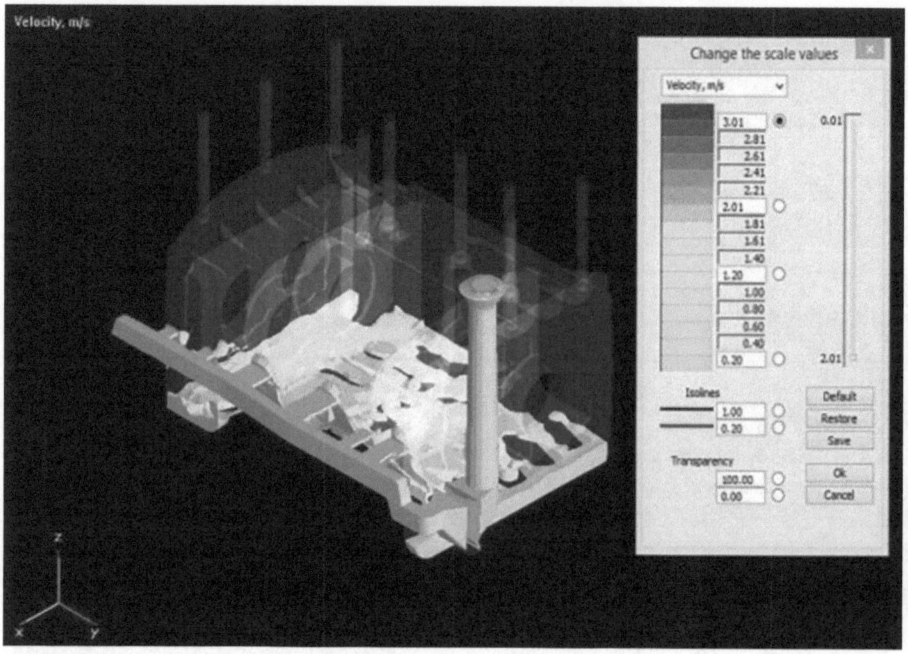

FIGURE 12.10 Velocity: filled volume, 20%.

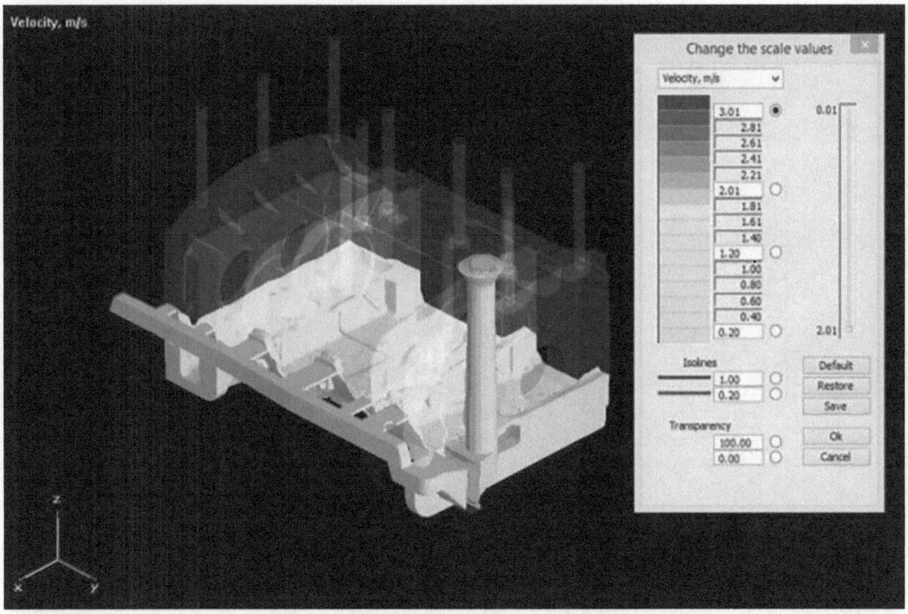

FIGURE 12.11 Velocity: filled volume, 40%.

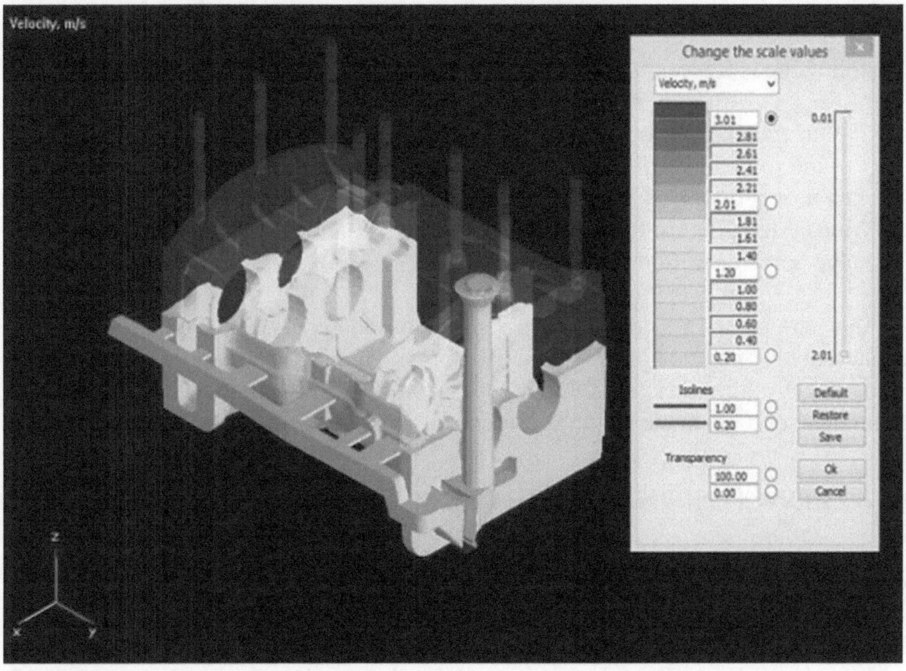

FIGURE 12.12 Velocity: filled volume, 70%.

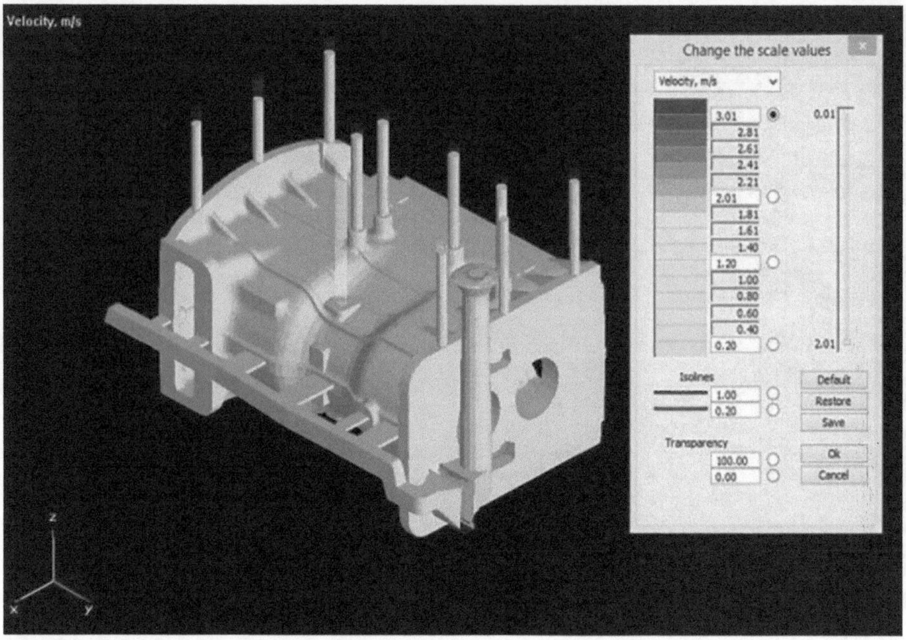

FIGURE 12.13 Velocity: filled volume, 100%.

12.6 RESULTS AND DISCUSSIONS

The following is the summary of various results observed:

- With the change in ingate sizes, the shrinkage porosity and hotspot were found to be shifted to sprue. Thus, the shrinkage was successfully shifted from functional area of gearbox housing to non-functional sprue of the gating system. Figure 12.14 shows the simulation picture of hotspot and shrinkage porosity shifted to sprue (new gating system).

Various parameters including steel scrap ration, inoculants, and alloy addition contributed to the reduction of the shrinkage porosity defects in areas. Gradual and uniform merger of the thick section with the thin section of the component profile helps in the reduction of shrinkage porosity tendency.

12.7 CONCLUSION AND FUTURE SCOPE

Gearbox housings made through the modified gating system have been found to be showing significant improvements in terms of drastic reduction of visible shrinkage porosity at the critical areas after machining. Modified castings can withstand dynamic loading conditions during working of the tractor in severe working conditions. There was substantial reduction in metal turbulence at both ingates and runners after carrying out the modifications in the ingates of the new

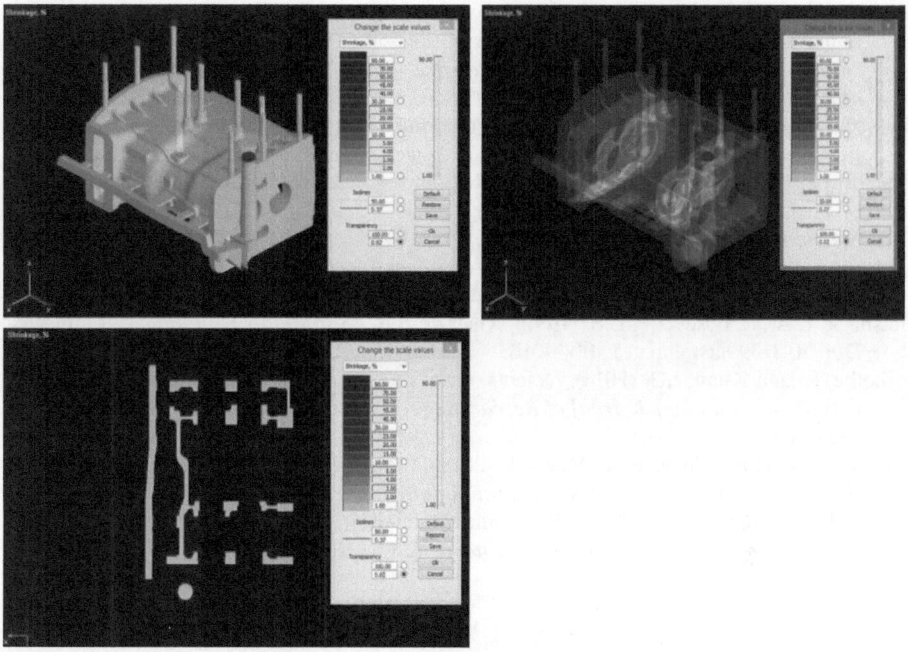

FIGURE 12.14 Simulation picture for shrinkage shifted in sprue (new gating system).

pouring system, and turbulence was reduced by decreasing the molten metal velocity at both runners and ingates. The new pouring system also shows near-laminar flow instead of high turbulent flow in the bulk fluid. MAGMASoft simulation study can be horizontally deployed to analyze and then improve defects such as shrinkage, leakage in different profiles, sizes, and material grades of gray iron such as SG iron.

ACKNOWLEDGMENTS

The authors acknowledge the support provided by the senior team of M/s International Tractors Limited (Punjab, India) during the fieldwork at various foundries and for the machining results of gearbox castings. The authors are also thankful for the foundry support and facilities provided by M/s DCM (Punjab, India) in carrying out the simulation work and pouring of the gearbox castings for the research work.

REFERENCES

Ahmad, N. (2015).Riser Feeding Evaluation Method for Metal Castings Using Numerical Analysis. Ohio State University, Permalink: http://rave.ohiolink.edu/etdc/view?acc_num=osu1447845668

Girish, N., Kidambi, T., Abhishek, A., & Hariths, A. (2013). Controlled solidification process in sand mould casting (carbon steel) to achieve uniform ferrite structure and increase the riser efficiency. *International Journal of Scientific & Technology Research*, *2* (10), 92–95. ISSN 2277-8616.

Hardin, R., & Beckermann, C. (2002, November). Effect of shrinkage on service performance of steel castings. In proceedings of the *56th Steel Founders' Society of America National Technical & Operating Conference, Paper No. 4.5, Steel Founders' Society of America,* Chicago, IL.

Havlicek, F., & Elbel, T. (2011). Geometrical modulus of a casting and its influence on solidification process. *Archives of Foundry Engineering, 11*(4), 170–6.

Mocellin, F., Melleras, E., Guesser, W. L., & Boehs, L. (2004). Study of the machinability of compacted graphite iron for drilling process. *Journal of the Brazilian Society of Mechanical Sciences and Engineering, 26*, 22–27.

Paknikar, S. K. (2014). Innovative simulation of castings–A technology to improve quality of castings as per global specifications with case studies. *Indian Foundry Journal, 60*(8), 28–32.

Plutshack, L.A., & Foseco, A.L.S. (1992). Riser design. *ASM Metals Handbook*, Volume 15. Doi: 10.31399/asm.hb.v15.a0009016.

Rajkolhe, R. and Khan, J.G. (2014). Defects, causes and their remedies in casting process: a review. *International Journal of Research in Advent Technology, 2* (3), 375–383. ISSN: 2321-9637.

Renukananda, K.H., Chavan, A., Ravi, B. (2012). Flow rates in multi-gate systems: experimental and simulation studies. *Indian Foundry Journal, 58* (4), 23–27.

Umezurike C., Onche W.O., (2010). Experimental analysis of porosity in grey iron castings. *Global Journal of Researches in Engineering*, 10(7), 65–70.

13 An Integrated Approach Based on Structural Modeling for Development of Risk Assessment Framework for Drivers Involved in Green Supply Chain Management in India

Pawan Koul and R.S. Rai

CONTENTS

13.1 INTRODUCTION

It was studied that society in today's times is facing a major challenge in the years to come regarding damage to the environment. Green supply chain management (GSCM) is about the prevention of deterioration of the environment, reducing the availability of raw materials, waste reduction, and increasing the level of pollution.

DOI: 10.1201/9781003158165-13

It has been found that multinational companies are taking advantage of weak environmental laws in developing countries such as India and thus are leaving a huge carbon imprint. There is a need to ethically engineer the products with the green element and by making strong laws to save the environment (Boy & Kuruba, 2015). There will be a positive effect on the environment by adopting GSCM practices. The green information helps in recording data related to the emission of greenhouse gases, and recording supply chain data regarding the use of energy, carbon emission, water management, and labor practices (Abbas, Mohamed, & Judi, 2014). It was found from the study that firms can achieve superior performance by managing green supply chains. China was the biggest waste importer of recyclable material (Song & Choi, 2018).

It was found that customers are showing concern for the environment. The incorporation of responsibility toward the environment as well as social responsibility has become increasingly relevant for the success of organizations in today's times. It was found from the study that there are external factors that impact actions related to supply chain management in organizations. These are regulatory pressures like government legislation, certifications, market pressures due to customer pressure, competitive advantage, and competitor pressure. There are societal pressures due to consumer organizations and due to media and the press. The internal pressures comprise corporate strategy, which involves top management commitment; the conduct of the organization; sociocultural responsibility; and skills and capabilities of human resources (Saeed & Kersten, 2019).

It was also found that there are six drivers to supply chain management. Facility decision is most important as this contributes to the success or failure of an organization. The inventory decisions help in managing the relationship between the suppliers and customers. Transportation plays a pivotal role in delivering products to customers. Information flow is also critical to supply chain management. Sourcing from the best suppliers is also critical to the supply chain. In order to gain competitive advantage, pricing is an important factor (Doan, 2020). It was found from the study that there are some drivers as well as non-drivers in the manufacturing supply chain of the Ukrainian industry.

Due to the lack of government support and slow demand from the customer, the implementation of GSCM is becoming difficult. The export market and customers are helping in driving the GSCM in India. The research has supported GSCM impact on achieving profit and lowering the firm's environmental costs (Kormych, Averochkina, Savych, & Pivtorak, 2019). It was found from the study that greening the supply chain is important in today's times. They also found that customer involvement is necessary for greening the supply chain. Customers have forced the management to adopt green procurement (Masudin, 2019).

The major objectives of this proposed work are to implement the concept of the GSCM, and the various drivers need to be identified. The identified variables are then prioritized. This leads to the establishment of a contextual relationship. The interpretive structural modeling (ISM) technique is used for developing a framework, and finally, MICMAC analysis is used for the classification of risks as independent enablers, linkage enablers, autonomous enablers, and dependent enablers.

13.2 LITERATURE REVIEW

Since the available literature on the subject is scarce in our country, the review of the literature of research work has been done for other regions in the world. How green practices can be adopted by producers was studied. In Pakistan, the management of adopting the green process in the supply chain is in the nascent stage today. The green supply chain means taking care of the environment. The study also shows that there is a positive impact of regulation of environmental management as well as green innovation policies adopted by stakeholders (Sarfraz & Abdullah, 2018). It was studied that an increase in greenhouse gas is one of the serious threats. There are seven economic factors related to GSCM. Some barriers are related to resource limitation, supply chain participants, perception of whether green initiatives are widely accepted or not, and motivation that powerful individuals resist change (Tihomir & Kolar, 2019). The design of green policy is to prevent the impact on the environment. The environmental taxation policy is still in the nascent stage. Adopting the green taxes will lead to an increase in the price of energy, and economies will look for alternate measures. The current atmosphere is getting degraded, and stringent laws are required to be made for enhancing compliance from various industries and people (Qayum, Gupta, Gupta, & Arya, 2016). How GSCM can enhance environmental stability was studied. Green supply chain application helps us to gain a competitive edge over the competitors and helps to enhance the quality of human living (Weeratunge & Herath, 2017).

The important factors for the successful implementation of GSCM were also studied in the automobile battery circuit in Brazil. Information management is done across the whole GSCM from suppliers to production to distribution network and finally to customers. There should be the involvement of all employees for good synergy, innovation, and good decision-making. The performance measures like monitoring environmental management systems should be implemented to ensure the quality of overall outcomes. Also, there is a need for top management commitment to raise environmental issues. Supplier management is also necessary to reduce waste. Training is required to promote environmental awareness among employees (Mauricio & De Sousa Jabbour, 2017). The implementation of GSCM in the FMCG industry in Nigeria was also studied. Quantitative analysis was done with 41 respondents from this industry. They reported that the adoption of GSCM has led to enhanced organizational performance and enhanced competitive advantage. The study also shows that there is a reduction in consumption of energy as well as waste emission using GSCM (Ogunlela, 2018).

A study was made for assessing the emission of greenhouse gases in the transportation sector. The study states that vehicle fuel efficiency is a key measure to reduce carbon emissions. Similarly, fuel taxes have to be implemented so that people go for alternate green fuels in their use (Wang & Pan, 2014). How to reduce the carbon footprint was studied. In the transportation area, it is possible by intermodal transportation and the use of hybrid vehicles. Improvements can also be made by using renewable sources of energy such as wind and biomass, and also in manufacturing new technologies such as carbon-capturing technology and in warehousing, maximum daylight should be used (Taborga, Lusa, & Coves, 2018). The buyer–supplier relationship in South African FMCG industry was studied. In South Africa, buyer organizations include environmental issues. The companies are not interested to use substitute products instead of using plastics. There is the cost of going green. It was also found from the study that every organization is not interested to implement GSCM (Meyer et al., 2019). GSCM application in the Selangor region of Malaysia in various manufacturing industries was studied. It was found that there is sustainable performance in the green supply chain (Rasit, Zakaria, Hashim, Ramli, & Mohamed, 2019).

It was studied that the green supply chain makes you bring all things in the supply chain which do not impact the environment. It starts with the selection of green suppliers, using green technology in manufacturing, logistics, reverse logistics, and recycling. Using green supply chain methods saves your money and time, and ensures timely delivery. There is a need for employing a green product, supply network compliance, and reverse logistics so that goods have to return to the supply chain and are properly disposed of (Bhattacharjee, 2017). It was studied that the GSCM has two dimensions, namely, inbound function and outbound function. Inbound function focuses upon techniques used in purchasing and green production, and outbound function emphasizes on activities related to marketing, delivery and recycling. Green purchasing is a set of policies of evaluation of suppliers and selection based on the criteria of GSCM. The suppliers have IS 14001 for environmental management systems. Purchasing of products is based on eco-labels. Besides, there are factors that relate to the impact of supplier establishment on the environment. The products should be able to get recycled and reused. The packaging should also be planned from an environmental perspective by reducing the weight of packaging and decreasing the amount of waste. Green transportation modes should be used to change the means of transportation. Green production is about planning the production activities that will ensure that raw material costs are decreased and productivity is increased, and there is less depletion in the environment. It can also be done by decreasing greenhouse emissions and preventing waste. The environmental effects of products are reduced by green design. The recycling of products and reprocessing of products should be done in the design stage. The design should enforce the use of renewable sources of energy. Reverse logistics should be used to reverse the products and materials for reuse and recycling, repair, and refurbishment (Sezen, 2016).

The supply chain is also influenced by customers as well as the government. Driver entities fulfill customer needs and expectations. There are coercive pressures on the government, and environmental interests and normative interests are social pressures and customer expectations. International environment management helps in developing the commitment and support of senior middle-level managers. Green purchasing involves firms that are implementing GSCM practices (Ali, Bentley, Cao, & Habib, 2017). Lean practices in the automobile sector were studied with the help of a literature review. The various practices found in the study are the elimination of waste and delivery of quality, reliable products, and environmentally safe products. Human values include leadership, teamwork, cooperation, and habits (Jadhav, Mantha, & Rane, 2014).

How inventories of deteriorating items can be controlled was studied. The solution to this problem is reverse logistics (Uthayakumar & Tharani, 2017). The management should have a cautious look at the services related to web and autonomic computing (Shukla, Joshi, Soni, & Kumar, 2019). How GSCM can be selected and developed was studied with the help of existing case studies. From the case studies, it was found that the main areas of focus are innovations and collaborations in supply chains. There should be minimal wastage and minimal energy usage in supply chains. The compliance in the supply chain should be as per the latest international standards. The supply chain should also use lean strategies by keeping in view environmental performance and waste reduction. The managers need to make identification of correct solutions for greening their product (Mutingi, 2013). In Indonesia, how in small-scale enterprises GSCM leads to green purchasing and further green marketing was found. Using SEM, it was found that green marketing is influenced by green purchasing (Sugandini, Muafi, Susilowati, Siswanti, & Syafri, 2020). It was found from the study that loss management ranks first in order, followed by production process improvement and interior environmental management, using the DEMETAL procedure in the railway industry in Iran (Hashemzadeh, Modiri, & Rahimi, 2014).

It was found from this study that the GSCM application is very poor in India. This study has used AHP for the ranking of various factors. The dimension international environment agreement has been found as the most important strategy as in the era of globalization, the companies have to follow the guidelines. The second dimension is the involvement of suppliers and vendors in the GSCM, which can be achieved by constantly educating them. The third dimension is the top management perspective, which is a necessary driver for the implementation of GSCM. The fourth perspective is implementing green supply chain practices. The fifth dimension is proper waste management. The sixth dimension is training employees (Luthra, Garg, & Haleem, 2013). It was also found from the study that a firm's performance is impacted by GSCM. The study was done in the automobile as well as shipping companies in Bahrain with the help of quantitative data analysis. It was concluded from the study that green production and marketing show a positive influence on firm performance (Jassim, Al-Mubarak, & Hamdan, 2020).

A study was also done to understand institutional pressure on GSCM. It was found that there are three factors influencing supply chain management: the support of top management, competitor pressure, and customer pressure. The research work was carried in 286 firms in Thailand that followed the ISO 14000 standard. Quantitative data analysis was done with 286 questions framed, and responses were asked on the Likert scale. The structural equation modeling technique was used, and it was found that the top management supports GSCM (Wongthongchai & Saenchaiyathon, 2019). How customers can get involved in GSCM was studied. GSCM integrates environment and supply chain management. The main factor involved in GSCM implementation is the help of top management, which further provides guidelines to the production, sales, distribution, recycling and IT functions in an organization (Kumar, Luthra, & Haleem, 2013). As per the review of the literature, it was found that there are twelve drivers found important in implementing the GSCM process. These are shown in Table 13.1.

TABLE 13.1
List of Green Drivers in Green Supply Chain Management

S. No.	Driver	ID	Brief Description
1	Top management	R1	Support of top management
2	Environmental management	R2	Following the IS 14000 standard
3	Material sourcing	R3	Sourcing from green suppliers
4	Product design	R4	Green design
5	Product manufacturing	R5	Use of green energy
6	Delivery of material	R6	Use of green transportation
7	Recycling of material	R7	Use of renewable materials
8	Training of manpower	R8	Training on green supply chain
9	Waste reduction	R9	Reduction by recycling
10	Customer pressure	R10	Customer asking for green product
11	Competitor pressure	R11	Competitors manufacturing green product
12	Information communication	R12	Communication in SCM stakeholders

13.3 INTERPRETIVE STRUCTURAL MODELING

The ISM technique analyzes complex relationships and was proposed by J. Warfield in 1974. The ideology used in this technique is the construction of a multilevel structural model. Many researchers have used this technique in various environments in supply chain management. The researchers used this approach to understand the root cause of problems due to the development of suppliers in manufacturing (Kumar, Luthra, & Haleem, 2013). This technique was also used to understand barriers in implementing GSCM techniques in the automobile sector (Luthra, Kumar, Kumar, & Haleem, 2011). It was used to study the challenges faced by the logistics sector in India (Gupta, Singh, & Suri, 2018). It was also used to study the relationship between barriers impacting various industries (Jayant & Azhar, 2014). Interpretive structural modeling also studied the interrelationship between the various risks which are enablers for sustainable supply chain management (Hussain, Awasthi, & Tiwari, 2016). The relationship between the buyer and seller and their interrelationship were studied using this technique (Mabrouk, 2020). Interpretive structural modeling technique was also used to study the relationship among GSCM performance indicators for the palm oil industry in Indonesia (Primadasa, Sokhibi, & Tauhida, 2019).

The flow process of this technique is as follows:

- Understanding the drivers of GSCM. The drivers are found from sourcing of materials, manufacturing, delivery, and recycling of products.
- The examination of factors as well as the interrelationship between them.
- Formation of structural self-interaction matrix of elements.
- Developing the initial and final matrices for reachability.
- Formulating the level partitions.
- Drawing of digraph showing the relationship between the elements.
- MICMAC analysis for grouping of factors.
- Finally, reviewing the model and looking for inconsistencies if any.

13.4 CASE STUDY

Three experts from the various facets of the industry were consulted. The experts chosen are having 20 years' industry experience and showed their interest to participate in the research. The experts' view was taken on the twelve factors, and the ISM technique was used to show their interrelationships. The same analysis was done using the MICMAC analysis.

13.4.1 Structural Self-Interaction Matrix

The interrelationship among the factors was formulated after consultation with experts from the industry. Contextual relationship checks relationships between complex factors. The interrelationship between these variables is developed based on this.

The relationship between any two factors is denoted by four symbols:

V: risk factor i will aggravate j factor.
A: risk factor j will aggravate i factor.
X: risk factors i and j will aggravate each other.
O: risk factors i and j are interrelated.

13.4.2 Reachability Matrix

The reachability matrix is made by substituting the V, A, X, and O with 1 and 0. The substitution rules are as follows:

If i and j entry in SSIM is V, then the (i, j) entry in reachability matrix becomes 1 and (j, i) entry becomes 0.
If i and j entry in SSIM is A, then the (i, j) entry in the reachability matrix becomes 0 and (j, i) entry becomes 1.
If i and j entry in SSIM is X, then the (i, j) entry in reachability matrix becomes 1 and (j, i) entry becomes 1.
If i and j entry in SSIM is O, then the (i, j) entry in the reachability matrix becomes 0 and (j, i) entry becomes 0.

After following the above-mentioned rules, we get the reachability matrix as shown in Table 13.2.

13.4.3 Final Reachability Formation Matrix

The final reachability matrix means that if a variable A is related to B and B in turn is related to C, then A is a subset of C.

13.4.4 Level Partitions

The level partitioning is done by finding the antecedent set as well as reachability set. The intersection set is the intersection of both the sets. This leads to the identification

TABLE 13.2
Structural Self-Interaction Matrix Risk Factor (i)

Code	Driving Factors (J)	R12	R11	R10	R9	R8	R7	R6	R5	R4	R3	R2
R1	Top management	V	V	V	V	V	V	V	V	V	V	V
R2	Environmental management	V	V	V	V	V	V	V	V	V	V	
R3	Material sourcing	V	V	V	V	V	V	V	O	O		
R4	Product design	V	V	V	V	V	V	V	X			
R5	Product manufacturing	V	V	V	V	X	O	V				
R6	Delivery of material	V	V	V	V	O	V					
R7	Recycling of material	V	V	V	O	O						
R8	Training of manpower	V	V	V	X							
R9	Waste reduction	V	V	A								
R10	Customer pressure	V	X									
R11	Competitor pressure	O										
R12	Information communication											

of top-level risk. The process is repeated until the risk factor of each level is not found. ISM model is developed on this basis. Table 13.3 gives the initial reachability matrix, and Table 13.4 gives the final reachability matrix. The level identification is done in Table 13.5.

Based on the iterations, the drivers that impact the implementation of GSCM have been shown in a hierarchical form as shown in Figure 13.1. The drivers that are obtained from iteration 1 are at the top of the framework, and the drivers obtained after iteration 6 are at the bottom of the framework. The drivers at the bottom are independent drivers as they are not followed by any other risks, and going from the bottom to the top of the framework, the dependency of risks increases.

TABLE 13.3
Initial Reachability Matrix

Code	Driving Factor (I)	R1	R2	R3	R4	R5	R6	R7	R8	R9	R10	R11	R12
R1	Top management	1	1	1	1	1	1	1	1	1	1	1	1
R2	Environmental management	0	1	1	1	1	1	1	1	1	1	1	1
R3	Material sourcing	0	0	1	0	0	1	1	1	1	1	1	1
R4	Product design	0	0	0	1	1	1	1	1	1	1	1	1
R5	Product manufacturing	0	0	0	1	1	1	0	1	1	1	1	1
R6	Delivery of material	0	0	0	0	0	1	0	1	1	1	1	1
R7	Recycling of material	0	0	0	0	0	0	1	0	0	1	1	1
R8	Training of manpower	0	0	0	0	1	0	0	1	1	1	1	1
R9	Waste reduction	0	0	0	0	0	0	0	1	1	0	1	1
R10	Customer pressure	0	0	0	0	0	0	0	0	1	1	1	1
R11	Competitor pressure	0	0	0	0	0	0	0	0	0	1	1	0
R12	Information communication	0	0	0	0	0	0	0	0	0	0	0	1

TABLE 13.4
Final Reachability Matrix

Code	Driving Factor (I)	R1	R2	R3	R4	R5	R6	R7	R8	R9	R10	R11	R12	Driving Power
R1	Top management	1	1	1	1	1	1	1	1	1	1	1	1	12
R2	Environmental management	0	1	1	1	1	1	1	1	1	1	1	1	11
R3	Material sourcing	0	0	1	0	1	1	1	1	1	1	1	1	9
R4	Product design	0	0	0	1	1	1	1	1	1	1	1	1	9
R5	Product manufacturing	0	0	0	1	1	1	0	1	1	1	1	1	8
R6	Delivery of material	0	0	0	0	1	1	0	1	1	1	1	1	7
R7	Recycling of material	0	0	0	0	0	0	1	0	1	1	1	1	5
R8	Training of manpower	0	0	0	0	1	0	0	1	1	1	1	1	6
R9	Waste reduction	0	0	0	0	1	0	0	1	1	1	1	1	6
R10	Customer pressure	0	0	0	0	0	0	0	1	1	1	1	1	5
R11	Competitor pressure	0	0	0	0	0	0	0	0	1	1	1	1	4
R12	Information communication	0	0	0	0	0	0	0	0	0	0	0	1	1
	Dependence power	1	2	3	4	8	6	5	9	11	11	11	12	

TABLE 13.5
Identification of Level

Factor	Reachability Set	Antecedent Set	Intersection	Level
		Iteration 1		
R1	1,2,3,4,5,6,7,8,9,10,11,12	1	1	
R2	2,3,4,5,6,7,8,9,10,11,12	1,2	2	
R3	3,5,6,7,8,9,10,11,12	1,2,3	3	
R4	4,5,6,7,8,9,10,11,12	1,2,4,5	4,5	
R5	4,5,6,8,9,10,11,12	1,2,3,4,5,6,8,9	4,5,6,8,9	
R6	5,6,8,9,10,11,12	1,2,3,4,5,6	5,6	
R7	7,9,10,11,12	1,2,3,4,7	7	
R8	5,8,9,10,11,12	1,2,3,4,5,6,7,8,9,10	8,9,10	
R9	5,8,9,10,11,12	1,2,3,4,5,6,7,8,9,10,11	5,8,9,10,11	
R10	8,9,10,11,12	1,2,34,5,6,7,8,9,10,11	10,11	
R11	9,10,11,12	1,2,3,4,5,6,7,8,9,10,11	9,10,11	
R12	12	1,2,3,4,5,6,7,8,9,10,12	12	i
		Iteration 2		
R1	1,2,3,4,5,6,7,8,9,10,11	1	1	
R2	2,3,4,5,6,7,8,9,10,11	1,2	2	
R3	3,5,6,7,8,9,10,11	1,2,3	3	
R4	4,5,6,7,8,9,10,11	1,2,4,5	4,5	
R5	4,5,6,8,9,10,11	1,2,3,4,5,6,8,9	4,5,6,8,9	
R6	5,6,8,9,10,11	1,2,3,4,5,6	5,6	
R7	7,9,10,11	1,2,3,4,7	7	
R8	5,8,9,10,11	1,2,3,4,5,6,7,8,9,10	5,8,9,10	
R9	5,8,9,10,11	1,2,3,4,5,6,7,8,9,10,11	5,8,9,10,11	ii
R10	8,9,10,11	1,2,34,5,6,7,8,9,10,11		
R11	9,10,11	1,2,3,4,5,6,7,8,9,10,11		
		Iteration 3		
R1	1,2,3,4,6,7	1	1	
R2	2,3,4,6,7	1,2	2	
R3	3,6,7	1,2,3	3	
R4	4,6,7	1,2,4,5	4	
R5	4,6	1,2,3,4,6	4,6	iii
R6	6	1,2,3,4,6	6	
R7	7	1,2,3,4,7	7	iv
		Iteration 4		
R2	2,3	1,2	2	
R3	3	1,2,3	3	v
		Iteration 5		
R2	2	1,2	2	vi
		Iteration 6		
R1	1	1	1	vii

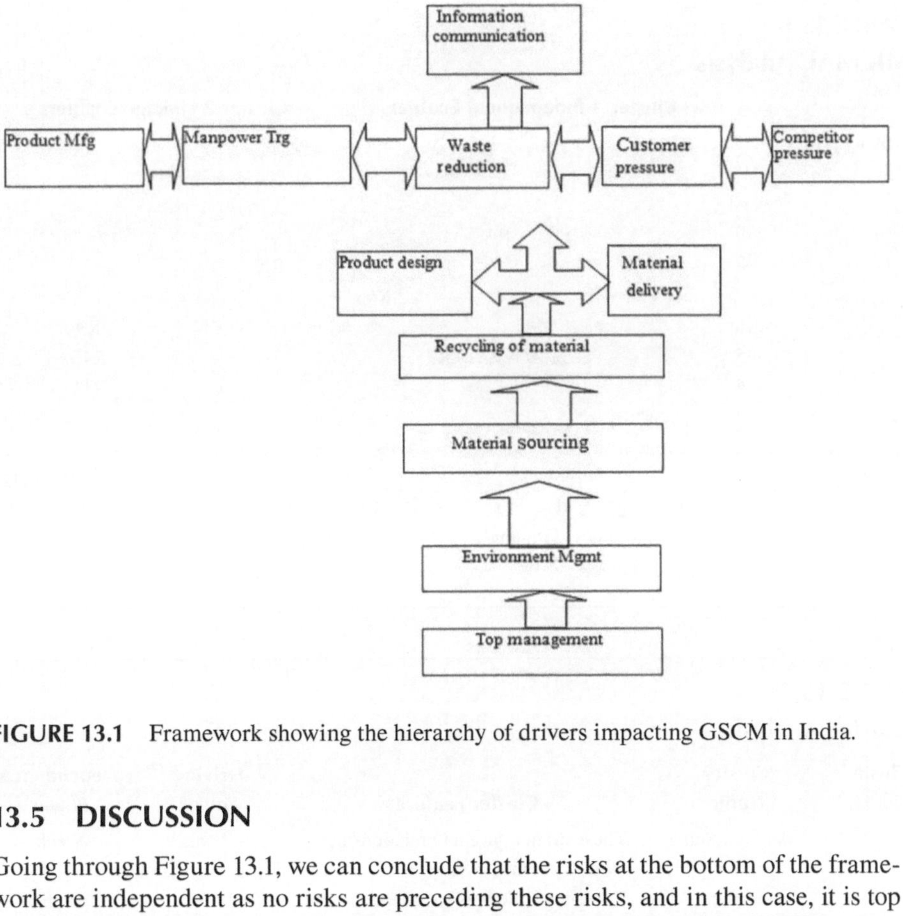

FIGURE 13.1 Framework showing the hierarchy of drivers impacting GSCM in India.

13.5 DISCUSSION

Going through Figure 13.1, we can conclude that the risks at the bottom of the framework are independent as no risks are preceding these risks, and in this case, it is top management that is the driving power, and without their intervention, GSCM cannot be implemented. Moving upward through the framework, the risks such as product design and delivery of material are the linkage between the independent risk and completely dependent risks such as information communication, product manufacturing, manpower training, waste reduction, customer pressure, and competitor pressure. This framework gives us an understanding of which drivers are more important and which are less important for the implementation of GSCM. MICMAC analysis is shown in Table 13.6, which gives the drivers R1, R2, R3, and R4, namely top management, environmental management, material sourcing, and product design, which are placed in cluster 4, meaning that these are independent enablers and no other variables impact them and hence management has to take the utmost care of these drivers for effective GSCM implementation. The drivers R6, R8, R9, R10, R11, and R12, namely delivery of material, training of manpower, waste reduction, customer pressure, competitor pressure, and information communication, are dependent enablers and are dependent on independent enablers. R5, product manufacturing, is a linkage enabler, which is the linkage between independent and dependent enablers. Driver R7, recycling of material, is an autonomous enabler and relatively disconnected from

TABLE 13.6
MICMAC Analysis

Cluster 4 Independent Enablers | **Cluster 3 Linkage Enablers**

Driving power	12	R1											
	11		R2										
	10												
	9			R3	R4								
	8								R5				
	7						R6						
	6									R8		R9	
	5					R7						R10	
	4											R11	
	3												
	2												
	1												R12
	0	1	2	3	4	5	6	7	8	9	10	11	12

Cluster 1 Autonomous Enablers | Cluster 2 Dependent Enablers

Dependence Power

TABLE 13.7
Cluster Analysis

Cluster No.	Cluster Name	Cluster Features	Driving Power	Dependence Power
1	Autonomous	These do not have any relationship with systems.	Weak	Weak
2	Dependent	These are followers to independent variables.	Weak	Strong
3	Linkage	These have risks on other risk factors.	Strong	Strong
4	Independent	Organizations have to pay extreme attention to these enablers.	Strong	Weak

the system. Table 13.7 shows the interrelationship between various enablers as well as their features.

13.6 CONCLUSION

This research has been used to find, analyze, and define the priority for implementing GSCM in India. The salient contributions of this research are as follows:

Twelve drivers that impact GSCM in India have been found from the literature.

The drivers such as top management, environmental management, material sourcing, and product design are important enablers with the highest independence. Thus, to get the green supply chain implementation done, top management involvement is of utmost importance, followed by environmental management, material sourcing, and recycling of material.

This study can be useful to the industry as well as academia, and while implementing GSCM, they should take the utmost care of independent enablers.

13.7 LIMITATIONS OF THE STUDY

The experts were selected from only a few selected organizations in the automobile sector which were not representing all the companies where supply chain management is implemented in India. The experts were based in Delhi/NCR region.

REFERENCES

Abbas, A. M. H., Mohamed, H., & Judi, H. M. (2014). The adoption of green ICT practices as a driver in supply chain management. *International Journal of Digital Content Technology and Its Applications*, 8(6), 75–82.

Ali, A., Bentley, Y., Cao, G., & Habib, F. (2017, September). Green supply chain management – food for thought? Doi: 10.1080/13675567.2016.1226788.

Bhattacharjee, K. (2017, June). Green supply chain management-challenges and opportunities. *Asian Journal of Technology and Management Research* (ISSN:2249-0892), 05(01).

Boy, R., & Kuruba, D. G. (2015). The drivers of green supply chain management: a theoretical framework. *International Journal of Logistics & Supply Chain Management Perspectives*, 4(4). Retrieved from https://www.researchgate.net/publication/288975623.

Doan, T. T. T. (2020). Supply chain management drivers and competitive advantage in manufacturing industry. *Uncertain Supply Chain Management*, 8(3), 473–800. Doi: 10.5267/j.uscm.2020.5.001.

Gupta, A., Singh, R. K., & Suri, P. K. (2018). Analysis of challenges faced by Indian logistics service providers. *Operations and Supply Chain Management*, 11(4), 214–225. Doi: 10.31387/oscm0350215.

Hashemzadeh, G., Modiri, M., & Rahimi, Z. (2014). Identification and ranking effective factors on establishment of green supply chain management in railway industry. *Uncertain Supply Chain Management*, 2(4), 293–302. Doi: 10.5267/j.uscm.2014.5.007.

Hussain, M., Awasthi, A., & Tiwari, M. K. (2016). Interpretive structural modeling-analytic network process integrated framework for evaluating sustainable supply chain management alternatives. *Applied Mathematical Modelling*, 40(5–6), 3671–3687. Doi: 10.1016/j.apm.2015.09.018.

Jadhav, J. R., Mantha, S. S., & Rane, S. B. (2014). Development of framework for sustainable Lean implementation : an ISM approach. Doi: 10.1007/s40092-014-0072-8.

Jassim, S., Al-Mubarak, M., & Hamdan, A. (2020). The impact of green supply chain management on firm's performance. *Journal of Information and Knowledge Management*, 19(1). Doi: 10.1142/S0219649220400262.

Jayant, A., & Azhar, M. (2014). Analysis of the barriers for implementing green supply chain management (GSCM) practices: an interpretive structural modeling (ISM) approach. *Procedia Engineering*, 97, 2157–2166. Doi: 10.1016/j.proeng.2014.12.459.

Kormych, B., Averochkina, T., Savych, O., & Pivtorak, H. (2019). Barriers and drivers of green supply chain management: a case study of Ukraine. *International Journal of Supply Chain Management*, 8(5), 305–313.

Kumar, S., Luthra, S., & Haleem, A. (2013). Customer involvement in greening the supply chain: an interpretive structural modeling methodology. *Journal of Industrial Engineering International*, 9, 6.

Luthra, S., Garg, D., & Haleem, A. (2013). Identifying and ranking of strategies to implement green supply chain management in Indian manufacturing industry using analytical hierarchy process. *Journal of Industrial Engineering and Management*, 6(4), 930–962. Doi: 10.3926/jiem.693.

Luthra, S., Kumar, V., Kumar, S., & Haleem, A. (2011). Barriers to implement green supply chain management in automobile industry using interpretive structural modeling technique-an Indian perspective. *Journal of Industrial Engineering and Management*, 4(2), 231–257. Doi: 10.3926/jiem.2011.v4n2.p231-257.

Mabrouk, N.B. (2020). Interpretive structural modeling of critical factors for buyer-supplier partnerships in supply chain management. *Uncertain Supply Chain Management*, 8(3), 613–626. Doi: 10.5267/j.uscm.2020.2.002.

Masudin, I. (2019). A literature review on green supply chain management adoption drivers. *Jurnal Ilmiah Teknik Industri*, 18(2), 103–115. Doi: 10.23917/jiti.v18i2.7826.

Mauricio, A. L., & De Sousa Jabbour, A. B. L. (2017). Critical success factors for GSCM adoption: case studies in the automotive battery industry. *Gestao e Producao*, 24(1), 78–94. Doi: 10.1590/0104-530X2267-16.

Meyer, A., Niemann, W., van Pletzen, P. R., & Smit, D. (2019). Environmental initiatives: a study of dyadic buyer and supplier relationships in the South African fast-moving consumer goods industry. *Journal of Transport and Supply Chain Management*, 13, 1–10. Doi: 10.4102/jtscm.v13i0.448.

Mutingi, M. (2013). Developing green supply chain management strategies A taxonomic approach Mutingi. *Journal of Industrial Engineering and Management*, 6(2), 525–546.

Ogunlela, G. O. (2018). Green supply chain management as a competitive tool in the fast-moving consumer goods manufacturing industry. *Journal of Business and Retail Management Research*, 12(4), 167–176. Doi: 10.24052/jbrmr/v12is04/art-17.

Primadasa, R., Sokhibi, A., & Tauhida, D. (2019). Interrelationship of green supply chain management (GSCM) performance indicators for palm oil industry in Indonesia. *IOP Conference Series: Materials Science and Engineering*, 598(1). Doi: 10.1088/1757-899X/598/1/012034.

Qayum, A., Gupta, A., Gupta, A., & Arya, R. (2016). Environmental taxation based integrated modeling towards sustainable environmental conservation approach. *Environmental Systems Research*, 5(1), 1–10. Doi: 10.1186/s40068-016-0074-1.

Rasit, Z. A., Zakaria, M., Hashim, M., Ramli, A., & Mohamed, M. (2019). Green supply chain management (GSCM) practices for sustainability performance: an empirical evidence of Malaysian SMEs. *International Journal of Financial Research*, 10(3), 371–379. Doi: 10.5430/ijfr.v10n3p371.

Saeed, M. A., & Kersten, W. (2019). Drivers of sustainable supply chain management: identification and classification. *Sustainability (Switzerland)*, 11(4). Doi: 10.3390/su11041137.

Sarfraz, M., & Abdullah, M. I. (2018). LogForum, (June 2019). Doi: 10.17270/J.LOG.2018.297.

Sezen, B. (2016, December). Ethics and sustainability in global supply chain management. Doi: 10.4018/978-1-5225-3909-4.ch006.

Shukla, O. J., Joshi, A., Soni, G., & Kumar, R. (2019). Analysis of critical drivers affecting implementation of agent technology in a manufacturing system. *Journal of Industrial Engineering International*, 15(2), 303–313. Doi: 10.1007/s40092-018-0293-3.

Song, B., & Choi, D. (2018). Dynamic capability of the firm as driver of green supply chain management implementation. *Sustainability (Switzerland)*, 10(7). Doi: 10.3390/su10072539.

Sugandini, D., Muafi, M., Susilowati, C., Siswanti, Y., & Syafri, W. (2020). Green supply chain management and green marketing strategy on green purchase intention: SMEs cases. *Journal of Industrial Engineering and Management*, 13(1), 79–92. Doi: 10.3926/jiem.2795.

Taborga, C. P., Lusa, A., & Coves, A. M. (2018). A proposal for a green supply chain strategy. *Journal of Industrial Engineering and Management*, 11(3), 445–465. Doi: 10.3926/jiem.2518.

Tihomir, O., & Kolar, D. (2019). Green supply chain management in Croatian companies, (December 2018). Doi: 10.31803/tg-20180207144632.

Uthayakumar, R., & Tharani, S. (2017). An economic production model for deteriorating items and time dependent demand with rework and multiple production setups. *Journal of Industrial Engineering International*, 13(4), 499–512. Doi: 10.1007/s40092-017-0202-1.

Wang, X., & Pan, Y. (2014). Research on supplier evaluation and selection of green procurement. *International Conference on Logistics, Engineering, Management and Computer Science*, LEMCS 2014, (Lemcs), 351–354. Doi: 10.2991/lemcs-14.2014.83.

Weeratunge, R. A. D. D., & Herath, R. (2017). The dimensions of green supply chain management practices. 8(6), 123–132. Doi: 10.17501/wcosm.2017.2111.

Wongthongchai, J., & Saenchaiyathon, K. (2019). The key role of institution pressure on green supply chain practice and the firm's performance. *Journal of Industrial Engineering and Management*, 12(3), 432–446. Doi: 10.3926/jiem.2994.

14 Human Resource Intelligent Systems
Rewriting the DNA of HR Function

Puja Sareen, Shikha Mishra, Vishal Kamra, and Amit Gupta

CONTENTS

14.1 INTRODUCTION

We find ourselves in the most transformational times of human history. The business environment is increasingly becoming volatile, dynamic, global, and complex. As a result, the pressure on organizations to excel is highly multifold and requires the organizations to introspect, rethink, and reimagine the ways to excellence. The key to success lies in developing the human resource management (HRM) function in the organization to yield competitive advantage. The current disruptions, challenges

DOI: 10.1201/9781003158165-14

of the global economy, fast improvements in the field of technology, and the growth of knowledge economy are among the crucial forces that have led to a vital change in the way the HR leaders are working. HR leaders are witnessing the technology integration in HR function to move from administrative roles to strategic roles of HRM. Integration of technology with the HRM function not only means symphonic coexisting, but also lays down a strong footing of the organizational journey toward a robust business model. The HR intelligent systems have enabled the journey of HRM to e-HRM. These intelligent systems provide a complete and holistic framework to HR leaders for decision making in HR.

14.1.1 Adoption of HR Intelligent Systems: Journey of "HRM" to "e-HRM"

HRM in today's world is different and follows an approach of science throughout the organization functions. This new scientific approach focuses more on the organization rather than individuals independently that make up an organization. Therefore, as a practice today, HRM grew out of a trend of being an organizational science and truly combined learning from old and much-studied practiced movements such as scientific management and human relations, with the most up-to-date contemporary research in different sciences of the human behavior (Fisher, 2007). HRM is now no longer a cost center function but a strategic partner which works in tandem with other business functions for organizational effectiveness. HR function has gained vital importance and is increasingly responsible and sought after. HR professionals are now rightly considered as architects in the growth of viable organizational systems. They need to embed themselves in the way the business operates. Dave Ulrich gave the four HR roles that have been widely used in the literature:

1. In the role of **"strategic partner"**, the practices, policies, and processes as suggested and implemented by the HR teams must always be aligned with the larger company policies.
2. Change management is the key to any large-scale venture impacting organization with a view to make or do things in a different manner; then, they have always been dealt with in the past. As a **"change agent"**, HR helps organization make that change keeping in view both the long-term and short-term targets.
3. As an **"administrative expert"**, HR professionals look after the various transactional and administrative processes related to workforce.
4. As an **"employee champion"**, HR professionals focus on not only knowing the employee well but also supporting them and creating open communication channels (Figure 14.1).

The nature and role of HR managers has undergone a drastic shift. Today, HR professionals face numerous challenges to realize the maximum potential of their most competitive resource, i.e., people. Globalization of economy brings with it the challenges related to managing a diverse workforce, and highly competitive workspaces.

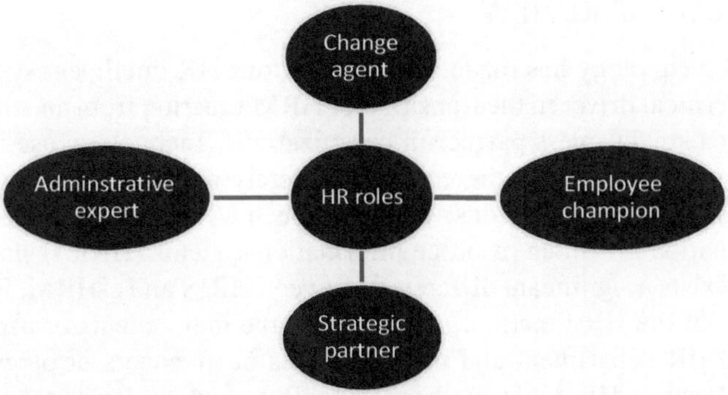

FIGURE 14.1 HR roles.

The other key challenges include increasing focus on quality of goods and services, emergent corporate governance issues, outsourcing developments happening in various industries, and foreseeable supremacy of information technology to gradually transform the overall function of HRM (Bernardin, 2007). The HR professionals must meet the growing demands of their profession. For this, it is imperative that they become strategic business partners, empower the business leaders by providing HR metrics, be change agents, and above all, effectively use the range of available intelligent systems available in the provision of HR services to its most important resource, i.e., employees. HR leaders have realized that to facilitate their strategic orientation for the business, gradually heading to adopt e-HRM systems and practices is the need of the hour. Such adoption will equip the HR leaders with better metrics for various HR-related activities and at the same time empower business heads.

To sum up the changing role of HR, to endure, HR function must establish its core competencies and prove its critical value, and adoption of the right technological tools can be the finest choice. HR must be equipped with IT knowledge and acquire technology acumen. HR leaders should utilize the various available intelligent systems. HR professionals must endorse and ensure availability of the right technology tools that not only provide easy access to employee-related crucial information but also facilitate in enhancing employee productivity and creating value for business. For a maximum outcome, it becomes essential that workforce-associated decisions be based on the real-time information so that there is a quick response to changes in business. It is apparent that e-HRM is earning significance in the current business scenario and the use of high-end technologies for HR activities is augmenting day by day within organizations (Cedar Crestone, 2005). The available literature mentions the term e-HRM or e-HR to refer to technology integration in the function of HR. Both the terms emphasize on technology adoption and integration in the HR function. e-HR is mainly involved with the use of cyberspace technology and mobile communication technology. Such technologies will impact the characteristics of relations among HRM staff, the business or line managers, and end-users (employees) from a purely direct relationship to interactions that are technology-enabled (Ruel et al., 2004).

14.2 LITERATURE REVIEW

The advent of technology has made available various HR intelligent systems and is regarded as a critical driver in the transition of HRM function from an administrative perspective to a key business partner in organizations. Technology use in the HRM domain has now gone beyond the context of merely managing the payroll system and other administrative operations. There has been a widespread usage of technology and automation in human resource information systems (HRISs) since the early 1980s. There exists a significant difference between HRIS and e-HRM. HRISs focus on automation of the HR function. In this sense, the main clients or users of HRIS have been the HR department and not employees or managers of other functional departments. Further, HRIS has not been capable enough to develop the HR virtual value chain. Since long, it has been effectively used to automate the various HR systems such as payroll processing and employee leave and attendance (Ball, 2001). Instead, e-HRM emphasizes the embracing and usage of Web or Internet-centered systems and mobile technologies. It has largely facilitated the interactions between HR department, business leaders, and employees. The interactions that were earlier face-to-face are now increasingly being facilitated by HR intelligent systems (Florkowski & Olivas-Lujan, 2006; Ruel & Bondarouk, 2008). Bondarouk and Ruel (2009) defined e-HRM as "An umbrella term covering all possible integration mechanisms and contents between HRM and information technologies, aiming at creating value within and across organizations for targeted employees and management".

Many researchers pointed out that HR leaders who primarily focused on administrative aspect of HR outlined transactional or efficiency objectives for e-HRM. In contrast, HR leaders who already act as a business/strategic partner to top management emphasized strategic goals behind e-HRM. Therefore, organization-level goals for e-HRM adoption embrace cost minimization of HRM operations (Marler, 2009; Bondarouk & Ruel, 2009) and increased effectiveness, which is a resultant of efficient and effective performance of HRM services and strategic orientation of HRM function (Ruel et al., 2004; Lepak & Snell, 1998).

The literature review has analyzed the research studies undertaken to understand how the HR processes have successfully established a set of interconnected, business service strategies, meeting long-range branding or reputational goals of the organizations (Reilly & Tamkin, 2006). A key internal strategy is the restructuring of the HRM function which combines the various HRM transformations. Recent times are witnessing new HRM delivery services that are an outcome of the available model of HR shared services, founded by Ulrich and Brockbank (2005) with outsourcing of HR processes, off-shoring the important HR activities, specifically the HRM shared service centers (Reddington et al., 2005; CIPD, 2005; Cooke, 2006). Recent research has highlighted the manner in which intelligent systems are being used to combine with HRM transformations and process outsourcing. The main objective of this is to radically change the way HR's internal operations are being carried out (Shrivastava & Shaw, 2004; Martin, 2005; Gueutal & Stone, 2005; Masum et al., 2018). It is found that business process re-engineering (BPR) and the technology tools are largely interconnected and co-dependent (Keebler, 2001). The virtual HRM is focusing on the quality of HR services. It also facilitates the development of new HR business models that lead to the emergence of better and more innovative organizations (Kavanagh & Thite, 2009). The technology enabled information facilitates the various HR transformation models (Ulrich, 2005). The role of intelligent systems and artificial intelligence is manifold (Johnson et al., 2020). Therefore, it can be concluded that e-HRM enables business process re-organization because automation helps to lower both the physical and positional hierarchy distance by centralization of key HR functions. Further, it also helps to create this distance by leading technology facilitation in the earlier face-to-face people relationships in organizations.

14.3 RESEARCH METHODOLOGY

The chapter details the research study that is carried out in selected Indian and multinational companies to identify key factors of employees' perception toward e-HRM systems being used in their respective organization and then to do a comparative analysis sector-wise between these organizations. The study could prove to be useful for organizations using e-HRM systems and help them to improve employees' experience toward the same.

14.3.1 Objectives of Research

To identify the factors that constitute perceptions of "End-users" (employees) to their organization's e-HRM practices/systems.

To compare selected Indian and multinational organizations in terms of perceptions of "End-users" (employees) to their e-HRM system.

The primary focus of the research study is to conduct a comparative study on perceptions of "End-users" (employees) toward the e-HR systems used in their organizations.

After a rigorous literature review and consultations with various HR leaders, it can be established that the e-HRM adoption is mounting at an exponential rate.

14.3.2 Procedure and Participants

The research is limited to the National Capital Region of India. Companies using high-end e-HRM systems were sent requests for participation. Indian companies were selected from the ET 500 ("List of top Indian companies") in the years 2018 and 2019. The Indian companies, for the purpose of this research, are those whose "country of origin" is India. Similarly, Fortune Global 500 companies' data were screened and requests were sent to the top multinational companies, which were a part of annual ranking of world's largest corporations for the years 2018 and 2019. These companies have the "country of origin" other than India and have been operating worldwide.

The multinational companies, from the above-mentioned list, whose operation was in NCR were shortlisted. Thus, the study compares Indian companies (3) with multinational companies (3).

Additionally, a sector-based comparison is done for Indian and multinational companies (banking and finance, automobile, and information technology). The research design is descriptive cum exploratory.

The population in this research includes employees who are the "End-users" of e-HR system. A total of 371 respondents (end-users) are part of the current research.

Employees of an organization have been referred to as "End-users" for the purpose of this research. Employees selected for the study are those who have a least work tenure of 4 years in the selected companies. This shall imply that they have considerable hands-on experience in e-HRM practices. Primary data are collected by means of a structured scale, following the non-probabilistic judgmental sampling (Malhotra, 2004).

Item generation for the formulation of various statements has been done through the following steps:

- An extensive review of literature regarding goals of e-HRM, HRM transformations, outcomes of e-HRM, limiting factors, and self-service applications formed the basis of statements for the perception scale.
- Informal discussions with HR managers of sampled companies took place to understand the changing scenario and the extent of technology integration in the HRM function.

For studying the perceptions of "End-users" toward e-HRM system and "HR professionals", a five-point Likert scale having 33 items was developed.

For "End-users" of e-HRM, the following set of hypotheses is framed for three sectors to compare the perception of "End-users" of Indian and multinational companies toward the e-HRM system:

H_1: There exists a significant difference in the perception of "End-users" of Indian and multinational IT companies toward the e-HRM system.

H_2: There exists a significant difference in the perception of "End-users" of Indian and multinational automobile companies toward the e-HRM system.

H_3: There exists a significant difference in the perception of "End-users" of Indian and multinational banking and finance companies toward the e-HRM system.

For testing the above-mentioned hypotheses, independent-samples t-test is used.

14.4 RESULTS AND DISCUSSION

The "item validation" was done by conducting exploratory factor analysis for the set of 33 items of the scale (Pattanayak et al., 2002).

Based on factor analysis, four factors were obtained. The researchers used principal component analysis for the extraction of the factors that are built on the latent root criterion (i.e., eigenvalue >1) (Chadha & Saini, 2014). The adequacy of sample was established through Kaiser–Meyer–Olkin (KMO) (Kaiser, 1960). The obtained value for KMO is 0.958, which is meritorious (Kaiser, 1960; Hair et al., 2013).

The results of principal component analysis are shown in Appendix Tables 14.A1–14.A3. The main factors identified for the "End-users" perception toward e-HRM system are as follows:

1. Empowers Employees
2. Operational Efficiency
3. User support and Ease of use
4. Constraints

Factor 1: Empowers Employees

This factor has a value of total variance explained of 26.464%. It has been interpreted so as to include scale items that have been taken from the existing literature of e-HRM perceptions relating to relational aspects of HR (Table 14.1).

Factor 2: Operational Efficiency

This factor has 22.656% as the value of total variance and has been interpreted because of its inclusion of various scale items that are taken from the existing literature of e-HRM perceptions relating to operational/administrative efficiency (Table 14.2).

Factor 3: User Support and Ease of Use

This factor has the total variance value of 13.170%. Based on the existing literature and the scale items, the factor is termed User support and Ease of use (Table 14.3).

TABLE 14.1
Factor Items and Loadings – Empowers Employees

Items	Factor Loading
Gives more control over employee's personal information.	0.639
Is responsive to the requirements for (real-time) information.	0.716
Gives the benefit of working as per the best practices.	0.693
ESS/MSS leads to a quick access to information.	0.745
ESS//MSS leads to self-efficacy.	0.790
Generates reports that improve decision making	0.750
The flow of knowledge is facilitated.	0.821
Helps in managing team more efficiently.	0.805
Get informed about the developments in organizations.	0.829
Get updated news of HR initiatives/activities.	0.793
ESS/MSS gives a sense of empowerment.	0.780
Quick response to changes with help of relevant information.	0.795
Manage team globally.	0.775

Cronbach's alpha = 0.958 (13 items).

TABLE 14.2
Factor Items and Loadings – Operational Efficiency

Items	Factor Loading
Time saving on administrative tasks.	0.594
Improves overall productivity.	0.828
Reduces HR-related paperwork.	0.801
Simplification of HR processes.	0.814
Brings improvement in overall HR processes.	0.814
Speeds up transaction processing related to HR activities.	0.802
Optimizes the workflow between employees, management, and HR department.	0.843
Availability of 24/7 HR information.	0.841
Is a reputational driver.	0.849
Transparency in HR systems and processes.	0.866

Cronbach's alpha = 0.960 (10 items).

TABLE 14.3
Factor Items and Loadings – User Support and Ease of Use

Items	Factor Loading
Has user friendly interface.	0.734
It is easy to work on e-HR system.	0.789
Receive support from seniors for using e-HR systems.	0.654
Proper training given to use e-HR systems.	0.762
Receive prompt help for problems faced.	0.763
Refresher training is provided to use e-HR system.	0.781

Cronbach's alpha = 0.919 (6 items).

Factor 4: Constraints

This factor has the value of total variance explained of 7.983%. It is interpreted as *Constraints*. Table 14.4 shows the various scale items that load onto Factor 4.

Hypothesis Testing

Four factors were derived based on exploratory factor analysis. These factors were further used to test research hypotheses. Data from end-users of e-HRM were taken from three different sectors – IT, automobile, and banking and finance. Respondent details from each sector are given in Table 14.5.

TABLE 14.4
Factor Items and Loadings – Constraints

Items	Factor Loading
Feel overburdened "doing HR's" job.	0.771
Losing "People touch".	0.748
Do not want to spend time on web-enabled "HR tools"	0.808
Concerns over data security.	0.680

Cronbach's alpha = 0.815 (4 items).

TABLE 14.5
Respondents' Details

Companies	End-Users
Indian automobile	65
Indian banking and finance	61
Indian IT	55
Multinational automobile	60
Multinational banking and finance	64
Multinational IT	66

14.4.1 Sector-wise Comparisons

14.4.1.1 IT sector

Table 14.6 shows the testing variables, which are factors of e-HRM perceptions for the "End-users", and the significance value of Levene's test for equality of variances and significance (2-tailed) at the 95% confidence level. The following interpretation is made based on the results:

For the factor "Empowers Employees", the Levene test to measure equality of variances shows a *p*-value of 0.005. It establishes that the difference between the variances is not statistically insignificant; therefore, equal variances are not assumed. Further, the *p*-value of 0.007, which is less than 0.05, shows that there is a significant

TABLE 14.6
Independent-Samples Test: "End-Users" (IT Sector)

Testing Variable (Factors)	Sig. Value of Levene's Test for Equality of Variances	Sig. (2-Tailed) at the 95% Confidence Level
Empowers Employees	0.005	0.007
Operational Efficiency	0.593	0.001
User support and Ease of use	0.025	0.236
Constraints	0.772	0.073

difference in the perceptions of "End-users" of Indian and multinational IT companies for the factor "Empowers Employees" at the 95% confidence level.

The Levene test to measure equality of variances for the factor "Operational Efficiency" yields a *p*-value of 0.593. The *p*-value of 0.001, less than 0.05, indicates that there is a significant difference in the perceptions of "End-users" of Indian and multinational IT companies for the factor "Operational Efficiency" at the 95% confidence level.

Levene's test shows a *p*-value of 0.025 for the factor "User support and Ease of use". It establishes that equal variances are not assumed and that the difference between the variances is not statistically insignificant. The *p*-value of 0.236, more than 0.05, suggests that there is no significant difference in the perceptions of "End-users" of Indian and multinational IT companies for the factor "User support and Ease of use" at the 95% confidence level.

For the factor "Constraints", the Levene test to measure equality of variances displays a *p*-value of 0.772. It can be interpreted that the difference among the variances is statistically insignificant and so equal variances are assumed. The *p*-value of 0.073, more than 0.05, establishes that there is no significant difference in the perceptions of "End-users" of Indian and multinational companies IT for the factor "Constraints" at the 95% confidence level.

Hypothesis Testing (H_1)

H_1: There exists a significant difference in the perceptions of "End-users" of Indian and multinational IT companies toward the e-HRM system. (*Partially accepted*)

$H_{1.1}$: There exists a significant difference in the perceptions of "End-users" of Indian and multinational IT companies for the factor "Empowers Employees". (*Accepted*)

$H_{1.2}$: There exists a significant difference in the perceptions of "End-users" of Indian and multinational IT companies for the factor "Operational Efficiency". (*Accepted*)

$H_{1.3}$: There exists a significant difference in the perceptions of employees who are "End-users" of Indian and multinational IT companies for the factor "User support and Ease of use". (*Not accepted*)

$H_{1.4}$: There exists a significant difference in the perceptions of employees who are "End-users" of Indian and multinational IT companies for the factor "Constraints". (*Not accepted*)

14.4.1.2 Automobile Sector

Table 14.7 shows the testing variables, which are factors of e-HRM perceptions for the "End-users", and the significance value of Levene's test for equality of variances and significance (2-tailed) at the 95% confidence level. The following interpretation is made based on the results:

For the factor "Empowers Employees", the Levene test displays a p-value of 0.370. It can be interpreted that the difference between the variances is statistically insignificant; therefore, equal variances are assumed. The *p*-value

TABLE 14.7
Independent-Samples Test: "End-Users" (Automobile Sector)

Testing Variable (Factors)	Sig. Value of Levene's Test for Equality of Variances	Sig. (2-Tailed) at the 95% Confidence Level
Empowers Employees	0.370	0.000
Operational Efficiency	0.135	0.155
User support and Ease of use	0.053	0.039
Constraints	0.571	0.362

of 0.000, less than 0.05, indicates that there is a significant difference in the perceptions of "End-users" of Indian and multinational automobile companies for the factor "Empowers Employees" at the 95% confidence level.

For the factor "Operational Efficiency", the Levene test to measure equality of variances displays a *p*-value of 0.135. It can be interpreted that the difference between the variances is statistically insignificant, and accordingly, equal variances are assumed. The *p*-value of 0.155, more than 0.05, indicates that there is no significant difference in the perceptions of "End-users" of Indian and multinational automobile companies for the factor "Operational Efficiency" at the 95% confidence level.

For the factor "User support and Ease of use", the Levene test to measure equality of variances displays a p-value of 0.053. It can be interpreted that the difference between the variances is statistically insignificant, and therefore, equal variances are assumed. The *p*-value of 0.039, less than 0.05, indicates that there is a significant difference in the perceptions of "End-users" of Indian and multinational automobile companies for the factor "User support and Ease of use" at the 95% confidence level.

For the factor "Constraints", the Levene test to measure equality of variances displays a *p*-value of 0.571. It can be interpreted that the difference between the variances is statistically insignificant and so equal variances are assumed. The *p*-value of 0.362, more than 0.05, indicates that there is no significant difference in the perceptions of "End-users" of Indian and multinational automobile companies for the factor "Constraints" at the 95% confidence level.

Hypothesis Testing (H_2)

H_2: There exists a significant difference in the perceptions of "End-users" of Indian and multinational automobile companies toward the e-HRM system. (*Partially accepted*)

$H_{2.1}$: There exists a significant difference in the perceptions of "End-users" of Indian and multinational automobile companies for the factor "Empowers Employees". (*Accepted*)

$H_{2.2}$: There exists a significant difference in the perceptions of "End-users" of Indian and multinational automobile companies for the factor "Operational Efficiency". (*Rejected*)

$H_{2.3}$: There exists a significant difference in the perceptions of employees who are the "End-users" of Indian and multinational automobile companies for the factor "User support and Ease of use". (*Accepted*)

$H_{2.4}$: There exists a significant difference in the perceptions of employees who are the "End-users" of Indian and multinational automobile companies for the factor "Constraints". (*Rejected*)

14.4.1.3 Banking and Finance Sector

Table 14.8 shows the testing variables, which are factors of e-HRM perceptions for the "End-users", and the significance value of Levene's test for equality of variances and significance (2-tailed) at the 95% confidence level. The following interpretation is made based on the results:

For the factor "Empowers Employees", equality of variances displays a *p*-value of 0.000. It can be interpreted that the difference between the variances is not statistically insignificant; therefore, equal variances are not assumed. The *p*-value of 0.001, less than 0.05, indicates that there is a significant difference in the perceptions of "End-users" of Indian and multinational banking and finance companies for the factor "Empowers Employees" at the 95% confidence level.

Levene's test for the factor "Operational Efficiency" displays a *p*-value of 0.153. It can be interpreted that the difference between the variances is statistically insignificant, and accordingly, equal variances are assumed. The *p*-value of 0.951, more than 0.05, shows that there is no significant difference in the perceptions of "End-users" of Indian and multinational banking and finance companies for the factor "Operational Efficiency" at the 95% confidence level.

For the factor "User Support and Ease of Use", the Levene test to measure equality of variances displays a *p*-value of 0.000. It can be interpreted that the difference between the variances is not statistically insignificant, and therefore, equal variances are not assumed. The *p*-value of 0.001, less than 0.05, indicates that there is a significant difference in the perceptions of "End-users" of Indian and multinational banking and finance companies for the factor "User support and Ease of use" at the 95% confidence level.

TABLE 14.8
Independent-Samples Test: "End-Users" (Banking and Finance Sector)

Testing Variable (Factors)	Sig. value of Levene's Test for Equality of Variances	Sig. (2-Tailed) at the 95% Confidence Level
Empowers Employees	0.000	0.001
Operational Efficiency	0.153	0.951
User support and Ease of use	0.000	0.001
Constraints	0.765	0.553

For the factor "Constraints", the Levene test to measure equality of variances displays 0.765 as *p*-value. It can be interpreted that the difference among the variances is statistically insignificant and so equal variances are assumed. The *p*-value of 0.553, more than 0.05, shows that there is no significant difference in the perceptions of "End-users" of Indian and multinational banking and finance companies for the factor "Constraints" at the 95% confidence level.

Hypothesis Testing (H_3)

H_3: There exists a significant difference in the perception of "End-users" of Indian and multinational banking and finance companies toward the e-HRM system. (*Partially accepted*)

$H_{3.1}$: There exists a significant difference in the perception of "End-users" of Indian and multinational banking and finance companies for the factor "Empowers Employees". (*Accepted*)

$H_{3.2}$: There exists a significant difference in the perception of "End-users" of Indian and multinational banking and finance companies for the factor "Operational Efficiency". (*Rejected*)

$H_{3.3}$: There exists a significant difference in the perception of "End-users" of Indian and multinational banking and finance companies for the factor "User support and Ease of use". (*Accepted*)

$H_{3.4}$: There exists a significant difference in the perception of "End-users" of Indian and multinational banking and finance companies for the factor "Constraints". (*Rejected*)

14.5 CONCLUSION

The present research work attempts to do a comparative study of Indian and multinational companies, with respect to e-HRM. There have been several studies on e-HRM earlier which have identified the importance and uses of e-HRM systems. Researches have been conducted to study the implementation of e-HRM systems but very few comparative studies have been undertaken in this subject. There are still a larger number of studies on e-HRM outside India. These include a study on attitudes toward e-HRM in Philips (Electronics), the Netherlands (Voermans & van Veldhoven, 2007); a study of e-HRM in Europe (Strohmeier & Kabst, 2009); research on e-HRM involving the top 1000 German firms (Laumer et al., 2010); a case-based study done in the context of Mexican firms to study the implementation aspects of the e-HRM strategy (Olivas-Lujan et al., 2007); a study to discuss the role of intelligent HRIS in HR excellence (Masum et al., 2018); and many more studies that have been discussed in the literature review by the researchers. Very few studies have been done in the Indian perspective. The present research has explored the factors involved in the perceptions of "End-users" toward the e-HRM systems and applications. These factors and the variables studied within them give a clear picture of the various benefits, limitations, and facilitating conditions as perceived by the respondents. The comparative study between Indian and multinational companies regarding these perceptions highlights the differences between them as far as e-HRM adoption and practice is concerned. The research outlines the four factors, including Operational Efficiency, Empowers

Employees, User support and Ease of use, and e-HRM Constraints that form the perception of "End-users" for e-HRM systems.

After comprehensively studying e-HRM systems of the selected companies and by statistically analyzing the data, it can be laid down that there are major differences in selected Indian and multinational companies in terms of the usage of technology-enabled HR intelligent systems, favorable conditions, and perceptions of "End-users" working in these selected companies. The companies need to not only keep on continuously updating the e-HRM technologies but also ensure that continuous and usage training is given to employees. Most importantly, a well-defined e-HRM strategy is the need of the hour.

14.6 FUTURE SCOPE

The current research contributes to the future scope of research on HR intelligent systems, particularly e-HRM systems. It provides a reference to HR managers and top management on what forms the perceptions of employees toward e-HRM systems and what critical steps could contribute to achieving the e-HRM success. There can be further research on implementation of e-HR systems, success factors, change management, and case studies which could study the e-HRM experiences of organizations.

14.7 LIMITATIONS OF THE STUDY

The research presented has its own limitations. Firstly, the sample size is not large. Further, the research is only confined to the National Capital Region. The findings of the research may be different in case the participants and geographical scope of study are varied. Furthermore, this study is limited to perceptions toward the e-HRM systems, and the results might substantially vary if the aspects of traditional HR processes are a part of the study. Future research can include a comparative analysis of traditional vs. electronic HR systems. As the study is undertaken with reference to only three sectors, the results may vary if more sectors are taken into consideration. Research may be conducted to compare the perceptions toward e-HRM, in case of A-class cities and B-class cities. However, the present study can be taken as a base to conduct further comparative studies in the e-HRM domain. The scale developed for measuring employees' perception toward e-HRM system can be used for future studies.

REFERENCES

Ball, K. S. (2001). The use of human resource information systems: a survey. *Personnel Review, 30* (6), 667–693.

Bernardin, H. J. (2007). *Human Resource Management: An Experiential Approach*. New York: McGraw-Hill.

Bondarouk, T. V., & Ruel, H. J. M. (2009). Electronic human resource management: challenges in the digital era. *The International Journal of Human Resource Management, 20* (3), 505–514.

Cedar Crestone. (2005). *HCM Survey: Workforce Technologies and Service Delivery Approaches–Eighth Annual Edition*. Retrieved May 16, 2008 fromhttp://www.cedarcrestone.com/serv_research_pr-100912.php.

Chadha, S., & Saini, R. (2014). Information technology support to knowledge management practices: a structural equation modeling approach. *IUP Journal of Knowledge Management, XII*(1), 39–52. Retrieved from http://web.a.ebscohost.com/ehost/detail/detail?sid=6503695b-83c5-49d2-9d4b-6deeaf7720b4@sessionmgr4004&vid=0&hid=4201&bdata=JkF1dGhUeXBlPWlwLHNoaWImc2l0ZT1laG9zdC1saXZl#db=bah&AN=95087463.

CIPD. (2005). *HR Outsourcing: The Key Decisions.* London: Chartered Institute of Personnel and Development.

Cooke, F. L. (2006). Modeling an HR shared services center: experience of an MNC in the United Kingdom. *Human Resource Management, 45,* 211–228.

Fisher, C. D., Schoenfeldt, L. F., & Shaw, J. B. (2007). *Human Resource Management.* Boston, MA: Houghton Mifflin Co.

Florkowski, G. W., & Olivas-Lujan, M. R. (2006). The diffusion of human resource information technology innovations in US and non-US firms. *Personnel Review, 35* (6), 684–710.

Gueutal, H. G., & Stone, D. L. (Eds.). (2005). *The Brave New World of eHR: Human Resource Management in the Digital Age.* San Francisco, CA: Josey Bass.

Hair, J., Black, W. C., Babin, B., & Anderson, R. E. (2013). *Multivariate Data Analysis.* 7th ed. Upper Saddle River, NJ: Pearson Education Inc.

Johnson, R. D., Stone, D. L., & Lukaszewski, K. M. (2020), The benefits of eHRM and AI for talent acquisition, *Journal of Tourism Futures.* Doi: 10.1108/JTF-02-2020-0013.

Kaiser, H. F. (1960). The application of electronic computers to factor analysis. *Educational and Psychological Measurement, 20,* 141–151.

Kavanagh, M. J., & Thite, M. (Eds.). (2009). *Human Resource Information Systems: Basics, Applications & Directions.* Thousand Oaks, CA: Sage.

Keebler, T. (2001). HR outsourcing in the internet era. In A. J. Walker (Ed.), *The Technologies and Trends that are Transforming HR: Web-Based Human Resource* (pp. 259–276). New York: McGraw Hill/Towers Perrin.

Laumer, S., Eckhardt, A., & Weitzel, T. (2010). Electronic human resource management in an E-business environment. *Journal of Electronic Commerce Research, 11* (4), 240–250.

Lepak, D. P., & Snell, S. A. (1998). Virtual HR: strategic human resource management in the 21st century. *Human Resource Management Review,* 8 (3), 215–234.

Malhotra, N. K. (2004). *Marketing Research: An Applied Orientation* (4th ed.). London: Prentice-Hall.

Marler, J. (2009). Making human resources strategic by going to the net: reality or myth? *International Journal of Human Resource Management, 20* (3), 515–527.

Martin, G. (2005). *Technology and People Management: The Opportunity and Challenges.* London: Chartered Institute of Personnel and Development.

Masum, A. K. M., Beh, L. S., Azad, M. A. K., & Hoque, K. (2018), Intelligent human resource information system (i-HRIS): a holistic decision support framework for HR excellence, *International Arab Journal of Information Technology, 15* (1). 121–130.

Olivas-Lujan, M. R., Ramirez, J., & Zapata-Cantu, L. (2007). e-HRM in Mexico: adapting innovations for global competitiveness. *International Journal of Manpower, 28* (5), 418–434.

Pattanayak. B., Misra, R. K., Niranjana, P., & Pethe, S. (2002), Corporate excellence: a conceptual framework and scale development. *Indian Journal of Industrial Relations, 37*(3), 420–433. Sri Ram Centre for Industrial Relations, New Delhi.

Reddington, M., Williamson, M., & Withers, M. (2005). *Transforming HR: Creating value through People.* Oxford: Butterworth Heinemann.

Reilly, P., & Tamkin, P. (2006). *The Changing HR Function: The Key Questions.* London: Chartered Institute of Personnel and Development.

Ruel, H. J. M., & Bondarouk, T. V. (2008). Exploring the relationship between e-HRM and HRM effectiveness: lessons learned from three companies. In G. Martin, M. Reddington, & H. Alexander (Eds.), *Technology, Outsourcing and HR Transformation* (pp. 161–192). Oxford: Butterworth Heinemann.

Ruel, H. J. M., Bondarouk, T., & Looise, J. C. (2004). E-HRM: innovation or irritation. An explorative empirical study in five large companies on web-based HRM. *Management Review, 15* (3), 364–381.

Shrivastava, S., & Shaw, J. B. (2004). Liberating HR through technology. *Human Resource Management, 42* (3), 201–222.

Strohmeier, S., & Kabst, R. (2009). Organizational adoption of e-HRM in Europe: an empirical exploration of major adoption factors. *Journal of Managerial Psychology, 24* (6), 482–501.

Ulrich, D. (1997). *Human Resource Champions: The Next Agenda for Adding Value and Delivering Results.* Boston MA: Harvard Business School Press.

Ulrich, D., & Brockbank, W. (2005). *The HR Value Proposition.* Boston, MA: Harvard Business School Press.

Voermans, M., & van Veldhoven, M. (2007). Attitude towards E-HRM: an empirical study at Philips. *Personnel Review, 36* (6), 887–902.

APPENDIX

TABLE 14.A1
KMO and Bartlett's Test

Kaiser–Meyer–Olkin Measure of Sampling Adequacy		0.958
Bartlett's test of sphericity	Approx. chi-square	1,0556.156
	df	528
	Sig.	0.000

TABLE 14.A2
Communalities

	Initial	Extraction
E1 User-friendly interface.	1.000	0.758
E2 Gives control over personal information.	1.000	0.497
E3 Working on e-HR system is easy.	1.000	0.752
E4 Saves time spent on administrative tasks.	1.000	0.495
E5 Support from seniors for using e-HR systems.	1.000	0.565
E6 Responsive to the needs for (real-time) information.	1.000	0.626
E7 Improves productivity.	1.000	0.751
E8 Reduces HR-related paperwork.	1.000	0.764
E9 Simplification of process steps of various HR process.	1.000	0.727
E10 Benefit of benchmarked HR practices	1.000	0.591
E11 Improvement in overall HR business processes.	1.000	0.750
E12 Speeds up transaction processing related to HR activities.	1.000	0.767
E13 Optimizes the workflow between employees, management, and HR department.	1.000	0.820

(Continued)

TABLE 14.A2 *(Continued)*
Communalities

	Initial	Extraction
E14 ESS/MSS leads to a quick access to information.	1.000	0.619
E15 ESS/MSS leads to self-efficacy.	1.000	0.692
E16 Generates reports that improve decision making.	1.000	0.683
E17 24/7 availability of information related to HR processes.	1.000	0.794
E18 Facilitates the flow of knowledge in the organization.	1.000	0.725
E19 Helps me in managing team more efficiently.	1.000	0.712
E20 Feel overburdened "doing HR's" job.	1.000	0.746
E21 Is a reputational driver.	1.000	0.789
E22 Get updated about the organizational developments.	1.000	0.728
E23 Get updated news of HR initiatives/activities.	1.000	0.706
E24 Losing "People touch".	1.000	0.650
E25 Transparency in HR systems and processes.	1.000	0.804
E26 ESS/MSS gives a sense of empowerment.	1.000	0.732
E27 Quick response based on HR information.	1.000	0.749
E28 Do not want to spend time on Web-enabled "HR tools".	1.000	0.663
E29 Manage team globally.	1.000	0.711
E30 Concerns over data security.	1.000	0.614
E31 Usefulness of training provided to work on e-HR systems.	1.000	0.723
E32 Get immediate help for problems faced.	1.000	0.747
E33 Usefulness of refresher training programs.	1.000	0.738

Principal component analysis.

TABLE 14.A3
Total Variance Explained

	Initial Eigenvalues				Extraction Sums of Squared Loadings			Rotation Sums of Squared Loadings		
Component	**Total**	**% of Variance**		**Cumulative %**	**Total**	**% of Variance**	**Cumulative %**	**Total**	**% of Variance**	**Cumulative %**
Dimension	1	15.045	45.590	45.590	15.045	45.590	45.590	8.733	26.464	26.464
	2	4.174	12.649	58.238	4.174	12.649	58.238	7.476	22.656	49.119
	3	2.126	6.443	64.681	2.126	6.443	64.681	4.346	13.170	62.289
	4	1.845	5.591	70.272	1.845	5.591	70.272	2.634	7.983	70.272

Extraction method: principal component analysis.

15 Role of Servitization in Society 5.0

Rahul Gupta, Vishal Kamra, and S. K. Roy

CONTENTS

15.1 INTRODUCTION

High-tech machineries, complex restructuring, and Internet services have changed customers' expectations and stresses upon service supports from manufacturers. In this dynamic business environment manufacturing businesses are reinventing their business models. Manufacturers are implementing the latest technological innovations into equipment and serve customers with real-time analysis of complex data available through the Internet (AMR, 2017). Users are experiencing service management, data analytics and machine-to-machine communication digitally with the use of Internet of things (IoT) and data mining softwares. Technology is amalgamated with equipment to enhance its service offerings. Support software is installed into equipment, and customers are suggested and served with useful analysis. The new manufacturing model enables to change focus from product selling to product-service system selling (Mathe and Shapiro, 1993). Businesses have evolved with service offerings. Worldwide, roughly 63% of the GDP contribution was through services.

DOI: 10.1201/9781003158165-15

Practicing servitization infuses a strategic shift in customer services. Vandemerwe and Rada (1988) revealed that manufacturers are adding service component with their offerings, just to increase competitiveness, gain market power, move beyond "servicing" to "moment of truth" and increase business turnover. Cusumano, Kahl and Suarez (2015) and Vandermerwe and Rada (1988) support the transition from "goods or services" through "goods and services" to "goods services, support, knowledge, and self-service". This relationship-based approach helps firms to generate a new revenue stream, have a better customer relation, reduce cost, and gain competitive advantages over competitors. Crozet and Milet (2017) found that servitization offers an opportunity to grow and to sustain long-term competitiveness to local manufacturing units. Bustinza et al. (2018) suggested that manufacturer gets benefitted by defining their ability to incorporate value-added services into their operations.

Manufacturers are overextending their service with their equipment; various examples are cited in the literature, from Xerox, a photocopier machine company, to Rolls-Royce engine manufacturer. Raval (2014) found that the shift from manufacturer to service-led organization has shown significant increase in return on investments. Xerox, after facing revenue losses and seeing growth through services, created an independent, service-centric business process outsourcing firm, and retains hardware-centric operations for high-quality color printers. Rolls-Royce initiated a total care program for their airplane engines. Customers need to pay based on the number of hours the airplane flies. Rolls-Royce will take care of repair and broken parts and modify and repair airplane engines remotely by analyzing real-time data with the help of IoT. This shift to servitization along with long-term relationships with air carrier has ensured more than 50% higher revenues as compared to earlier ones.

Manufacturers are creating a whole new system by offering value and earning long-term relationships and creating an entry barrier to competition. From a strategic perspective, they are getting competitive advantage, with blurred boundaries of product differentiation, and increasing customers' expectations for product and service performance. Increasing demand for services along with product offerings has forced manufacturers to think out of the box.

15.2 REVIEW OF THE LITERATURE

Services need to be an integrated part of manufacturer core offerings (Gadiesh & Gilbert, 1998; Quinn et al., 1990). The rationale behind integration lies in economic suggestions. The products with long life cycles help to generate extra revenue and services associated with those products ensure higher margins (Knecht et al., 1993; The Economist, 2000). Demand for products with associated services is higher among users (Guillot, 2017). Competition has forced the manufacturer to be flexible to develop core competencies by adopting technological innovations (Lojo, 1997). Heskett et al. (1997) studied that it is difficult to imitate services, they are intangible and inseparable, and the provider enjoys competitive advantage.

Most of the manufacturers are opting for services strategies with a slow pace except few companies such as GE, Xerox, Rolls–Royce, and Caterpillar just to overcome the hurdles in transition process. Firms might doubt economic computability for services. Oliva and Kallenberg (2003) suggested through an interview that "it is difficult for an engineer to get excited for service contract worth $10,000 for a multi-million equipment". Firms may find that providing services is beyond their scope of work (e.g., an independent dealer network restricted Ford Motor Co.'s offerings into after-sales services).

Manufacturer's transition to services provider comes with various challenges. Services require varied capabilities, principles, metrics, and incentives and a changed model from transition to relationship. These capabilities demand managerial and financial resources for new product development and to gain competitive advantages.

Mathe and Shapiro (1993) studied that services complement the sales growth and success in competitive market. Earlier, services were neglected, and now, better services enhance sales. Oliva and Kallenberg (2003) acknowledged service/product interface as sales function, and their fieldwork yielded focuses on services provided by manufacturers with installation of products, which was in accordance with Patton and Bleuel (2000).

The existing literature is sparse in studying the challenges in transition from product to services. It is not clear even at the strategic level that what extent of services should be offered or what factors should be considered while deciding service mix.

15.3 SHIFT TO SERVITIZATION

Earlier manufacturers considered the services of their equipment as an extra cost. With the adoption of servitization and implementation of technological advancements, field service technicians now retain all important information about the users. Analyzing information fetches new products and customers. Servitization gets success with immediate customer support, with efficient data management, monitoring equipment remotely, and managing work complexities through a central service management system. The recent phenomenon has forced manufacturers to compete strategically with a portfolio of integrated services with products.

15.4 TECHNOLOGY

Servitization's success rests on the use of technology, to connect, inform, and improve maintenance and repairs. A higher level of delivery suiting to an individual customer's requirements depends on the implication of technological advancements. A few mapped processes are as follows:

- **Monitor**: The critical product system is continuously sensed through transducers, data storage, and fault code generation.
- **Transmit**: Data for fault code is transmitted through cell phone, Internet, radio, GPRS, or satellite.
- **Store**: The manufacturer stores data on soft or hard storage systems.
- **Analyze**: Diagnosis and analysis are performed to predict future behavior and trends.
- **Respond**: The manufacturer intervenes in replacement/repair offering, modifying designs, drafting a contingency plan, and contacting the customers.

15.4.1 PR Actioners for Servitization

Rolls-Royce, the famous engine manufacturer practices a new concept called "power by the hour" program. The airlines who buy power of aero engine from Rolls-Royce get full support and maintenance. The company ensures continuous power delivery to their clients. With the Total Care service package, engines are rented, and Rolls-Royce monitor data to predict any potential maintenance problem, saving cost for unnecessary and unplanned maintenance downtime. Forty-nine percent of their revenue was generated by adopting servitization.

Caterpillar, a famous manufacturer of heavy earth moving machines practices a concept called "Cat Product Link". The company optimizes performance of its products by analyzing the data tracked remotely from each location, preventive maintenance, regular components monitoring and reducing unnecessary downtime.

Alstom's TrainLife Services package and "lost customers hours" opt for reducing delays by train failures. Services such as condition-based maintenance, technical and managerial support, and performance improvement provide reliable and quality services.

MAN Truck and Bus, a famous transport company in United Kingdom practices a concept called "Trucknology". With its fleet management services by using sensors to identify inefficiencies and analyzing data, the company reduces the cost of service operations. New technology helps the company to monitor whether engine is running while stoppage and technology helps to check inefficient and fast deriving by the users. This type of monitoring helps the customer to improve fuel efficiency, lower down the maintenance cost, lessen the accidents, and improve total performance.

Xerox, initially known for photocopiers, is now known as an enterprise for business process and management and is now earning half of its revenue from document publishing, document management, and BPO. Xerox provide curriculum content and learner care services as learning solutions to Hertz. For Siemens, they have created a digital archive and interface to improve document control and decrease paper waste. For Apple, 15% of their revenue and gross margin is through services. General Electric (GE) projects are expecting to generate $12 billion digital revenue.

15.5 SERVITIZATION DRIVERS AND MODELS

Enabling an organization as servitization organization, it needs to identify the key components, major challenges, and considerations that help in strategizing in developing programs. Firms such as chemical, aerospace, defense, energy, and technological firms practice servitization. A survey done by Infosys Consulting suggests 5%–15% increased revenue, 10%–20% increased utilization of production capabilities, 15%–30% reduction in logistical costs, 10%–25% savings in cost due to small batch size, 20%–30% reduced lead times and inventories, and 10%–25% increased flexibility and responsiveness.

These savings can be achieved by adopting a model for servitization. A few suggested models are discussed here. **Operating services** include operations and life cycle of products. Services include maintenance, replacements, design and development of products, installation, and upgrade services. **Value-added services** include consulting and trainings and product customization. **Service contract** includes

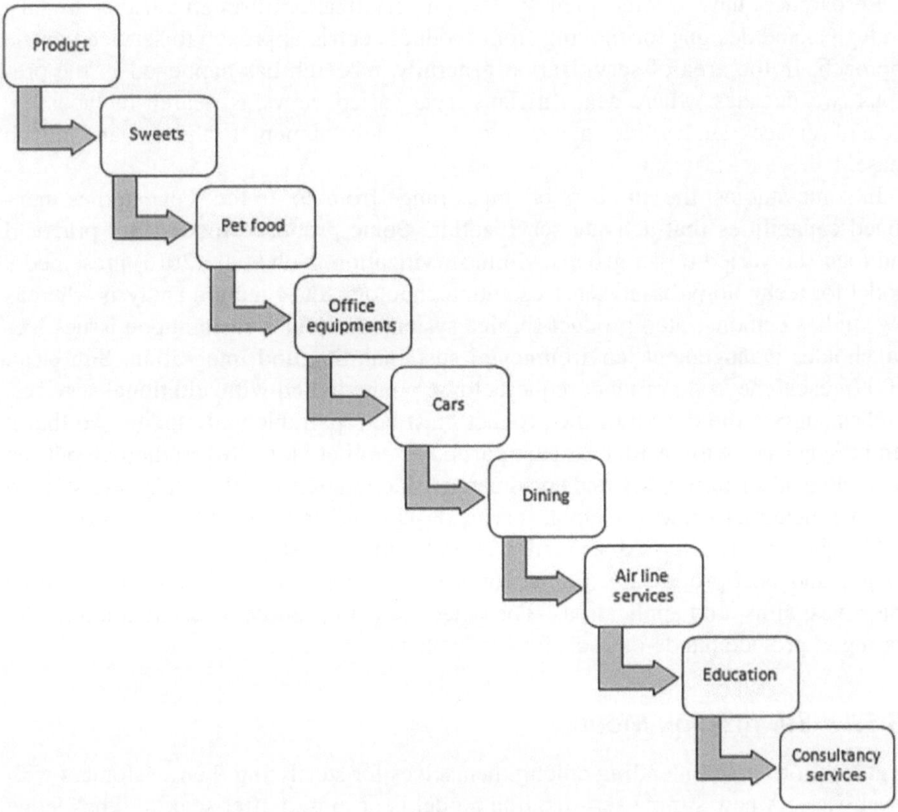

FIGURE 15.1 Differentiation of services and products.

contract management, extended warranties, and claims management. Financial services include financial assistance and leasing services (Figure 15.1).

Differentiation in products and services is done based on their understanding on a continuum, ranging from tangible to intangible dominant. Physical attributes are considered as tangibility aspect and physiological attributes are considered as intangible aspect. Differentiation may be possible by looking into these aspects. For example, computer software is considered as the tangible side on a continuum. This assumption is based on absolute service that means only services can be felt or product does not exist. They can only be categorized on intangible and tangible dominance. Tangibility is associated with physical aspect and intangibility as services.

From the users' perspective, risk is reduced, and the level of uncertainty is low with servitization. Liabilities, level of competences, and knowledge of processes for various resources get reduced with servitization, as services are taken care of by the provider. Servitization should be viewed from the buyer's perspective, not from the seller's perspective, on the tangible continuum as it has physiological dimensions. A taxi owner views the business as tangible while taking vehicle as physical product, but from the riders' perspective, it is considered as intangible. It becomes difficult to overtly ghettoize products and services in various categories; hence, they are considered as an offering.

Researchers have offered their studies on servitization through various models, strategies, and designs for moving from product-centric approach to service-centric approach. In the area of servitization generally, research has happened in the previous two decades, where academicians have varied views. Research includes the cycle of services, such as idea generation, design development, test phase, and launch phase.

In some studies, the numbers of stages range from six to ten. Few studies mentioned capabilities that activate servitization. Some proposed models are practical and focus on varied target groups within servitization. Noh et al. (2016) presented a model for technology-based services with technology and function analysis whereas few studies demonstrated product-service systems models focusing upon issues like stakeholder management, environmental sustainability, and integration. Sun et al. (2009) researched into product requests to be premeditated with additional services. Studies suggest the design of the product must be repairable and modular so that it can be serviced, with available spare parts. Fargnoli et al. (2018) studied an outline for ecological sustainability and customer satisfaction. Kim et al. (2013) have studied a model including value modeling, service design, integration, and experience.

Marques et al. (2013) discussed a model with four steps: readiness, planning, design, and post-processing. Frameworks and models were studied with specific objectives, aims, and applications. The objective is to propose a servitization model aiming to provide hands-on tools for the industry.

15.5.1 Servitization Model

Organizations are contending among themselves for sterilizing their customers with novel ways. A new simple servitization model is proposed after seeking knowledge from various published studies. The proposed model is a time-efficient tool, following seven steps for servitizing of products, practically. The evaluation is completed with the help of qualitative judgments and the performer's knowledge. Models find it difficult to generate excellent ideas for services, but instead lead to develop services by small and medium firms.

An offering in the form of product with services is selected for servitization. Various customer groups are selected; this helps in understanding offerings as utilization. These offerings are decomposed into its barriers and enablers for enhanced understanding. This leads to get reasons for the buyer to opt for the services. The discussed barriers and enablers are then synchronized with their belongingness for major buyer groups. Servitization is then applied for changes to implement barriers and enablers, so tangibility is reduced (Figure 15.2).

Step 1: Selecting Offerings

Technological progresses and business models gave way to servitize many products and services. Innovative avenues are opened due to augmented technological advances. Earlier, few services were added, such as diesel generator maintenance, but it was difficult to add services to products such as furniture. New advance technology and trends have powered servitization in various product categories. For example, furniture can

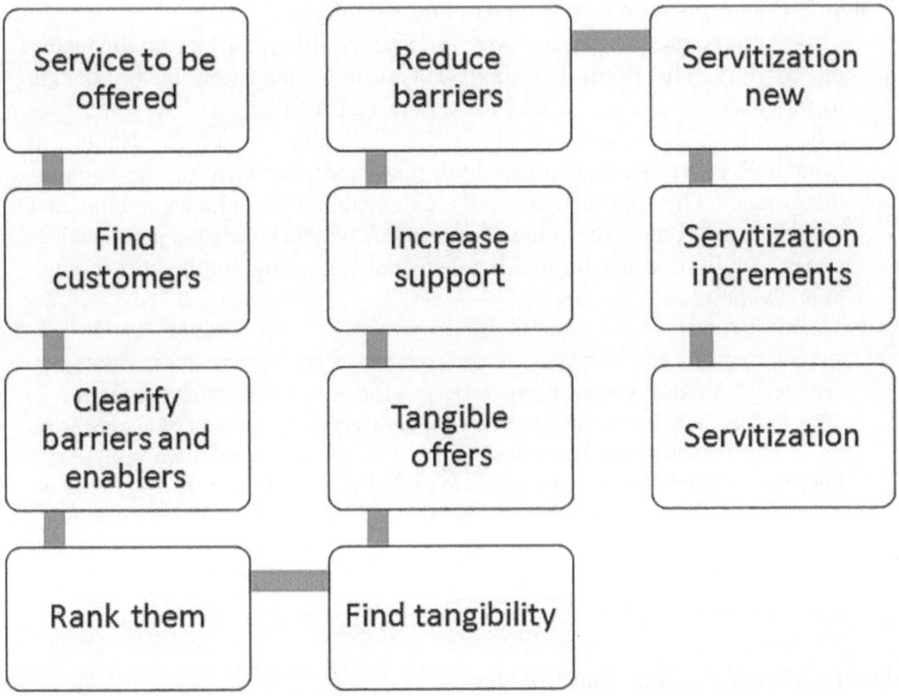

FIGURE 15.2 Servitization models.

be servitized by installing temperature sensors to track the well-being of the users. Being a continuous process, servitization can be executed in many areas.

Step 2: Identifying Customer Segment/Group

Product utility can be enhanced with respect to the customer segment or group. By following segmentation processes, various customers can be identified and grouped as per their similarities. Marketing discipline uses Segmenting Targeting and Positioning for identifying customers' groups for selected product or service offerings. In our case, the present product may be servitized. It gets assistance in barriers and enablers for the next stage.

Any purchaser may be pertinent to the model, and the suggested parameters are psychographic, geographic, monetary, and demographics.

Psychographic	Openness, Innovation	Personal interests, socially responsible, lifestyle, quality conscious
Demographics	Type, size, industry, employees	Age, occupation, education, gender
Monetary	Revenue, assets, net profit	Income, homeowner
Geographic	Office locations, presence	Location, mobility

Step 3: Decomposition into Barriers and Enablers

A lot of research has been done on product utility, still it remains unexplored fully. Therefore, it is important to understand how buyers decide to purchase products and which barriers restrict them to buy some specific products. Sheth et al. (1991) emphasize "value", which corresponds to "enabler". Barriers and enablers both take credit for forming the value of the product. The approach proceeds with enablers for enhanced value, and barriers compromise the value. The tradeoff between barriers and enablers serves as a floor seater for making decisions regarding purchase or to postpone the purchase.

Specific	**Product-Specific Barriers**: Unique hindrances in different customer groups that stop them from buying the offering?	**Subutility**: Utilities that different customer groups are looking for
Common	**Barriers:** Hindrances related to ownership/financing, operation, and risk that stop customers	**Utility:** Utility that most of the customer groups are looking for
	Barriers	**Enablers**

Identification of Barriers and Enablers

Barriers and enablers can be qualitatively discussed as per the table. Various customer segments depict common utilities and subutilities for a product. The core enabler is the most significant and recursive utility. For an excavation machine, the customer segment would be miner, and construction companies, infrastructural development companies, etc., have a common utility for these machines for excavating earth ore excavating mines. The sub utilities for these machines would be excavating leaps and humps of overburden. These utilities and sub utilities are enablers for excavating machines.

Barriers for Various Products

Procurement time and cost
Maintenance hassles and expenses
Redundancy/downtime
Operational expenses
Depreciations
Clearances and certifications for operations
Rapid technological advancements
Skilled labor for operations
Business liabilities/accidents
Regular taxes and premiums
Handling and storage cost
Hassel of reselling

The impression of barriers in these utilities to enhance their intangibility for an offering is increased by Servitiztion. While buying a car, customers look for enablers such as commutation from one position to another, flexibility in commutation, personalized availability, and status, and common barriers while purchasing a car are high purchase cost, taxation/insurance liability, recurring maintenance, legal obligations such as license to drive and acquiring driving skills or hiring a skilled driver for the car. When these enablers are offered in an intangible way they may enhance service opportunities. Likewise, when barriers are disconnected, newer opportunities for services can be explored.

Step 4: Ranking

Barriers and enablers might fail in identifying the true potential for servitization. So, it becomes imperative to prioritize them to ensure high-market-demand enablers are prioritized during market segmentation and modeling. The given figure will place them as per customers' priorities. The sternness of enabler will be measured qualitatively between "nice to have" and "essential". Essential enabler increases the satisfaction of consumers, and in the absence of them, buyers may postpone or may not make purchase decision. Accordingly, barriers will also be qualitatively determined among "moderate" and "extreme" barriers, from the buyers' perspective. Highly ranked barriers and enablers are preferred chances for servitization. Here, one point is also to be noted that barriers and enablers placed on the lower side of the scale must not be ignored.

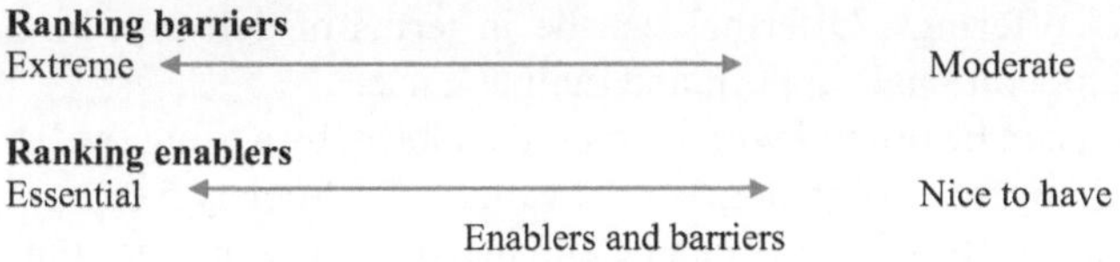

Step 5: Locating Offering on Tangibility Scale

Once barriers and enablers of offering are identified, then their qualitative judgment of tangibility is done. Here, the presentation of the model is tracked, to comprehend that tangibility of offering has compromised after placing it in the model. Different assignees may form different qualitative judgments for the purpose of evaluation of its tangibility and comparability with present offerings.

Step 6: Checking Option

The model systematically checks the key objectives of selected offerings thorough the first five steps. Then, servitization options are present and can be functional to select offerings for servitization.

Radical servitization and incremental servitization are the two subcategories used for servitization. They are comparable to radical and incremental innovations. The focus here is to decrease tangibility, instead of innovation of the aids for servitization. Radical innovation and incremental innovation are defined as "doing what is not done earlier" and "doing better what is already done", respectively. The servitization approach is referred to an incremental servitization, where a steady decrease in tangibility in the offering takes place. For example, maintenance is added as a value-added service. Radical servitization refers to a situation where existing offering is significantly changed. For example, any product offering has substantially reduced its tangible offerings. Uber is a classic example for disruption in taxi services. Nowadays, servitization can be considered as an instrument for innovation or a subset of modernization.

Step 7: Checking Tangibility

The model at the end compares the tangibility of the offering after and before the intervention of numerous options in the model. Servitization is successfully introduced if tangibility in the offerings is abridged.

15.5.2 Incremental Servitization

Incremental servitization can be defined as reduced tangibility by growing enablers or lowering barriers. Here, the utility is gradually increased for users, and barriers are removed which hinder buyers for making purchase. It becomes imperative how enablers are increased and barriers are curtailed.

Easy way to servitize a product is to magnify its enablers either by enhancing or simplifying the complexities of the product or value-added services can be added, with additional offerings. Offerings can be in terms of delivery, installation, customer support, operational support, and maintenance.

Servitization can be offered with increased enablers by adding the latest technologies. Various innovative technologies are available for Society 5.0.

For simple products, offerings can be enhanced by growing the complexity of the product (Figure 15.3).

15.6 REDUCING BARRIERS

Succeeding in the model barrier is an integral part of servitization. These barriers are generally common, unlike enablers. Various types of barriers are product specific barriers, risk barriers, operational barriers, ownership barriers, and financing barriers. All these barriers can be addressed as follows: ownership barriers such as high cost of procurement, operational barriers such as shortage of skilled staff, inventory barriers, and risk barriers such as rapid advancement in know-hows. These barriers need to be checked and addressed before they stop buyers from buying (Figure 15.4).

15.6.1 Ownership Barriers

Ownership barriers have maximum tangibility associated with the products. Oliva and Kallenberg (2003) studied various firms who were transiting from product to

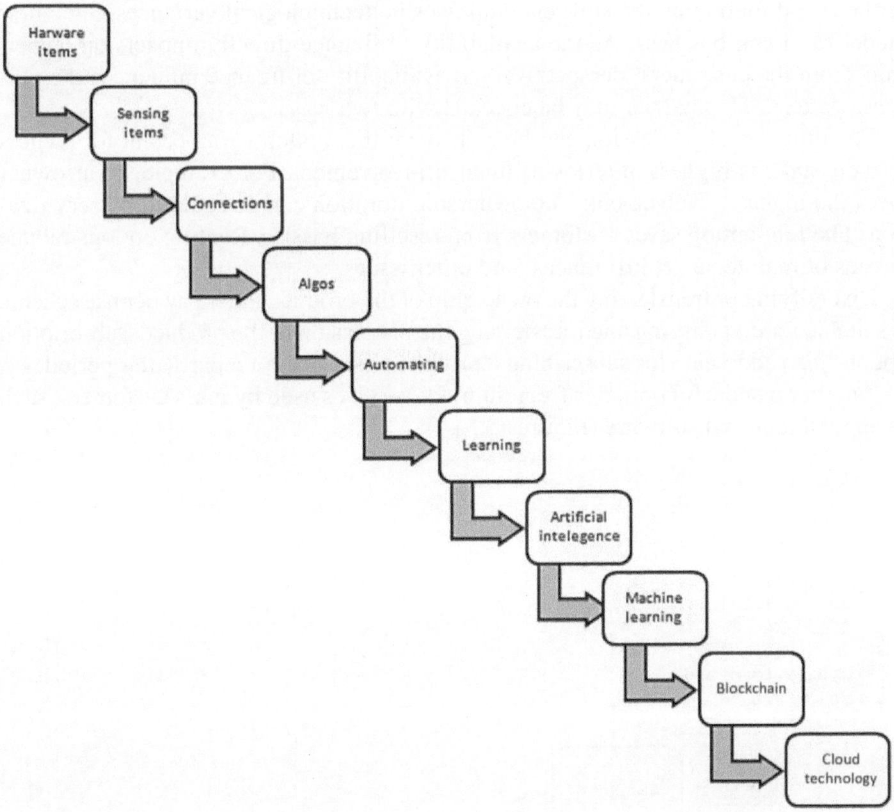

FIGURE 15.3 Increasing the enablers.

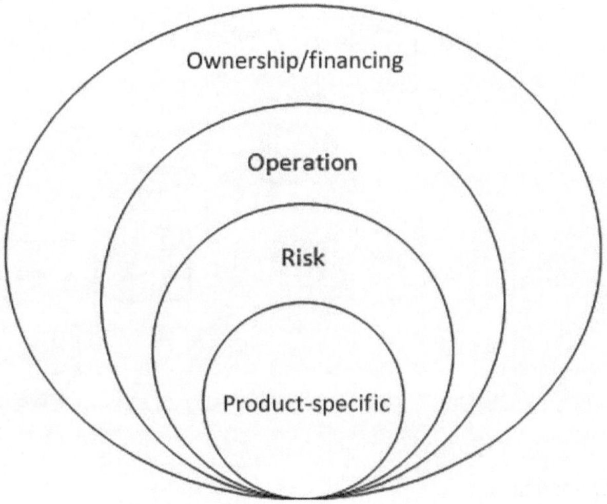

FIGURE 15.4 Types of barriers.

services and found that the shift encompasses no technological variances, but a new model for doing business. As the availability of finance directly impacts on ownership, from the customers' perspective, the availability of finance influences the purchase pattern. This barrier may be checked with two ways.

The duration of possession can be restricted; this option is important for owners as their stake is highest in terms of finance involvement. For example, a car owner faces the highest involvement. The ownership duration can be reduced by servitization. The rent option saves customers from reselling hassles. Renting options reduce hassles of maintenance, insurances, and other issues.

Diversifying or transferring the ownership of the product under pay per use scheme is suitable for diversifying and transferring the possession of the product. Subscription options also add value for subscribing a product or service for a particular period.

Another wonderful option is freemium, for services used by mass customers, such as applications and software (Figure 15.5).

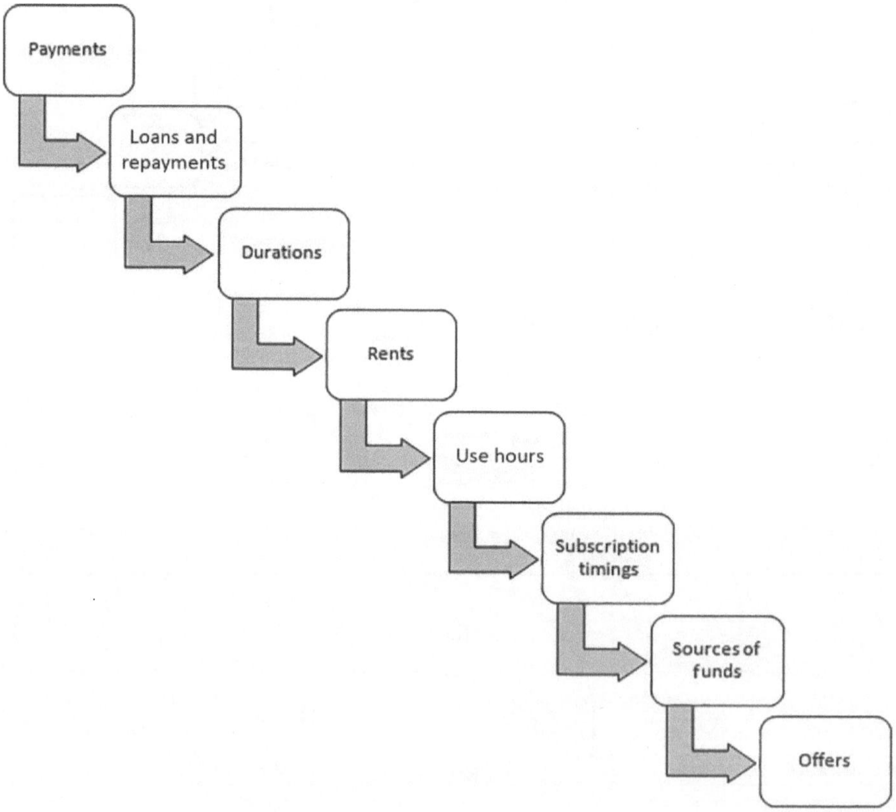

FIGURE 15.5 Ownerships.

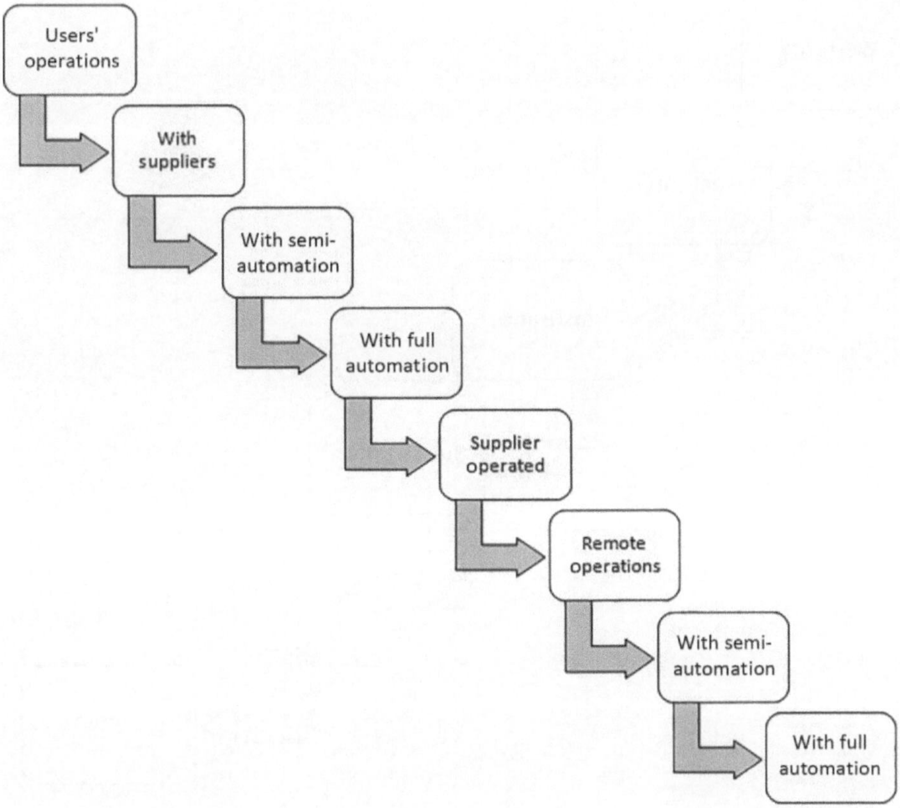

FIGURE 15.6 Operational barrier.

15.6.2 Operational Barrier

By transferring many business activities such as store operations, maintenance, and handlings offer to the suppliers, buyers felt relieved with this risk (Figure 15.6).

15.6.3 Risk Barrier

Risk barrier is associated with the use of the product. This barrier demotivates the buyers to opt for purchasing a new product. This risk can be reduced by taking care for breakdown risk, liability risk, risk associated with the change in technological advancements, and depreciation risk.

Independent product-related risk also needs to be addressed. For example, environmental concerns may discourage a buyer to purchase a particular vehicle, due to emission norms. These kinds of risks are difficult to address (Figure 15.7).

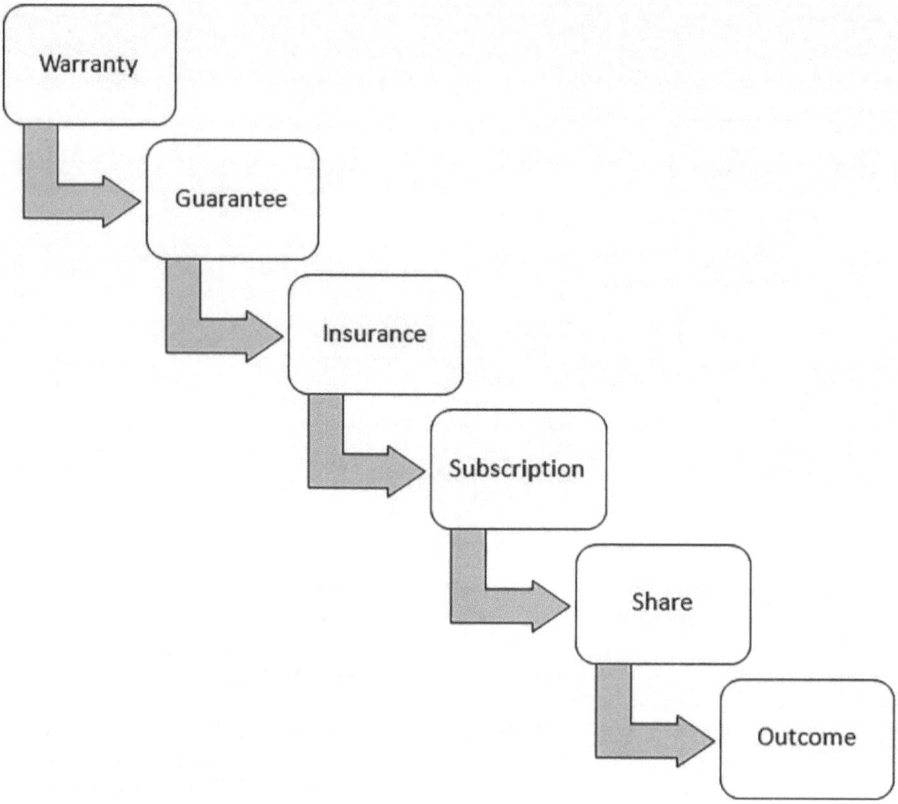

FIGURE 15.7 Risk barrier.

15.7 CONCLUSION

With a shift in focus to services, firms have secured a good customer base. They monitor, transmit, and analyze usage patterns and accordingly offer their services and charge. Customers feel they are getting required services and paying for the same. With advancement in Internet speed and availability, we also enhance usages of servitization in our country, especially in agriculture and other industries. Increased expectations and abridged product life cycle demand sustainable alternatives for survival. Competition added fuel to this burning fire of sustainability. The existing literature suggests that many approaches are explored. Numerous models are suggested for servitization. Business owners find it difficult to survive and seek sustainable options to compete. The study offers a conceptual model, which needs thorough investigation and verification. Studying servitization with advanced technological developments, with service options, requires further research directions.

REFERENCES

Annual Manufacturing Report. (2017). https://www.barclayscorporate.com/content/dam/corp-public/corporate/Documents/Industry-expertise/annual-manufacturing-report-2017.pdf.

Index

Bustinza, O. F., Herrero, F. V., Gomes, E., Lafuente, E., Basáez, M. P., Rabetino, R., & Vaillant, Y. (2018). Product–service innovation and performance: unveiling the complexities. *Int. J. Business Environ.* 10(2), 95–111. doi:10.1504/IJBE.2018.095819.

Crozet, M., & Milet, E. (2017). Should everybody be in services? The effect of servitization on manufacturing firm performance. *J. Econ. Manage. Strat.* 26(4), 820–841. doi:10.1111/jems.12211.

Cusumano, M. A., Kahl, S. J., & Suarez, F. F. (2015). Services, industry evolution, and the competitive strategies of product firms. *Strategic Manage. J.* 36(4), 559–575. doi:10.1002/smj.2235.

Daimler Key Figures. (2017). (Divisions): https://www.daimler.com/investors/key-figures/divisions.html-(The).

Economist. (2000). In search of Fiat's soul. *The Economist*, 3 June, pp. 69–70.

Fargnoli, M., Haber, N., & Sakao, T. (2018b). PSS modularisation: a customer-driven integrated approach. *Int. J. Prod. Res.* 1–17.

Gadiesh, O., & Gilbert, J.L. (1998). Profit pools: a fresh look at strategy. *Harvard Business Rev.* 76(3), 139–47.

Guillot, C. (2017). *Servitization is a Growing Manufacturing Model.* Chief Executive magazine: https://chiefexecutive.net/servitization-growing-manufacturing-model.

Heskett, J. L., Sasser, W. E., & Schlesinger, L. A. (1997). *The Service Profit Chain.* Free Press, New York.

Kim, Y. S., Lee, S. W., Jeong, H., Kim, S. R., Kim, J. H., Noh, J. H., & Won, J. H. (2018). A systematic design framework for product-service systems and its implementation. In *Proceedings of the Fifth International Conference on Service Science and Innovation (ICSSI),* Kaohsiung, Taiwan, pp. 59–66.

Knecht, T., Leszinski, R., & Weber, F. (1993). Memo to a CEO. *The McKinsey Quarterly*, Vol. 4, pp. 79–86.

Lojo, M. (1997). *Contracting of High-Technology Industrial Services*, unpublished PhD dissertation, Sloan School of Management, Massachusetts Institute of Technology, Cambridge, MA.

Marques, P., Cunha, P. F., Valente, F., & Leitão, A. (2013). A methodology for product-service systems development. *Procedia CIRP*, 7, 371–376.

Mathe, H., & Shapiro, R. D. (1993). *Integrating Service Strategy in the Manufacturing Company.* Chapman & Hall, London.

Noh, H., Song, Y., Park, A. S., Yoon, B., & Lee, S. (2016). Development of new technology-based services. *Serv. Ind. J.* 36, 200–222.

Oliva, R., & Kallenberg, R. (2003). Managing the transition from products to services. *Int. J. Serv. Ind. Manag.* 14: 160–172.

Patton, J. D., & Bleuel, W. H. (2000). *After the Sale.* Solomon Press, New York.

Quinn, J. B., Doorley, T. L., & Paquette, P. C. (1990). Beyond products: services-based strategy. *Harvard Bus. Rev.* 68(2), 58–67.

Raval, A. (2014, January 24). Xerox says shift to services is paying off. *Financial Times.* Retrieved from: www.ft.com/cms/s/0/bac264c8-662e-11e2-bb67-00144feab49a.

Rolls-Royce Holdings plc Annual Report 2017: https://www.rolls-royce.com/~/media/Files/R/Rolls-Royce/documents/annual-report/2017/2017-full-annual-report.pdf - Servitization-leading-manufacturing-new-business-models-infosys-consulting.

Sun, H. B., Mo, R., & Chang, Z. Y. (2009). Study on product service-oriented enterprise servitization methods. In *Materials Science Forum*; Trans Tech Publications, Stafa-Zurich, Switzerland, Vol. 626, pp. 747–752.

Vandermerwe, S., & Rada, J. (1988). Servitization of business: adding value by adding services. *Eur. Manage. J.* 6(4), 314–324. doi:10.1016/0263-2373(88)90033-3.